Dynamic Econometrics

"Very bold and confident advance in dynamic econometric theory and practice. While quite formal in maths, the book is quite exciting to read".

—I. Litvine, *Professor in statistics, Nelson Mandela University, Port Elizabeth, South Africa*

"This book is an outstanding contribution to econometrics, coming at a crucial time to fill a significant gap in the field. With a rigorous approach and a rich array of models and applications, this book is an essential resource for students, educators and professionals in economics".

—Maria do Rosário Grossinho, *Professor of Analysis and Mathematical Finance, ISEG—University of Lisbon, Portugal*

"An impressive and outstanding book presenting pedagogically the most recent developments in dynamic econometrics by combining theoretical and empirical analyses".

—Valérie Mignon, *Professor at the University of Paris Nanterre, France*

Francis J. Bismans · Olivier Damette

Dynamic Econometrics

Models and Applications

palgrave
macmillan

Francis J. Bismans
BETA (Bureau of Applied and Theoretical Economics)
University of Lorraine
Nancy, France

Olivier Damette
BETA (Bureau of Applied and Theoretical Economics)
University of Lorraine
Nancy, France

ISBN 978-3-031-72909-6 ISBN 978-3-031-72910-2 (eBook)
https://doi.org/10.1007/978-3-031-72910-2

Translation from the French language edition: "Économétrie dynamique - Modèles et applications" by Francis J. Bismans and Olivier Damette, © Éditions Ellipses 2023. Published by Éditions Ellipses. All Rights Reserved.

The original submitted manuscript has been translated into English. The translation was done using artificial intelligence. A subsequent revision was performed by the author(s) to further refine the work and to ensure that the translation is appropriate concerning content and scientific correctness. It may, however, read stylistically different from a conventional translation.

Cover illustration: eStudio Calamar

This Palgrave Macmillan imprint is published by the registered company Springer Nature Switzerland AG
The registered company address is: Gewerbestrasse 11, 6330 Cham, Switzerland

Preface

Econometrics, in its broadest sense, is the combination of economic theory, probability theory and mathematical statistics. Nowadays, it is as essential to the basic training of an economist as microeconomics or macroeconomics. Ultimately, the *Principles of Econometrics* can be seen as an essential component of the *Principles of Economics.*

Dynamic econometrics, on the other hand, is nothing other than time series econometrics. In other words, it is the product of the coming together of general econometrics and that sub-discipline of inferential statistics known as time series analysis. To tell the truth, we now need to provide a few historical details on how this latest symbiosis came about.

A Brief History...

Let's go back to the seventies of the previous century (for more details, see Bismans and Grossinho, 2021). The discipline of econometrics had made great strides since its first steps, which really dated back to the years of the Great Depression, marked in particular by the work of the first two Nobel Prize winners in economics, Jan Tinbergen and Ragnar Frisch. Johnston's *Econometric Methods* (1963, 1972), probably the most widely used textbook during the *Trente Glorieuses*, is fully representative of this trend towards deepening and broadening econometric theory.

In particular, this theory was enriched by the analysis of systems (or models) of simultaneous equations (SSE), a development that led to the construction of large-scale macro-models. The most significant of these was the American *Brookings* model, designed in the 1960s.

To judge its size, we need only to mention that it contained no fewer than 176 endogenous variables and 89 purely exogenous variables. In its smallest version, it contained 176 equations divided into recursive blocks.

Two major developments were to challenge the dominance of econometric macro-models: firstly, the development of time series analysis; and secondly, the recognition of the non-stationarity characteristic of economic series.

Firstly, since the end of the *Trente Glorieuses*, the analysis of time series has been considerably enriched, particularly following the work of Box and Jenkins (1976). Their methodology was based on the use of ARMA or ARIMA-type processes (*AutoRegressive* (*Integrated*) *Moving* Average).

The explicit aim of this methodology was to predict the future behaviour of a series based solely on knowledge of its past.

Such an approach obviously had no basis in economic theory, unlike macro-models which were based on a strong theoretical structure, essentially Keynesian in nature. But both approaches ultimately shared the same central objective: *forecasting*.

From this point of view, the fatal blow to large models was dealt by Nelson (1972), who compared the predictive performance of the FMP (Federal Reserve Board-MIT-PENN) macro-model with that of ARIMA-type models. The comparison was indisputable: the latter provided better forecasts in seven out of fourteen cases, and the macro-model in only two. This begs the question: what's the point of using systems of several dozen, or even several hundred, equations, when just one of these equations often gives a better predictive result?

Secondly, it is almost self-evident that economic series are not stationary. We need only think, for example, of Gross Domestic Product, which, despite cyclical fluctuations, incorporates a marked trend or tendency.

However, this simple observation has had significant theoretical consequences. Granger and Newbold (1974) pointed out, using Monte Carlo simulations, that non-stationarity could lead to what they called *spurious* regressions. Phillips (1986) was to provide a theory of such spurious regressions. This led to the opening of an enormous theoretical field devoted to the identification and study of the properties of non-stationary series. In fact, it would be no exaggeration to speak of a veritable revolution in this area, which led to the creation of dynamic econometrics.

In the end, the forecasting failure of macro-models will have two major consequences: on the one hand, their relative decline; on the other, the birth of a genuine "time series econometrics", which will take full account of the non-stationary nature of the main economic series.

Purpose and Outline of the Book

The aim of this book is to report on the most recent developments in dynamic econometrics. It therefore focuses on the analysis of the consequences of non-stationarity.

It does not therefore cover the entire range of econometric subjects, although it should be noted that it is intended to cover the various dynamic aspects, from techniques associated with time series to those based on panel data models or qualitative variables.

This book is divided into ten chapters. The first is a sort of elementary presentation of the simple linear regression model. The next six chapters are devoted to the study of dynamic linear models, whether univariate or multivariate. The next two chapters deal with panel data models, characterised by the dual dimension of time and *cross-section*. Finally, the tenth and last chapter deals with the vast field of qualitative models, of the probit and logit type, restricting itself to their dichotomous variety, i.e. with two qualitative characteristics.

So much for the general structure of the book. Let us now look at the organisation of each of the chapters, adding that they all end with one or more applications, generally taken from the work and research of the two authors. These applications are then followed by appropriate exercises.

Chapter 1 begins by defining econometrics as it is understood today. It goes on to analyse the techniques for estimating the linear regression model between two variables. It also presents some statistical tests that apply in this elementary case. It may, of course, be overlooked by students who have already taken a first course in econometrics.

Chapter 2 begins by introducing the main stationary time series models, before moving on to study non-stationary processes. This is followed by a discussion of unit root tests for assessing the stationary or non-stationary nature of a time series. Finally, it looks at stochastic simulations, better known as Monte Carlo simulations.

Chapter 3 is devoted to the estimation of the classical linear regression model in its usual form. The two classical methods of least squares and maximum likelihood are developed, as are the criteria for selecting a model, which are based on the maximum likelihood technique. This chapter also studies the optimality properties of linear estimators before presenting the general linear dynamic model.

Given that inferential statistics comprises two main areas—estimation and test theory—Chapter 4 develops all the tests that can be applied to the general linear dynamic model studied in the previous chapter, with the aim of validating the hypotheses concerning the random term of the model. It also develops the three commonly used asymptotic tests.

Chapter 5 addresses the problem of spurious regressions, a problem caused by the presence of non-stationarity in the model variables. It introduces the concept of cointegration and details the different methods for estimating cointegrating relationships: Engle-Granger, DOLS, FM-OLS and, finally, Johansen's multivariate method. It also shows how the ARDL model is associated with an error correction mechanism and provides details of Pesaran's specific approach to estimating cointegrating relationships.

The aim of Chapter 6 is to detail the procedure to be followed in order to specify a dynamic model as correctly as possible, a procedure that can be summed up by the formula: "from the general to the specific". It also discusses the various specification tests available.

Chapter 7 looks in detail at dynamic systems of the VAR (vector autoregression) type, from their formalisation to their estimation. It also discusses VAR models in their structural form and impulse response functions, in particular on the basis of local projections.

Chapter 8 presents the two main models of panel data, data with a dual dimension, time and cross-section. It shows which methods to use to estimate and test them. It also looks at stationary dynamic models, while Chapter 9—it should be noted—is entirely devoted to the study of non-stationary panels.

Finally, Chapter 10 first introduces the two main non-linear models—logit and probit—used when the dependent variable is qualitative. It then focuses on the dynamic probit model as a forecasting tool.

Two appendices, to which we will return shortly, complete this handbook.

Prerequisites and Advice

As you will have realised, this manual is not intended to be introductory. On the contrary, for a fruitful approach, it assumes that the reader has taken a first introductory course in econometrics. However, Chapter 1 presents a few preparatory elements which may guide the student in reading and assimilating this work.

In addition, there are a number of prerequisites for studying this book. Knowledge of the basics of linear algebra, and more specifically matrix calculus, is imperative. The same applies to the theory of optimisation of functions of real variables. However, to help the reader, we have grouped together a number of relevant results in Appendix B. Nevertheless, some more specific mathematical developments appear in several chapters, where a good understanding of the econometric subject requires it.

Another prerequisite is at least an elementary knowledge of the elements of probability theory that underpin econometrics. From this point of view, we would refer you to the work of one of the authors of this book—see Bismans (2016), non-translated—which covers this subject in all its complexity. In English, the undergraduate textbook by Chung (1979) provides the essential concepts of probability theory.

Finally, a word of advice, not a prerequisite: if you would like to know more about models and the analysis of time series, we can do no better than refer you to two important textbooks: one in French by Gouriéroux and Monfort (1995); the other, in English, by Harvey (1993).

The applications and exercises in each chapter of our book require the use of one or other econometric software package. In the first instance, we have used the free software GRETL—the term is not only a reference to the Brothers Grimm fairy tale but also an acronym for GNU *Regression, Econometrics and Time Series Library*. Appendix A provides more details on this software, as well as on the *Microfit* of Bahram and M. Hashem Pesaran.

However, we have endeavoured to provide the various application or exercise files in several software extensions such as R, Stata, PC-Give and several others. These multiple data files can of course be found on the publisher's website.

As for the rest, we hope you enjoy reading this and learning about dynamic econometrics.

Nancy, France

Francis J. Bismans
Olivier Damette

Contents

List of Figures

List of Tables

Chapter 1
General Introduction

1.1 What Is Econometrics?

The first Nobel Price winner, Ragnar Frisch, as early as 1926, coined the term econometrics, a discipline at the crossroads of mathematics, statistics and political economy. The founding Charter of the *Econometric Society*, as reproduced by its secretary Charles F. Roos (1933, p. 106), says no more than this: "The Econometric Society is an international society for the advancement of economic theory in its relations with statistics and mathematics. The Society will operate as a scientific organisation without political, social, financial or nationalistic bias. Its main object will be to promote studies which aim at a unification of theoretical-quantitative and empirical-quantitative approaches to economic problems".

However, a question immediately arises: what is meant by "statistics"? Frisch (1933, p. 2) responds to this in the editorial of the first issue of the Society's journal, *Econometrica*, dated January 1933, saying that it is the unification of economic statistics, economic theory and mathematics that constitutes econometrics. The term "economic statistics" is particularly precise; it clearly indicates that it does not in any way refer to inferential or mathematical statistics.

Another term must also be clarified is that of mathematics. Indeed, econometrics as it is conceived and practiced today does not use mathematics in general, but this branch of the discipline consists of axiomatised probability theory. (Note that the decisive step in the axiomatisation of probability was taken by Andrey Kolmogorov in his brochure published in German in 1933.)

Haavelmo will play a major role in the evolution of econometrics, through two publications published in 1943 and 1944. What he proposes is nothing less than

Supplementary Information The online version contains supplementary material available at https://doi.org/10.1007/978-3-031-72910-2_1.

F. J. Bismans and O. Damette, *Dynamic Econometrics*,
https://doi.org/10.1007/978-3-031-72910-2_1

establishing the econometric approach on probabilistic foundations, all making full use of the resources of mathematical statistics.

Overall, econometrics has undergone a conceptual evolution which has placed it at the intersection of economics, inferential statistics and probability theory. Such an evolution is parallel but had been anticipated, by biometrics.

In *Biometrika*, whose name clearly served as a model for the journal of the Econometric Society, Davison (2001, p. 13) describes the purpose of the latter as follows: "*Biometrika* will serve not only as a means (...) of publishing biological data (...), but also of extending the knowledge of statistical theory that may be required for their scientific treatment".

1.2 Regression Between Two Variables

We first consider the linear model linking the two variables x and y, before discussing the methods that allow it to be estimated, in particular ordinary least squares and maximum likelihood.

The Classical Linear Model

At this stage, we will consider the simplest possible model, i.e.

$$y_t = \beta_1 + \beta_2 x_t + u_t, \qquad t = 1, \ldots, T. \tag{1.1}$$

The variables y and x are said to be respectively dependent and independent (alternatively, explained and explanatory variables) or again, y is the regressand and x the regressor. For its part, u_t is an error term, a disturbance, while β_1 and β_2 are parameters and T is the number of observations in the sample. (We used the index t, because the series we will consider are time series.)

Of the five quantities present in (1.1), two are observed, namely x_t and y_t, and three (i.e. β_1, β_2 and u_t) are not; two, y_t and u_t, are random, while the other three are not. If we involve the perturbation u_t in (1.1), this is essentially because we are unable to specify all the determining factors of y_t and as a result, we model our (relative) ignorance by introducing randomness, that is to say chance or stochasticity, in the equation.

The task of the econometrician is to estimate (secondly, test) the parameters β_1 and β_2 of the regression model. However, we cannot progress in this direction without formulating some hypotheses on the unobserved random variable u_t. From this point of view, it is convenient to interpret (1.1) as a model of the expectation of y_t conditionally on x_t:

$$E(y_t|x_t) = \beta_1 + \beta_2 x_t + E(u_t|x_t).$$

If we consider that x_t is a deterministic variable, we can then simply assume that $E(u_t) = 0$. Such a hypothesis is, however, not very realistic on an applied level, because the explanatory variable x_t can turn out to be just as random as y_t.

It is therefore indicated to fall back on a weaker hypothesis, compatible with reality: that of independence between the variable x_t and the random term u_t. Formally, it translates into

$$E(u_t|x_t) = 0. \tag{1.2}$$

Formulation (1.2) allows a certain dose of stochasticity in x_t, but on the other hand, since the conditional expectation is zero, it rules out the possibility that the expectation (the mean) of u_t.

This is a so-called predeterminity hypothesis. It also implies that $E(u_t) = 0$, because

$$\begin{aligned} E(u_t) &= E_x(E(u_t|x_t)) \\ &= E_x(0) \\ &= 0. \end{aligned}$$

It follows from this hypothesis and (1.2), that the model (1.1) can then be written

$$E(y_t|x_t) = \beta_1 + \beta_2 x_t. \tag{1.3}$$

The Ordinary Least Squares (OLS)

It is now a matter of estimating the parameters β_1 and β_2, the first being the regression constant and the second the slope of the regression line. They are obtained by minimising the sum of the squares of the errors in (1.1), i.e.

$$S(\beta_1, \beta_2) = \sum_{t=1}^{T} u_t^2 = \sum_{t=1}^{T} (y_t - \beta_1 - \beta_2 x_t)^2. \tag{1.4}$$

Assuming that function (1.4) is differentiable, we obtain the first-order conditions for an extremum:

$$\begin{cases} \frac{\partial}{\partial\beta_1} S(\beta_1, \beta_2) = \frac{\partial}{\partial\beta_1}\left[\sum_{t=1}^{T} (y_t - \beta_1 - \beta_2 x_t)^2\right] = \sum_{t=1}^{T} (-2)(y_t - \beta_1 - \beta_2 x_t) = 0 \\ \frac{\partial}{\partial\beta_2} S(\beta_1, \beta_2) = \frac{\partial}{\partial\beta_2}\left[\sum_{t=1}^{T} (y_t - \beta_1 - \beta_2 x_t)^2\right] = \sum_{t=1}^{T} (-2x_t)(y_t - \beta_1 - \beta_2 x_t) = 0. \end{cases}$$

From these two conditions, we draw the following equations, called *normal equations*

$$\begin{cases} T\hat{\beta}_1 + \hat{\beta}_2 \sum_{t=1}^{T} x_t = \sum_{t=1}^{T} y_t \\ \hat{\beta}_1 \sum_{t=1}^{T} x_t + \hat{\beta}_2 \sum_{t=1}^{T} x_t^2 = \sum_{t=1}^{T} x_t y_t. \end{cases} \tag{1.5}$$

We thus obtained a system of two linear equations which can be solved by substitution. From the first of the equations of (1.5), it comes

$$\hat{\beta}_1 = \overline{y} - \hat{\beta}_2 \overline{x}, \tag{1.6}$$

where $\overline{y} = (1/T) \sum_t y_t$ and $\overline{x} = (1/T) \sum_t x_t$.

Taking into account that $\sum_t x_t = T\,\overline{x}$, we obtain by substituting (1.6) into the second normal equation

$$\hat{\beta}_2 = \frac{\sum_{t=1}^{T} x_t y_t - T\,\overline{x}\,\overline{y}}{\sum_{t=1}^{T} x_t^2 - T\,\overline{x}^2}. \tag{1.7}$$

The ordinary least squares estimators thus obtained provide the estimation residuals, denoted $\hat{u}_t$, i.e.

$$\hat{u}_t = y_t - \hat{\beta}_1 - \hat{\beta}_2 x_t,\ t = 1, \ldots, T. \tag{1.8}$$

To refine the interpretation of the regression between our two variables, it is interesting to plot on a graph the observed values of y_t and those estimated, $\hat{y}_t$, both based on the equation $\hat{y}_t = \hat{\beta}_1 + \hat{\beta}_2 x_t$ (Fig. 1.1).

If we take, for example, the point with coordinates (x_1, y_1), the actually observed value y_1 differs from the estimated value $\hat{y}_1$, so that $\hat{u}_1 = y_1 - \hat{y}_1$. The same goes for

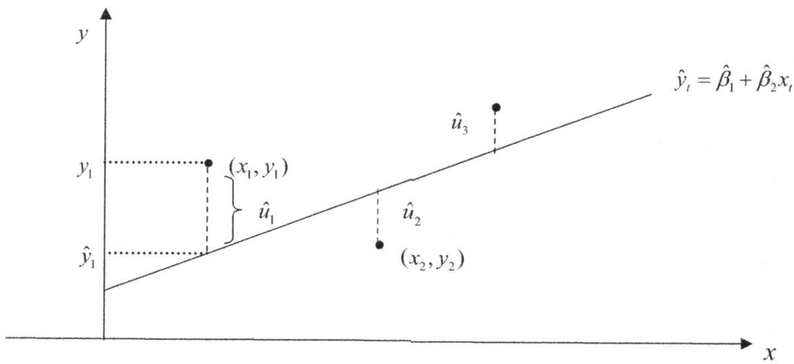

Fig. 1.1 A simple regression by OLS

the other observed values $y_2, \ldots, y_T$. In total, for all t,

$$\hat{u}_t = y_t - \hat{y}_t.$$

This graphic representation clearly illustrates the fact that the regression line $\hat{\beta}_1 + \hat{\beta}_2 x_t$ is drawn in such a way as to make the sum of the squares of the estimation residuals, that is to say $\sum_t \hat{u}_t^2$, is minimal.

Estimation by Maximum Likelihood

We return to model (1.1), namely

$$y_t = \beta_1 + \beta_2 x_t + u_t, \qquad t = 1, \ldots, T.$$

and we add an assumption relating to the distribution of disturbances u_t which are supposed to follow a Gaussian distribution:

$$u_t | \, x_t \sim N(0, \sigma^2), \quad t = 1, \ldots, T. \tag{1.9}$$

Given the normality of the errors, variable y is also normally distributed with mean $\beta_1 + \beta_2 x_t$, since

$$E(y_t | \, x_t) = E(\beta_1 + \beta_2 x_t | \, x_t) + E(u_t | \, x_t) = \beta_1 + \beta_2 x_t$$

and with variance equal to

$$\text{Var}\,(y_t | \, x_t) = \text{Var}\,(\beta_1 + \beta_2 x_t | \, x_t) + \text{Var}(u_t | \, x_t) = \sigma^2.$$

In summary, for any observation y_t, $t = 1, \ldots, T$, we have

$$y_t | \, x_t \sim N(\beta_1 + \beta_2 x_t, \, \sigma^2). \tag{1.10}$$

The likelihood function of the sample $(y_{1,\ldots,} \, y_{T})$ is then given by

$$L(\beta_1, \beta_2, \sigma^2; \, y_1, \ldots, y_T) = \prod_{t=1}^{T} \frac{1}{\sigma \sqrt{2\pi}} \, e^{-\frac{1}{2\sigma^2}(y_t - \beta_1 - \beta_2 x_t)^2}$$

$$= \left(2\pi\sigma^2\right)^{-T/2} e^{-\frac{1}{2\sigma^2}\sum\limits_{t=1}^{T}(y_t - \beta_1 - \beta_2 x_t)^2}. \tag{1.11}$$

From (1.11), we easily obtain the log-likelihood function:

$$\ln L(\beta_1, \beta_2, \sigma^2 ; \mathbf{y}) = -\frac{T}{2} \ln(2\pi) - \frac{T}{2} \ln(\sigma^2) - \frac{1}{2\sigma^2} \sum_{t=1}^{T} (y_t - \beta_1 - \beta_2 x_t)^2. \tag{1.12}$$

By calculating the partial derivatives of (1.12) with respect to each of the three parameters, we obtain the likelihood equations

$$\begin{cases} \frac{\partial \ln L}{\partial \beta_1} = \frac{1}{\sigma^2} \sum_{t=1}^{T} (y_t - \beta_1 - \beta_2 x_t) = 0 \\ \frac{\partial \ln L}{\partial \beta_2} = \frac{1}{\sigma^2} \sum_{t=1}^{T} x_t (y_t - \beta_1 - \beta_2 x_t) = 0 \\ \frac{\partial \ln L}{\partial \sigma^2} = -\frac{T}{2\sigma^2} + \frac{1}{2\sigma^4} \sum_{t=1}^{T} (y_t - \beta_1 - \beta_2 x_t)^2 = 0 \end{cases} \tag{1.13}$$

Note that the first two equations of (1.13) constitute a subsystem independent of the variance. From the first of them, we draw:

$$\hat{\beta}_1 = \frac{1}{T} \sum_{t=1}^{T} y_t - \hat{\beta}_2 \frac{1}{T} \sum_{t=1}^{T} x_t = \overline{y} - \hat{\beta}_2 \overline{x}, \tag{1.14}$$

where $\overline{y}$ and $\overline{x}$ represent respectively the empirical mean of the y_t and x_t variables.

Substituting (1.14) in the second equation of the system (1.13) yields

$$\hat{\beta}_2 \sum x_t^2 - \hat{\beta}_2 \overline{x} \sum x_t = \sum x_t y_t - \overline{y} \sum x_t,$$

which leads, given that $\sum x_t = T\overline{x}$, to

$$\hat{\beta}_2 = \frac{\sum_{t=1}^{T} x_t y_t - T\overline{x}\,\overline{y}}{\sum_{t=1}^{T} x_t^2 - T\overline{x}^2}. \tag{1.15}$$

Comparing (1.14) and (1.15) with the OLS estimators, we see that they coincide.

However, one of the advantages of the maximum likelihood method is that it allows in addition the estimation of the variance of the random term. From this point of view, it can be shown that

$$\hat{\sigma}^2 = \frac{1}{T} \sum_{t=1}^{T} \left(y_t - \hat{\beta}_1 - \hat{\beta}_2 x_t\right)^2. \tag{1.16}$$

The Method of Moments (MM)

Least squares and maximum likelihood are obviously not the only estimation procedures, even if their use is very widespread, especially in econometrics. Another procedure gives them real competition in this respect: it is the method of moments.

The method is simple in principle: it consists of equalising theoretical moments (i.e. of the population) and empirical moments (i.e. of a sample). Let's immediately apply it to the regression model (1.1)

$$y_t = \beta_1 + \beta_2 x_t + u_t, \qquad t = 1, \ldots, T.$$

The error term for observation t is given by

$$u_t = y_t - \beta_1 - \beta_2 x_t.$$

One of the hypotheses on disturbances is that their expectation is zero. Its empirical counterpart is that the mean of the sample residuals is also zero:

$$\frac{1}{T}\sum_{t=1}^{T} u_t = \frac{1}{T}\sum_{t=1}^{T} (y_t - \beta_1 - \beta_2 x_t) = 0.$$

To determine a second equation on the moments, we will appeal to the fact that it is the conditional expectation of the disturbances relating to the independent variable, which is zero. It follows therefore that

$$E(x_t u_t) = E_x(E(x_t u_t | x_t)) = E_x(x_t E(u_t | x_t)) = 0.$$

By substituting the sample mean for the expectation in the previous equation, it comes

$$\frac{1}{T}\sum_{t=1}^{T} x_t (y_t - \beta_1 - \beta_2 x_t) = 0,$$

which provides the second equation sought.

Now let's rewrite the two equations in the following form:

$$\begin{cases} T\hat{\beta}_1 + \hat{\beta}_2 \sum_{t=1}^{T} x_t = \sum_{t=1}^{T} y_t \\ \hat{\beta}_1 \sum_{t=1}^{T} x_t + \hat{\beta}_2 \sum_{t=1}^{T} x_t^2 = \sum_{t=1}^{T} x_t y_t. \end{cases}$$

On examination, this system is exactly identical to that of the normal Eq. (1.5). Accordingly, their solutions are also identical:

$$\begin{cases} \hat{\beta}_1^{MM} = \hat{\beta}_1 = \overline{y} - \hat{\beta}_2\overline{x} \\ \hat{\beta}_2^{MM} = \hat{\beta}_2 = \frac{\sum_{t=1}^{T} x_t y_t - T\overline{x}\,\overline{y}}{\sum_{t=1}^{T} x_t^2 - T\overline{x}^2}. \end{cases}$$

The three estimation methods considered therefore provide the same estimators for the two-variable regression model.

1.3 Correlation and Variance Decomposition

We first introduce the notion of correlation coefficient between the variables x and y, before showing how this coefficient is linked to the decomposition of the variance of the dependent variable y.

Pearson's Correlation Coefficient

This coefficient is of high quality because it is invariant whatever the units of measurement used. Here is what it is defined in probabilistic terms:

Definition 1.1 The correlation coefficient between the random variables X_1 and X_2, denoted by ρ_{X_1,X_2}, is defined as following

$$\rho_{X_1,X_2} = \frac{\text{Cov}(X_1, X_2)}{\sqrt{\text{Var}(X_1)\text{Var}(X_2)}}.$$

The value of ρ_{X_1,X_2} is always comprised between -1 and $+1$, knowing that the positive or negative sign is determined by the covariance, the denominator always being positive.

In the case of the linear model consisting of two variables that we consider, the covariance between y_t and x_t is given by

$$\text{Cov}(y, x) = \sum_{t=1}^{T} (x_t - \overline{x})(y_t - \overline{y})$$

and the variances by

$$\text{Var}(y) = \sum_{t=1}^{T} (y_t - \overline{y})^2 \text{ and } \text{Var}(x) = \sum_{t=1}^{T} (x_t - \overline{x})^2.$$

It follows that

$$\hat{\rho}_{y,x} = \frac{\sum_{t=1}^{T}(x_t - \overline{x})(y_t - \overline{y})}{\left[\sum_{t=1}^{T}(x_t - \overline{x})^2 \sum_{t=1}^{T}(y_t - \overline{y})^2\right]^{1/2}}. \tag{1.17}$$

Examination of (1.17) shows that the correlation coefficient between y and x is identical to the correlation coefficient between x and y:

$$\hat{\rho}_{y,x} = \hat{\rho}_{x,y}. \tag{1.18}$$

Furthermore, there is a specific relationship between $\hat{\rho}_{x,y}$ and the coefficients of the regression of y on x and the coefficients of the "inverse" regression of x on y. First, to reveal this relationship, we will need to give an alternative formulation of the regression coefficient as we derived it in (1.7) or (1.15). In this regard, we will need the two following transformations:

$$\sum_t (x_t - \overline{x})(y_t - \overline{y}) = \sum_t x_t y_t - 2T\,\overline{x}\,\overline{y} + T\,\overline{x}\,\overline{y}$$
$$= \sum_t x_t y_t - T\,\overline{x}\,\overline{y}$$

and similarly

$$\sum_t (x_t - \overline{x})^2 = \sum x_t^2 - T\overline{x}^2.$$

The coefficient of the regression of y on x can then be written in the alternative form

$$\hat{\beta}_2 = \frac{\sum_{t=1}^{T} x_t y_t - T\overline{x}\,\overline{y}}{\sum_{t=1}^{T} x_t^2 - T\overline{x}^2} = \frac{\sum_t (x_t - \overline{x})(y_t - \overline{y})}{\sum_t (x_t - \overline{x})^2}. \tag{1.19}$$

Let us then designate by $\hat{\beta}_{2\cdot y,x}$ the coefficient of the regression of y on x and by $\hat{\beta}_{2\cdot x,y}$ the coefficient of the regression of x on y. Now comes the relationship sought using (1.17)– (1.19)

$$\hat{\beta}_{2\cdot x,y}\hat{\beta}_{2\cdot y,x} = \frac{\left[\sum_{t=1}^{T}(x_t - \overline{x})(y_t - \overline{y})\right]^2}{\left[\sum_{t=1}^{T}(x_t - \overline{x})^2 \sum_{t=1}^{T}(y_t - \overline{y})^2\right]} = \hat{\rho}_{x,y}^2. \tag{1.20}$$

Several conclusions can be drawn from (1.20):

(i) If $\hat{\beta}_{2\cdot x,y} > 0$, then $\hat{\beta}_{2\cdot y,x} > 0$.

(ii) Since $\rho^2_{y,x} \leq 1$, if $\hat{\beta}_{2\cdot y,x} > 0$, then $\hat{\beta}_{2\cdot x,y} \leq 1/\hat{\beta}_{2\cdot y,x}$.

(iii) If $0 < \hat{\beta}_{2\cdot y,x} < 1$, then $\hat{\beta}_{2\cdot x,y} = \hat{\rho}^2_{x,y}/\hat{\beta}_{2\cdot y,x} > \hat{\rho}^2_{x,y}$.

Decomposition of the Variance of y

The decomposition of the variance is based on the fact that the variation of y can be decomposed into two parts, one explained by the bivariate regression and the other resulting from the variation of the estimation residuals.

From this viewpoint, we have

$$\begin{aligned}\sum (y_t - \bar{y})^2 &= \sum \left[(\hat{y}_t - \bar{y}) - (\hat{y}_t - y_t)\right]^2 \\ &= \sum (\hat{y}_t - \bar{y})^2 + \sum (y_t - \hat{y}_t)^2 - 2\sum (\hat{y}_t - \bar{y})(\hat{y}_t - y_t) \\ &= \sum (\hat{y}_t - \bar{y})^2 + \sum (y_t - \hat{y}_t)^2 + 2\sum \hat{u}_t(\hat{y}_t - \bar{y}). \end{aligned} \tag{1.21}$$

According to the normal equations grouped in (1.5), $\sum \hat{u}_t = 0$ and $\sum \hat{u}_t x_t = 0$, so that

$$\sum \hat{u}_t(\hat{y}_t - \bar{y}) = \sum \hat{u}_t(\hat{\beta}_1 + \hat{\beta}_2 x_t) - \bar{y}\sum \hat{u}_t = 0.$$

It follows that (1.21) is then written

$$\sum (y_t - \bar{y})^2 = \sum (\hat{y}_t - \bar{y})^2 + \sum (y_t - \hat{y}_t)^2, \tag{1.22}$$

which means that the total sum of squares (TSS) is equal to the addition of the explained sum of squares (ESS) and the residual sum of squares (RSS). From then on comes the practical formula:

$$TSS = ESS + RSS.$$

This decomposition of the total variation of y leads to the analysis of variance (abbreviated (ANOVA). Here is a table that shows how this analysis works for the case of a two-variable regression (Table 1.1).

Table 1.1 ANOVA for the bivariate regression

Source of variation	Sum of squares	Degrees of freedom	Mean squares
Explained by x_t	ESS	1	$ESS/1$
Residual	RSS	$T-2$	$RSS/(T-2)$
Total variation	TSS	T	TSS/T

Finally, it remains to show how the square of the correlation coefficient is related to the analysis of variance. This is the subject of the following proposition.

Proposition 1.1 The square of the correlation coefficient is equal to

$$\hat{\rho}_{x,y}^2 = 1 - \frac{RSS}{TSS} = 1 - \frac{\sum_t (y_t - \hat{y}_t)^2}{\sum_t (y_t - \overline{y})^2}.$$

(The demonstration will be left as an exercise.)

This proposition is very interesting. It will be generalised and will find its full usefulness when introducing the concept of multiple correlation coefficient when studying multivariate regression (see "Optimality Properties of Linear Estimators" section).

1.4 Properties of OLS Estimators

We will now examine two properties that the OLS estimators of x and y possess: firstly, they are unbiased, and secondly, on the other hand, their variance is minimal in the class of linear unbiased estimators.

Unbiased Estimators

Let's begin by grouping and stating the assumptions we have made so far regarding the error term in bivariate regression:

(H1) *Zero expectation*: the disturbances have zero mathematical expectations, i.e.

$$E(u_t) = 0.$$

(H2) *Uncorrelated error*: the disturbances are serially uncorrelated, i.e.

$$\forall t \neq s, \quad \text{Cov}(u_t, u_s) = 0.$$

(H3) *Homoscedasticity*: the variance of u_t conditionally to x_t is constant, i.e.

$$\forall t, s, \quad \text{Var}(u_t | x_s) = \sigma^2.$$

(H4) *Orthogonality*: conditionally to x_t, u_t has a zero expectation

$$\forall t, s, \quad E(u_t | x_t) = 0.$$

Here is now the general definition of the absence of bias for an estimator of the parameter θ.

Definition 1.2 Let us consider an unknown parameter θ to be estimated based on the random sample $(X_1, \ldots, X_T)$ of size T and $R(X_1, \ldots, X_T)$ an estimator of θ. The estimator $R(X_1, \ldots, X_T)$ is said *unbiased* if

$$E(R(X_1, \ldots, X_T)) = \theta.$$

The *bias* of the estimator, denoted b, is the number equal to

$$b(T, \theta) = E(R(X_1, \ldots, X_T)) - \theta.$$

We now need to prove that the two OLS estimators of the classical linear model are unbiased. However, we will only show it for the second of them, that is to say for the slope coefficient. Equation (1.19) allows us to write this estimator as:

$$\hat{\beta}_2 = \frac{\sum_t (x_t - \overline{x})(y_t - \overline{y})}{\sum_t (x_t - \overline{x})^2}. \tag{1.23}$$

The numerator of (1.23) can be written in the form

$$\sum_t (x_t - \overline{x})(y_t - \overline{y}) = \sum_t y_t(x_t - \overline{x}) - \sum_t y_t(x_t - \overline{x}).$$

The last sum in the right-hand side of the previous equation being zero, it comes

$$\sum_t (x_t - \overline{x})(y_t - \overline{y}) = \sum_t y_t(x_t - \overline{x}).$$

Consequently, (1.23) can be written in the form of a linear function weighted by the weights of the y_t's:

$$\hat{\beta}_2 = \sum_{t=1}^{T} w_t y_t, \tag{1.24}$$

where

$$w_t = \frac{(x_t - \overline{x})}{\sum_{t=1}^{T} (x_t - \overline{x})^2}. \tag{1.25}$$

Let us specify that these weights are fixed and that their sum is zero.

Proposition 1.2 The OLS estimator $\hat{\beta}_2$ is *unbiased*, i.e.

$$E(\hat{\beta}_2) = \beta_2.$$

Proof Taking into account the hypothesis (H4) and Eq. (1.3), we are allowed to write that

$$\begin{aligned} E(\hat{\beta}_2) &= E\left(\sum_t w_t y_t | \; x_1, \ldots, x_T\right) \\ &= \sum_t w_t E(y_t | \; x_t) \\ &= \sum_t w_t y_t (\beta_1 + \beta_2 x_t) \\ &= \beta_1 \sum_t w_t + \beta_2 \sum_t w_t y_t. \end{aligned} \tag{1.26}$$

By using (1.25), we have

$$\sum_t w_t x_t = \frac{\sum_t x_t (x_t - \overline{x})}{\sum_t (x_t - \overline{x})^2} \tag{1.27}$$

and also

$$\begin{aligned} \sum (x_t - \overline{x})^2 &= \sum_t (x_t - \overline{x})(x_t - \overline{x}) \\ &= \sum_t x_t (x_t - \overline{x}) - \overline{x} \sum_t (x_t - \overline{x}) \\ &= \sum_t x_t (x_t - \overline{x}). \end{aligned} \tag{1.28}$$

Substituting (1.28) into (1.27), we obtain

$$\sum_t w_t x_t = 1.$$

Since the sum of the weights is zero, it follows that (1.26) is equal to

$$E(\hat{\beta}_2) = \beta_2.$$

□

The Gauss–Markov Theorem

We will use (1.25) to determine the variance of the slope parameter:

$$\begin{aligned}\mathrm{Var}(\hat{\beta}_2) &= \sum_t w_t^2 \mathrm{Var}(y_t|\ x_t) \\ &= \sum_t w_t^2 \mathrm{Var}(u_t|\ x_t) \\ &= \sigma^2 \sum_t w_t^2.\end{aligned}$$

By introducing the sum of squares of (1.25) into the previous equation, it comes

$$\mathrm{Var}(\hat{\beta}_2) = \frac{\sigma^2}{\sum_{t=1}^{T} (x_t - \overline{x})^2}. \tag{1.29}$$

The Gauss–Markov theorem establishes that among all the linear unbiased estimators of β_2, the OLS estimator has the smallest variance. To this extent, the OLS estimator is said to be the best linear unbiased estimator (BLUE), which justifies the frequent use of least squares as an estimation technique.

We can now formulate the Gauss–Markov theorem in the form of a formalised proposition.

Proposition 1.3 Let be $\mathcal{L}$ the class of all linear unbiased of β_2. Then the OLS estimator is the *best linear unbiased estimator* (BLUE) in class $\mathcal{L}$, i.e.

$$\forall \tilde{\beta}_2 \in \mathcal{L}, \quad \mathrm{Var}(\tilde{\beta}_2) - \mathrm{Var}(\hat{\beta}_2) \geq 0.$$

Proof As $\tilde{\beta}_2 \in \mathcal{L}$, it is linear and unbiased. These conditions result in the fact that

$$\tilde{\beta}_2 = \sum_t \tilde{w}_t y_t,$$

where the weights are independent of y_t and satisfy

$$\sum_t \tilde{w}_t = 0 \text{ and } \sum_t \tilde{w}_t x_t = 1. \tag{1.30}$$

Let us formulate the hypothesis that the weights $\tilde{w}_t$ differ by a quantity q_t from those given in (1.25), so that

$$\tilde{w}_t = w_t + q_t, \quad t = 1, \ldots, T. \tag{1.31}$$

Given (1.30), we obtain the double condition

$$\begin{cases} \sum_t q_t = 0 \\ \sum_t q_t x_t = 0 \end{cases} \tag{1.32}$$

The variance of $\tilde{\beta}_2$ is then equal to

$$\begin{aligned} \text{Var}(\tilde{\beta}_2) &= \sum_t \tilde{w}_t^2 \text{Var}(y_t \mid x_t) \\ &= \sum_t \tilde{w}_t^2 \text{Var}(u_t \mid x_t) \\ &= \sigma^2 \sum_t \tilde{w}_t^2 \end{aligned}$$

and becomes taking account of (1.32)

$$\text{Var}(\tilde{\beta}_2) = \sigma^2 \left(\sum_t w_t^2 + 2 \sum_t w_t q_t + \sum_t q_t^2 \right). \tag{1.33}$$

Using (1.25) yields

$$\sum_t w_t q_t = \frac{\sum_t q_t (x_t - \overline{x})}{\sum_t (x_t - \overline{x})^2}. \tag{1.34}$$

The numerator of (1.34) is equal to zero, because using (1.32) again, it comes

$$\sum_t q_t (x_t - \overline{x}) = \sum_t q_t x_t - \overline{x} \sum_t q_t = 0.$$

From the previous equation, we deduce that

$$\sum_t w_t q_t = 0$$

and that consequently,

$$\text{Var}(\tilde{\beta}_2) = \sigma^2 \left(\sum_{t=1}^{T} w_t^2 + \sum_{t=1}^{T} q_t^2 \right) \geq \text{Var}(\hat{\beta}_2) = \sigma^2 \sum_{t=1}^{T} w_t^2, \tag{1.35}$$

which completes the demonstration. □

Let us specify, however, that the equality in (1.35) only occurs when $q_t = 0$ pour tout $t = 1, \ldots, T$.

The Special Case of Variance

As shown, for example, by Eq. (1.35), the variance of the OLS estimator depends on the unknown parameter σ^2. It is therefore necessary to estimate the latter. Note, however, that the hypotheses (H1), (H3) and (H4) make it possible to write the relationship

$$\mathrm{Var}(u_t | x_t) = E(u_t^2) = \sigma^2.$$

Based on this relationship, it is natural to take as the variance estimator the mean of the squares of the residuals, i.e.

$$\hat{\sigma}_{ML}^2 = \frac{\sum_{t=1}^{T} \hat{u}_t^2}{T} = \frac{\sum_{t=1}^{T} \left(y_t - \hat{\beta}_1 - \hat{\beta}_2 x_t\right)^2}{T}. \tag{1.36}$$

Note that this variance estimator is exactly the one that was obtained by maximum likelihood—see (1.16)—and that it is for this reason that we used the abbreviation ML.

The estimator (1.36), however, has the drawback of being biased. It can be shown, from this point of view, that

$$E(\hat{\sigma}_{ML}^2) = (T - 2)\sigma^2.$$

The problem is not serious when the sample size is large. On the other hand, when T is small, it is better to use the unbiased estimator given by

$$\hat{\sigma}^2 = \frac{\sum_{t=1}^{T} \left(y_t - \hat{\beta}_1 - \hat{\beta}_2 x_t\right)^2}{T - 2}, \tag{1.37}$$

where the number (− 2) appearing in the denominator designates the number of unknown parameters to be estimated.

1.5 Testing the Two-Variable Linear Model

Statistical inference essentially includes two main parts: estimation on the one hand; test theory on the other hand. It is the latter which will now interest us, but before going further, we will first present some generalities relating to hypothesis tests.

Table 1.2 Hypotheses and decision

	H_0 true	H_0 false
Rejecting H_0	Error of type I	Correct decision
Accepting H_0	Correct decision	Error of type II

Basic Concepts

Let be $(X_1, X_2, \ldots, X_T)$ a sample of size T, i.e. a set of random variables independently and identically distributed, which constitute as many observations taken from a parent population, of density $f(\cdot\,;\theta)$, where θ is a parameter such that $\theta \in \Theta \subset \mathbb{R}$.

Any assertion relating to the parameter $\theta \in \Theta$ is said to be a *statistical hypothesis.* The hypothesis being tested is called the *null hypothesis*, denoted by $H_0 : \theta \in \Theta_0 \subset \Theta$. The latter therefore induces a partition of the parameter space Θ into two subsets Θ_0 and Θ_1 such that $\Theta_0 \cap \Theta_1 = \emptyset$ and $\Theta_0 \cup \Theta_1 = \Theta$. The assertion that $\theta \in \Theta_1 \subset \Theta$ is called the *alternative hypothesis* to H_0 and denoted H_1.

Once a null hypothesis has been formulated, the whole problem boils down to either accepting or rejecting it. In doing so, there are two major risks:

1. rejecting H_0, even though it is true, which amounts to accepting it even though it is false; in this case, you commit the *error of type* I;
2. accepting H_0, even though it is false, which amounts to rejecting H_1, even though it is true; in this case, you commit an *error of type* II.

Table 1.2 summarises the four possible decisions and their results.

Formally, we adopt the following definitions for the specific probabilities of the two types of errors.

Definition 1.3 We call *probability of error of type* I, noted α, the probability of wrongly rejecting the null hypothesis and *probability of error of type* II, noted β, the probability of wrongly accepting the null hypothesis. The probability α is called the *level of significance* of the test or the *size of the critical region.* The probability β is called the *power* of the test.

A decision rule which leads, on the basis of the examination of data from a random sample, to select either the *acceptance* of the null hypothesis or its *rejection* is a test. More generally, if X denotes the set of possible sampling results, i.e. the set of values of the random variable X, we have the definition:

Definition 1.4 Consider a partition of X into two (disjoint) subsets A and R, called respectively *acceptance region* and *rejection region* of H_0. Then any decision rule such that if

$$(x_1, \ldots, x_T) \in A,$$

we accept H_0 and if

$$(x_1, \ldots, x_T) \in R,$$

we reject H_0, is said to be a test of H_0 against H_1.

Fig. 1.2, based on a standardised normal distribution, under H_0, illustrates this testing procedure

This is a right-sided test. We took α equal to 0.05, so that the ***critical region*** represents 95% of values of the distribution. The critical value is given by $z_\alpha = 1.645$. If the test statistic z is lower than 1.645, we are in the acceptance region of the null hypothesis. On the other hand, if $z > z_\alpha = 1.645$, we are in the reject region of H_0.

However, we can conduct this test in a slightly different way, which will allow us to introduce the alternative concept of p-value. Indeed, if the test statistic takes the value z_p, we can determine the empirical probability of the type I error associated with this value of the test statistic.

This empirical probability will be denoted α_{obs} and called the p-value associated with z_p. It then follows the alternative decision rule: if this p-value is greater than α, we accept H_0; if it is lower than α, we reject H_0.

Formally, we have

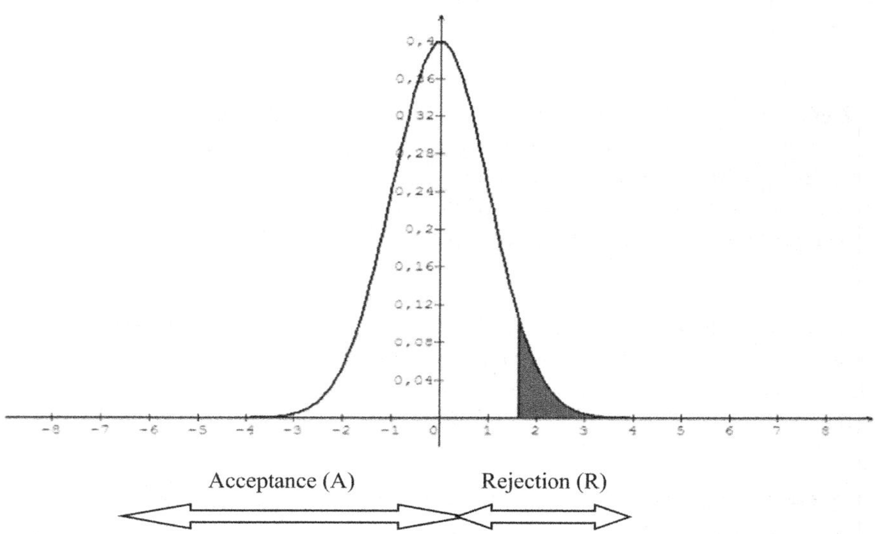

Fig. 1.2 A hypothesis test

$$\begin{cases} \alpha_{obs} > \alpha \Rightarrow z_p \in A \\ \alpha_{obs} < \alpha \Rightarrow z_p \in R. \end{cases}$$

Let us specify that the two procedures which have just been described are equivalent.

We can now, while remaining at an intuitive level, propose a method to determine the "optimal" tests. This method is based on a theorem known as the "Neyman-Pearson Lemma". To do this, consider the two simple hypotheses H_0 and H_1.

The essence of the procedure proposed by Neyman-Pearson to test $H_0 : \theta = \theta_0$ versus $H_1 : \theta = \theta_1$ lies in two main points:

(i) Give yourself the level of significance α, for example, 0.05;
(ii) There is then a so-called ***most powerful*** test of level α, such as it presents the maximum power among all tests of the same size α.

Test of Coefficients of the Regression Model

Our objective is now to test the coefficients of the model

$$\forall t, \qquad y_t = \beta_1 + \beta_2 x_t + u_t, \qquad u_t \sim N(0, \sigma^2). \tag{1.38}$$

First, we will need a reminder and a definition. So, the random variable $X = \sum_{t=1}^{T} X_t^2$, where for all t, $X_t \sim IN(0, 1)$, is distributed as a ***Chi-square*** with T degrees of freedom:

$$X \sim \chi^2(T).$$

Here is the announced definition:

Definition 1.5 Let be the independent random variables $X_1 \sim N(0, 1)$ and $X_2 \sim \chi^2(T)$. Then the random variable

$$Y = X_1/(X_2/T)$$

follows a ***Student distribution*** with m degrees of freedom:

$$Y \sim t\,(T).$$

We will now show that it is possible to implement a significance test of β_2 in the regression (1.38). The null hypothesis will be

$$H_0 : \beta_2 = \beta_0$$

against the two-side alternative hypothesis

$$\beta_2 \neq \beta_0.$$

An estimator of β_2 is given by (1.24), namely

$$\hat{\beta}_2 = \sum_t w_t y_t, \tag{1.39}$$

where the weights w_t are determined by (1.25), i.e.

$$w_t = \frac{(x_t - \overline{x})}{\sum_{t=1}^{T} (x_t - \overline{x})^2}.$$

Substituting (1.38) into (1.39) yields

$$\begin{aligned}\hat{\beta}_2 &= \sum_t w_t(\beta_1 + \beta_2 x_t + u_t) \\ &= \beta_1 \sum_t w_t + \beta_2 \sum_t w_t x_t + \sum_t w_t u_t.\end{aligned}$$

Considering—see Sect. Unbiased Estimators—that the weights are such that

$$\sum_t w_t = 0 \text{ and } \sum_t w_t x_t = 1,$$

it follows that the previous equation will be written

$$\hat{\beta}_2 = \beta_2 + \sum_t w_t u_t. \tag{1.40}$$

Since the $u_t's$ are normally distributed, it also follows that

$$\hat{\beta}_2 | \, x_t \sim N\left(\beta_2, \operatorname{Var}(\hat{\beta}_2)\right), \tag{1.41}$$

with by (1.29)

$$\operatorname{Var}(\hat{\beta}_2) = \frac{\sigma^2}{\sum_t (x_t - \overline{x})^2}. \tag{1.42}$$

Under H_0, we can obtain the corresponding test statistic by standardising (1.41), so that it comes

$$\frac{\hat{\beta}_2 - \beta_0}{\sqrt{\text{Var}(\hat{\beta}_2)}} = \frac{\hat{\beta}_2 - \beta_0}{SE(\hat{\beta}_2)} \sim N(0, 1),$$

where SE is the standard error of $\hat{\beta}_2$.

However, the variance in (1.42) is an unknown parameter. But we have shown—see Sect. The Gauss-Markov theorem—that an unbiased estimator of σ^2 was given by

$$\hat{\sigma}^2 = \frac{\sum_t \left(y_t - \hat{\beta}_1 - \hat{\beta}_2 x_t\right)^2}{T - 2}. \tag{1.43}$$

Equation (1.43) is the sum of the squared $\hat{u}_t$ divided by $(T - 2)$, so that

$$\hat{\sigma}^2 \sim \chi^2(T - 2).$$

Therefore, the statistic $t_{\hat{\beta}_2}$ follows a Student distribution with $(T - 2)$ degrees of freedom

$$t_{\hat{\beta}_2} = \frac{\hat{\beta}_2 - \beta_0}{\left(\text{Var}(\hat{\beta}_2)\right)^{1/2}} = \frac{\hat{\beta}_2 - \beta_0}{\left[\hat{\sigma}^2 \Big/ \sum_t (x_t - \overline{x})^2\right]^{1/2}} \sim t\,(T - 2), \tag{1.44}$$

since this statistic is the ratio of a distribution $N(0, 1)$ to a chi-square distribution with $(T - 2)$ degrees of freedom.

Correlation and Tests on Coefficients

We have shown that—see Eq. (1.20)—that the correlation coefficient between y and x could be estimated by

$$\hat{\rho}^2_{x,y} = \frac{\left[\sum_t (x_t - \overline{x})(y_t - \overline{y})\right]^2}{\sum_t (x_t - \overline{x}) \sum_t (y_t - \overline{y})}. \tag{1.45}$$

Furthermore, we have also established that

$$\hat{\beta}_2 = \frac{\sum_t (x_t - \overline{x})(y_t - \overline{y})}{\sum_t (x_t - \overline{x})^2} \tag{1.46}$$

and

$$\operatorname{Var}(\hat{\beta}_2) = \frac{\hat{\sigma}^2}{\sum_t (x_t - \overline{x})^2}. \tag{1.47}$$

From then on, (1.20) will be written using (1.45)

$$\begin{aligned}\hat{\rho}^2_{x,y} &= \hat{\beta}^2_2 \frac{\sum_t (x_t - \overline{x})^4}{\sum_t (x_t - \overline{x})^2 \sum_t (y_t - \overline{y})^2} \\ &= \hat{\beta}^2_2 \frac{\sum_t (x_t - \overline{x})^2}{\sum_t (y_t - \overline{y})^2}.\end{aligned} \tag{1.48}$$

If we test $H_0 : \beta_2 = 0$ against $H_1 : \beta_2 \neq 0$, the correspondent statistic $t_{\hat{\beta}_2}$—see (1.44)—will be given by

$$t_{\hat{\beta}_2} = \frac{\hat{\beta}_2}{\left[\operatorname{Var}(\hat{\beta}_2)\right]^{1/2}},$$

which, using (1.46), leads to

$$t^2_{\hat{\beta}_2} = \frac{\hat{\beta}^2_2}{\hat{\sigma}^2 \Big/ \sum_t (x_t - \overline{x})^2} = \frac{\hat{\beta}^2_2 \sum_t (x_t - \overline{x})^2}{\hat{\sigma}^2}. \tag{1.49}$$

Proposition 1.1—see Sect. Pearson's correlation coefficient—states that the square of the estimated correlation coefficient is equal to

$$\hat{\rho}^2_{x,y} = 1 - \frac{ESS}{TSS} = 1 - \frac{\sum_t (y_t - \hat{y}_t)^2}{\sum_t (y_t - \overline{y})^2}, \tag{1.50}$$

where *ESS* is the Explained Sum of Squares and *TSS* the Total Sum of Squares.

Given that by (1.43),

$$\hat{\sigma}^2 = \frac{\sum_t (y_t - \overline{y})^2}{T - 2},$$

the correlation coefficient (1.49) becomes

$$\hat{\rho}^2_{x,y} = 1 - \frac{(T-2)\hat{\sigma}^2}{\sum_t (y_t - \overline{y})^2},$$

which, after transformation, gives

$$\frac{(T-2)\hat{\sigma}^2}{\sum_t (y_t - \bar{y})^2} = 1 - \hat{\rho}^2_{x,y} \Rightarrow \hat{\sigma}^2 = \frac{\left(1 - \hat{\rho}^2_{x,y}\right) \sum_t (y_t - \bar{y})^2}{T-2}. \tag{1.51}$$

By introducing (1.50) into (1.48), we obtain

$$\begin{aligned} t^2_{\hat{\beta}_2} &= \frac{\hat{\beta}_2^2 \sum_t (x_t - \bar{x})^2}{\hat{\sigma}^2} \\ &= \frac{(T-2)\hat{\beta}_2^2 \sum_t (x_t - \bar{x})^2}{\left(1 - \hat{\rho}^2_{x,y}\right) \sum_t (y_t - \bar{y})^2} \\ &= \frac{(T-2)\hat{\rho}^2_{x,y}}{\left(1 - \hat{\rho}^2_{x,y}\right)}. \end{aligned} \tag{1.52}$$

From these results, we deduce that, in the context of a simple regression between two variables, the test of $H_0 : \hat{\rho}^2_{x,y} = 0$ against $H_1 : \hat{\rho}^2_{x,y} \neq 0$ is equivalent to that which takes as null hypothesis $\beta_2 = 0$ and as alternative hypothesis $\beta_2 \neq 0$.

1.6 Application

We will now estimate a "Keynes-style" consumption function. The Cambridge economist, in fact, presented a new approach to the said function in his *General Theory*, even if many other developments were recorded subsequently.

In the words of Keynes (1936, p. 96), "The fundamental psychological law (...) is that men are disposed (...) to increase their consumption as their income increases, but not by as much as the increase in their income. That is to say, if C is the amount of consumption and Y is income (...), dC/dY is positive and less than unity". (This derivative is precisely what Keynes called the "propensity to consume".)

The function that we will estimate will therefore look like this:

$$c_t = \beta_1 + \beta_2 y_t + u_t, \qquad u_t \sim N(0, \sigma^2),$$

while specifying that c and y represent *variations* in overall consumption and income.

The Used Data

We will use two-time series consisting of annual data on US aggregate income and consumption, 1950–1985. (These data are extracted from *The Economic Report of*

Table 1.3 The American business cycle (1950–1985)

Peak month	Trough month	Duration (peak to peak)
July 1953	May 1954	56
August 1957	April 1958	49
April 1960	February 1951	32
December 1969	November 1970	116
November 1973	March 1975	47
January 1980	July 1980	74
July 1981	November 1982	18

the President, 1986, and reprinted from William Greene, *Econometric Analysis*, 4th edition, Table A10.3.)

However, we will transform the series c_t and y_t by taking their first differences and this for reasons which will be fully explained in Chapter 2—see in particular Sect. 2.3: in short, the essential reason for working with the first differences is that both the series considered are then stationary series.

We will therefore present the respective graphs of aggregate consumption and income. But we will superimpose on each of these graphs the American recessions and their duration, so as to show, in parallel, the cyclical movement of the economy.

In the United States, it is the *Business Cycle Dating Committee* (BCDC) of the National Bureau of Economic Research (NBER) which establishes the diagram and dating of cyclical fluctuations. Table 1.3 presents the dating obtained by the BCDC for the period considered, i.e. 1950–1985.

Here is the graph of the two series coupled with the indication of the recession periods (Fig. 1.3).

The profile of the two series is very similar. Note that the longest period of expansion, even if there were fluctuations, took place during the Golden Sixties. Furthermore, the period from 1980 to 1985 is also the one which recorded two almost consecutive recessions.

Estimated Equation

We will now estimate by OLS the model

$$c_t = \beta_1 + \beta_2 y_t + u_t, \tag{1.53}$$

knowing that we have 35 data, i.e. 1951–1985.

Table 1.4 presents the results relevant to the estimation and testing of model (1.52). Two remarks can be made when reading the table:

1. The coefficients of the linear model are significantly different from zero. Indeed, if the level of significance α is equal to 0.05 and we test the null hypotheses

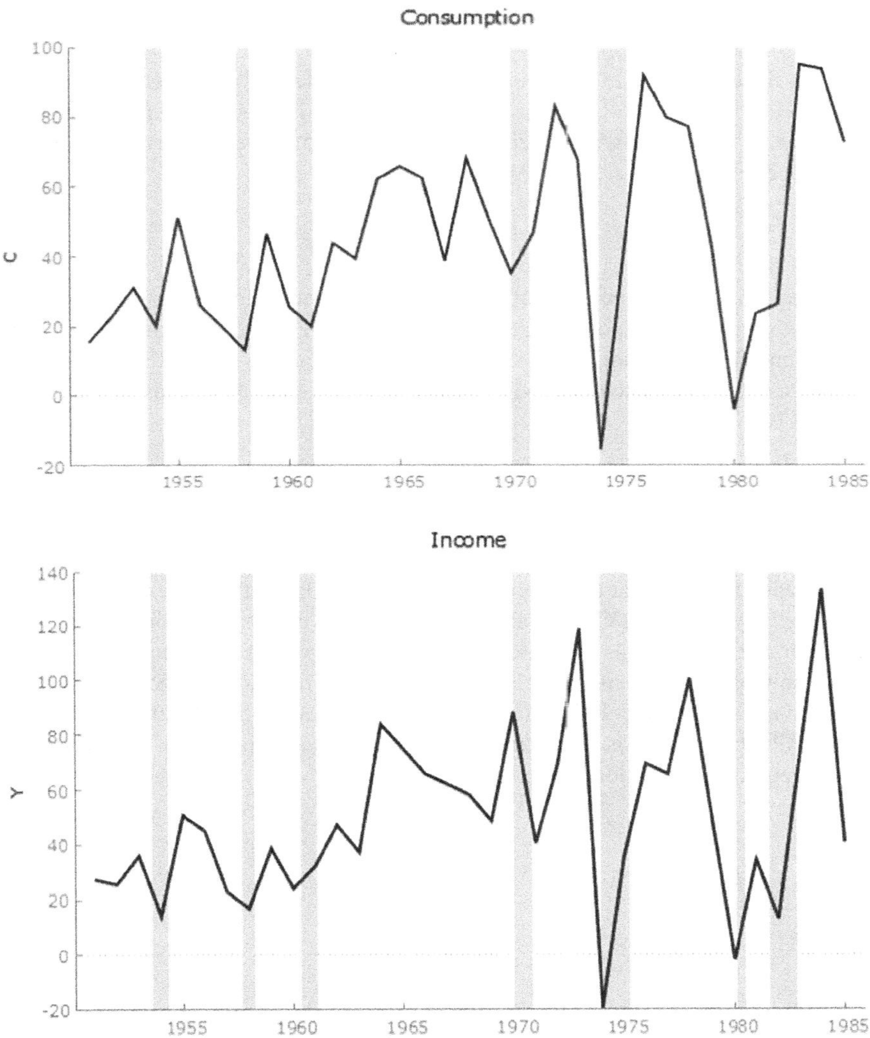

Fig. 1.3 Recessions in the US economy

Table 1.4 Estimation and significance of coefficients

	Coefficient	Standard error	t-ratio	p-value
Constant	10.642	5.095	2.089	0.0445
Income (y)	0.7028	0.087	8.049	0.0003

$$H_0 : \beta_1 = 0 \text{ and } H_0 : \beta_2 = 0,$$

then the critical value of the statistic $t_{0.95}(35) = 1.68$. Consequently, as $2.089 > 1.68$ and $8.049 > 1.68$, we reject the null hypothesis in both cases. Alternatively, since the two p-values are strictly less than $\alpha = 0.05$, we reject the null hypotheses and conclude that

$$H_1 : \beta_1 \neq 0 \text{ and } H_1 : \beta_2 \neq 0.$$

2. Regarding the interpretation of the results, it is immediately necessary to observe the existence of a positive relationship between aggregate consumption and income. The marginal propensity to consume is equal to + 0.7028, which means that a change equal to unity in income results in an increase in consumption equal to 0.7028.

The following graph shows the fitted and actual values for the regressand c_t (Fig. 1.4).

In view of the graph, the adjustment obtained with the estimated model seems relatively satisfactory, even if the deviations $(y_t - \hat{y}_t)$ appear to increase over time, which tends to show that the hypothesis of homoscedasticity of the residuals is not automatically verified. (We will see in Sect. Testing the two-variable linear model that there are several tests of homoscedasticity of the residuals variance.)

We now need to establish what is the value of the correlation coefficient associated with the estimated model. As Proposition 1.1 showed it, this coefficient is easily obtained from an analysis of variance based on the decomposition of the variance of the aggregate consumption c_t (Table 1.5).

Fig. 1.4 Actual and fitted values for consumption

Table 1.5 ANOVA

	Sum of squares	Degrees of freedom	Mean square
ESS	17,239.9	1	17,239.9
RSS	8,781.84	33	266.116
TSS	26,021.7	34	765.345

The correlation coefficient will then be given by

$$\rho_{c,y}^2 = 1 - \frac{RSS}{TSS} = 1 - \frac{8781.84}{26021.7} = 0.662519.$$

In the case that concerns us, this coefficient rounded to 0.62 can be interpreted as the proportion of the variation in consumption attributable to the linear regression on income.

A Prediction Exercise

The general problem is to obtain the *best* forecast of y_{T+1} while having the information set

$$I_t = \{y_1, \ldots, y_T; x_1, \ldots, x_T, x_{T+1}\}.$$

The solution to this problem is to take the mathematical expectation of y_{T+1} conditional on the information set I_t:

$$E(y_{T+1}|\, I_T) = E(y_{T+1}|\, x_{T+1}) = \beta_1 + \beta_2 x_{T+1}. \tag{1.54}$$

An OLS estimator of (1.53) is given by

$$\hat{y}_{T+1} = \widehat{E}(y_{T+1}|\, x_1, \ldots, x_T, x_{T+1}) = \hat{\beta}_1 + \hat{\beta}_2 x_{T+1}. \tag{1.55}$$

This is the approach to adopt to achieve an *optimal forecast* of y_{T+1}.

Adapting this procedure to our example will require carrying out the optimal forecast in two stages:

- firstly, we will estimate the linear model linking consumption to income over the period 1951–1980, i.e.

$$c_t = \beta_1 + \beta_2 x_t + u_t, \qquad t = 1, \ldots, 30. \tag{1.56}$$

- then we will predict future values of aggregate consumption for the years 1981–1985 using the estimates of the model (1.55):

$$c_{T+h} = \hat{\beta}_1 + \hat{\beta}_2 y_{T+h}, \qquad h = 1, \ldots, 5. \tag{1.57}$$

Here are the results of the estimation of the model (1.55), results which are presented in the now usual manner:

Reading the Table 1.6 calls for several remarks:

(i) the estimated coefficients are highly significant, since the p-values for the constant and even more for $\hat{\beta}_2$, are well below $\alpha = 0.05$, which leads to the rejection of the corresponding null hypotheses, that is to say, the nullity of the two estimated coefficients.

(ii) the comparison of the estimators used for forecasting with those obtained in the general model—see Table 1.4—leads to the conclusion that they are close to each other; in particular, the marginal propensity to consume is around 0.7 in both cases.

(iii) the correlation coefficient is in fact very close to 0.68, which indicates the proportion of the variation in consumption attributable to the linear regression carried out; such a value is satisfactory, even if its importance should not be exaggerated; please note that this value of the correlation coefficient is slightly higher than that obtained for the general model (0.66 in the latter case).

Table 1.7 presents the forecasts for the 5 years considered and compares them with the observed values for aggregate consumption. Also included are the 95% confidence intervals for these forecasts.

One thing that certainly stands out from this table is that the predicted values are included in the corresponding confidence intervals. For the rest, there is clearly a significant gap between forecasts and achievements.

Here is now a graphical representation of these results, which will allow, as we will see, to go further in the analysis (Fig. 1.5).

Table 1.6 Estimation of the predictive model ($T = 30$)

	Coefficient	Standard error	t-ratio	p-value
Constant	8.1868	3.5911	2.280	0.0304
Income (y)	0.7182	0.0863	8.321	0.0006
		$\hat{\rho}^2_{c,y} = 0.6799$		

Table 1.7 Forecasts (1981–1985)

	Consumption	Prediction	95% interval
1981	23.8	32.8	(2.1, 63.5)
1982	26.5	17.5	(-13.5, 48.4)
1983	95.2	60.7	(29.3, 92.1)
1984	94.0	104.3	(63.0, 139.5)
1985	72.7	37.3	(6.7, 68.0)

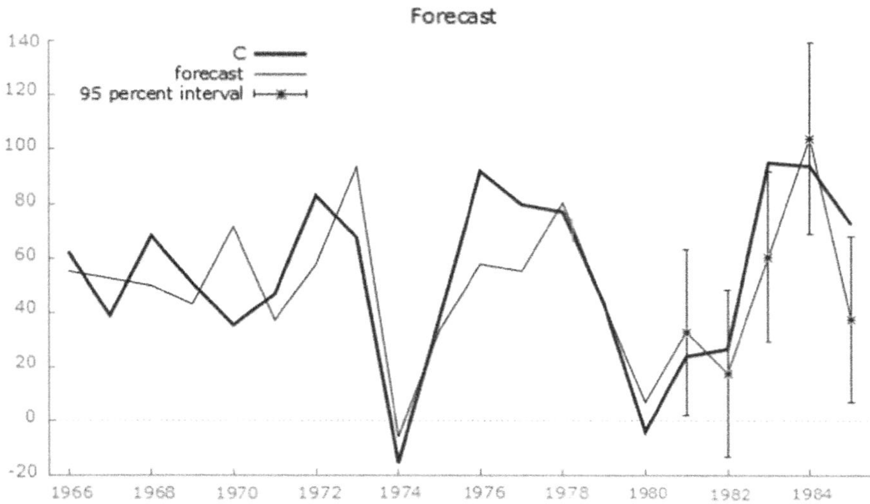

Fig. 1.5 An exercise in forecasting

The whole problem now is to evaluate the relevance and accuracy of these predictions compared to the consumption values actually observed.

From this point of view, there are several measures that allow us to assess the degree of accuracy of a forecast. To introduce them, we will adopt some writing conventions. Let y_t the value of a variable at time t and let $\hat{y}_t$ the corresponding forecast of y_t. The prediction error will be then given by

$$e_t = y_t - \hat{y}_t, \qquad t = 1, \ldots, h,$$

where h constitutes the forecast horizon.

We can now introduce two measures of forecast accuracy: the Mean Square Error (MSE) and the Root Mean Square Error ($RMSE$), defined respectively by

$$ME = h^{-1} \sum_{t=1}^{h} e_t^2 \text{ and } RMSE = \left(h^{-1} \sum_{t=1}^{h} e_t^2 \right)^{1/2}.$$

More interesting, however, is the Theil coefficient, denoted U_2 and defined as the positive square root of

$$U_2^2 = \frac{(1/h) \sum_{t=1}^{h-1} \left(\frac{\hat{y}_{t+1} - y_{t+1}}{y_t} \right)^2}{(1/h) \sum_{t=1}^{h-1} \left(\frac{y_{t+1} - y_t}{y_t} \right)^2}. \tag{1.58}$$

We can interpret the numerator of (1.57) as being the *RMSE* of the forecasts associated with the selected model. In case of perfect forecast, i.e. for all t, $\hat{y}_{t+1} = y_{t+1}$, Theil's coefficient takes the value 0.

On the other hand, if the forecast is "naïve", that is to say if simply, in the denominator, $y_{t+1} = y_t$, for all t, numerator and denominator in (1.57) are equal and consequently, the coefficient U_2 takes a value equal to the unity. The values above (resp. below) 1 indicate a prediction which is lesser (resp. better) than the naïve approach.

Regarding our forecast of aggregate consumption, Theil's coefficient is equal to

$$U_2 = 0.54211,$$

so we can conclude that our forecast is more accurate than the naïve approach.

1.7 Exercises

1.7.1 Let be the model of regression

$$y_t = \beta_1 + \beta_2 x_t + u_t, \qquad t = 1, \ldots, T.$$

We ask:

(i) to establish what are the second-order conditions on the parameters which make it possible to minimise the sum of the squares of the residuals $\hat{u}_t^2$;

(ii) to show that if $u_t \sim N(0, \sigma^2)$, the maximum likelihood estimator of σ^2 is given by

$$\hat{\sigma}_{ML}^2 = (1/T) \sum_{t=1}^{T} (y_t - \hat{\beta}_1 - \hat{\beta}_2 x_t)^2,$$

where $\hat{\beta}_1$ and $\hat{\beta}_2$ are the maximum likelihood estimators of the parameters of regression model;

(iii) to prove that $\hat{\sigma}_{ML}^2$ is a biased estimator of σ^2.

1.7.2 In the model of the simple regression, but without constant,

$$y_t = \beta x_t + u_t, \qquad t = 1, \ldots, T,$$

where the disturbances are $IID(0, \sigma^2)$, prove that the OLS estimator of β is given by

$$\hat{\beta}_{OLS} = \frac{\mathrm{Cov}(y_t, x_t)}{\mathrm{Var}(x_t)}.$$

1.7.3 Suppose a random sample $(X_1, \ldots, X_T)$ composed of variables all taken from a population normally distributed of parameters μ and σ^2. We ask to find by the method of moments the estimators of μ and σ^2.

1.7.4 Prove proposition 1.1, namely that the square of the correlation coefficient is equal to

$$\rho^2_{x,y} = 1 - \frac{RSS}{TSS}.$$

1.7.5 If $X_1, \ldots, X_T$ are T independent observations on X, normally distributed with density equal to

$$f_X(x\,;\,\mu) = \frac{1}{2\pi} e^{-\frac{(x-\mu)^2}{2}}, \qquad -\infty < x < +\infty,$$

determine the best linear unbiased estimator (BLUE) of μ.

(Indication: use $E(X^2) = 1 + \mu^2$.)

1.7.6 Let us assume a normal population of unknown expectation and variance equal to 25. We think, with good reason, that this average is equal to 30, but we cannot completely rule out the possibility that it is equal to 28.

We ask to

(i) formulate the hypothesis test corresponding to the problem posed;
(ii) determine the size of the sample and the decision rule which will make it possible to make a choice between the two hypotheses while knowing that the probabilities of type I and type II errors cannot exceed respectively 0.01 and 0.05.

1.7.7 Consider a population distributed according to a normal law with mean μ and standard deviation equal to 5. We wish to test: $H_0 : \mu = 22$ against $H_1 : \mu = 25$. We also have a sample of size 25, taken from this population and we apply the following decision rule:

$$\text{Accept } H_0 \text{ if } \overline{X} \leq 24 \text{ and reject it if } \overline{X} > 24,$$

where $\overline{X}$ is the sample mean.

Compute the appropriate level of significance α and power β.

1.7.8 Using annual data on US aggregate consumption and income (1950–1985)—see in this chapter, Sect. 6.1 -, estimate the parameters and the correlation coefficient of the regression

$$c_t = \beta_1 + \beta_2 x_{t-1} + u_t, \qquad t = 2, \ldots, 34,$$

and show that the obtained equation is not statistically satisfactory.

1.7.9 Let the estimated model

$$c_t = \underset{(2.3)}{8.19} + \underset{(8.3)}{0.72}\, y_t,$$

where the numbers in parentheses are the values of the two t-ratios.

The analysis of variance table is given by

	Sum of squares	Degrees of freedom	Mean square
ESS	13,132.3	1	13,132
RSS	6181.42	28	220.77
TSS	19,313.7	29	695.99

We ask, on the basis of the elements appearing in this table, to determine the estimated value of the correlation coefficient $\hat{\rho}^2_{c,y}$. How would you interpret this result?

1.7.10 Consider the two forecasts F1 and F2, whose characteristic statistics (*RMSE* and U_2) are as follows:

Statistic	F1	F2
RMSE	0.9810	2.0000
U_2	0.4208	1.2200

We are asked to evaluate each of these forecasts in relation to the available statistics, and then to compare them.

Reference

Davison, A.C. (2001), Biometrika centenary: Theory and general methodology, *Biometrika, 88*, 13–52.

Frisch, R. (1933), Propagation problems and impulse problems in dynamic economics. In *Economic essays in honour of Gustav Cassel* (pp.171–205). George Allen & Unwin.

Keynes, J.M. (1936), *The general theory of employment, interest and money, in Collected Writings of John Maynard Keynes*, Vol VII, Macmillan.

Roos, C.F. (1933), Constitution of the Econometric Society, *Econometrica, 1*, 106–108.

Chapter 2
Dynamics in Econometrics

2.1 The Multiple Regression Model

The linear model of multiple regression—with the restriction that it only includes a single equation—can be formulated as follows:

$$y_i = \beta_1 + \beta_2 x_{i2} + \cdots + \beta_k x_{ik} + u_i, \qquad i = 1, \ldots, n, \tag{2.1}$$

where y_i is the dependent variable on which we have n observations, the x_{ij} are the independent or explanatory variables and u_i is a random variable with the appropriate properties.

Equation (2.1) can easily be rewritten in the form:

$$y_i = (1, x_{i2}, \ldots, x_{ik}) \begin{pmatrix} \beta_1 \\ \beta_2 \\ \vdots \\ \beta_k \end{pmatrix} + u_i = \boldsymbol{x}' \boldsymbol{\beta} + u_i, \qquad i = 1, \ldots, n. \tag{2.2}$$

(Conventionally, a row vector will be denoted in bold and will be associated with the prime ($'$), while a column vector will simply be designated by a lowercase letter, also in bold.)

However, the model formulated in (2.1) or in (2.2) is not dynamic, the chronological dimension is not even present. From this point of view, we can initially date the variables of the model and introduce lagged explanatory variables, which gives

---

Supplementary Information The online version contains supplementary material available at https://doi.org/10.1007/978-3-031-72910-2_2.

F. J. Bismans and O. Damette, *Dynamic Econometrics*, https://doi.org/10.1007/978-3-031-72910-2_2

$$y_t = m + \beta_{10}x_{t1} + \beta_{11}x_{t-1,1} + \cdots + \beta_{1k_1}x_{t-k_1,1} + \beta_{20}x_{t1} + \beta_{22}x_{t-1,2} + \cdots + \beta_{2k_2}x_{t-k_2,2}$$
$$+ \cdots + \beta_{K0}x_{tm} + \beta_{K1}x_{t-1,m} + \cdots + \beta_{Kk_K}x_{t-k_K,K} + u_t, \quad t = 1, \ldots, T.$$

Using a double summation, the previous equation becomes

$$y_t = m + \sum_{j=0}^{k_i} \sum_{i=1}^{K} \beta_{ij} x_{t-j,i} + u_t, \quad t = 1, \ldots, T. \tag{2.3}$$

A further step in the dynamisation of the model will consist of adding p lagged dependent variables to (2.3), that is

$$y_t = m + \sum_{r=1}^{p} \alpha_r y_{t-r} + \sum_{j=0}^{k_i} \sum_{i=1}^{K} \beta_{ij} x_{t-j,i} + u_t, \quad t = 1, \ldots, T. \tag{2.4}$$

2.2 Models of Stationary Time Series

Time series models comprise several classes, including those known as ARMA or even those of purely financial models of the GARCH type.

Some Preliminary Results

Consider the **discrete-time stochastic process** $(X_t, 1, \ldots, T)$. Informally, one could define such a process as "a random variable immersed in time". A realisation of this process, denoted $(x_t, t = 1, \ldots, T)$, is called a **time series.**

The **autocovariance of order** k of the series $(x_t, t = 1, \ldots, T)$, which we will denote by γ_k, is the covariance between x_t and x_{t-k} :

$$\gamma_k = \text{Cov}(x_t, x_{t-k}) = E\big[(x_t - \mu_t)(x_{t-k} - \mu_{t-k})\big],$$

where μ_t is the mean of the series at t and $k = 0, \pm 1, \pm 2, \ldots$. (Of course, if $k = 0$, we then obtain the variance.) The **autocovariance function** is an application from $\mathbb{Z}$ to $\mathbb{R}$. As it is even, that is to say that for all k, $\gamma_k = \gamma_{-k}$, its graph is limited to non-negative values of k.

From the autocovariances, we introduce the notion of autocorrelation, which is obtained by dividing each autocovariance of the considered series by the corresponding variance. The **autocorrelation of order** k of $(x_t, t = 1, \ldots, T)$ is thus given by

$$\rho_k = \frac{\gamma_k}{\gamma_0}, \qquad k = 0, \pm 1, \pm 2, \ldots.$$

We can then define the autocorrelation function in parallel to the autocovariance function:

Definition 2.1 We call partial autocorrelation function the mapping ρ from $\mathbb{Z}$ to $\mathbb{R}$ such that

$$\rho_k = \frac{\gamma_k}{\gamma_0}, \qquad k \in \mathbb{Z}.$$

Just like the autocovariance function, the function ρ_k is even. However, it differs in that $\rho_k = \gamma_0/\gamma_0 = 1$. This function is graphically represented by limiting to non-negative values of k, taking into account its parity. It is then called **correlogram.**

Let's now introduce the concept of stationarity of a process or a discrete series.

Definition 2.2 A time series $(x_t,\ t \in \mathbb{Z})$ is said to be weakly stationary or second-order stationary if for all $t \neq s$ and arbitrary k

(i) $E(x_t) = E(x_s) = \mu < \infty$;
(ii) $\text{Var}(x_t) = \text{Var}(x_s) = \sigma^2 < +\infty$;
(iii) $\text{Cov}(x_t, x_{t-k}) = \text{Cov}(x_s, x_{s-k})$.

Example 2.1 Consider the **Gaussian white noise** process defined by $(\varepsilon_t) \sim IIN\left(0, \sigma^2\right)$, where for all t, the random variables are independently and identically normally distributed. This process is second-order stationary, since

(i) $\forall t,\ E(\varepsilon_t) = 0$;
(ii) $\forall t,\ \text{Var}(\varepsilon_t) = \sigma^2$;
(iii) $\forall t \neq s$ and $k \neq 0$, $\text{Cov}(\varepsilon_t, \varepsilon_{t+k}) = \text{Cov}(\varepsilon_s, \varepsilon_{s+k})$.

This last equality is verified given that the ε_t are independently distributed.

Let's now provide some details about the **lag operator** L defined by

$$L^j x_t = x_{t-j}, \quad j = 0, 1, \ldots.$$

Formally, L can be treated as a variable, which allows to write a polynomial of the p-th degree in L, denoted $\phi(L)$ and called **lag polynomial**, in the form

$$\phi(L) = \phi_0 + \phi_1 L + \phi_2 L^2 + \cdots + \phi_p L^p. \tag{2.5}$$

By multiplying both sides of (2.5) by x_t, we then get

$$\begin{aligned}\phi(L)x_t &= \left(\phi_0 + \phi_1 L + \phi_2 L^2 + \cdots + \phi_p L^p\right)x_t \\ &= \phi_0 x_t + \phi_1 x_{t-1} + \phi_2 x_{t-2} + \cdots + \phi_p x_{t-p} \\ &= \sum_{j=0}^{p} \phi_j x_{t-j}.\end{aligned}$$

As we will have the opportunity to see later, the lag operator allows for the multiplication and division of time series. In particular, in the following section, we will use the following results:

Proposition 1.1 *If* $|\phi| < 1$, *then*

$$(1 + \phi L + \phi^2 L^2 + \phi^3 L^3 + \cdots)x_t = (1 - \phi)^{-1} x_t. \tag{2.6}$$

Proof By multiplying both sides of (2.6) by $(1 - \phi L)$, we get.

$$(1 - \phi)\left(1 + \phi L + \phi^2 L^2 + \phi^3 L^3 + \cdots\right)x_t = x_t,$$

which gives after realising the product in the first side

$$(1 - \phi L + \phi L - \phi^2 L^2 + \phi^2 L^2 - \phi^3 L^3 + \cdots)x_t = \lim_{n \to +\infty} (1 - \phi^n L^n)x_t.$$

Given that $|\phi| < 1$, $\phi^n L^n$ tends to 0 when $n \to +\infty$, and therefore (2.6) is verified.

Autoregressive Processes

Let's start with autoregressive models, known as AR. (This acronym is the abbreviation of "AutoRegressive".)

Definition 2.3. An autoregressive process of order p, denoted by AR(p), is written

$$y_t = \phi_1 y_{t-1} + \cdots + \phi_p y_{t-p} + \varepsilon_t, \qquad t = 1, \ldots, T, \tag{2.7}$$

where $\varepsilon_t \sim WN(0, \sigma^2)$ is a white noise with variance σ^2. We then note

$$y_t \sim \text{AR}(p).$$

Consider, for illustration, the AR(1) model

$$y_t = \phi_1 y_{t-1} + \varepsilon_t, \qquad t = 1, \ldots, T. \tag{2.8}$$

It is easily verified that this process is stationary. Using the lag operator, it comes.

$$
\begin{aligned}
y_t &= \phi L y_t + \varepsilon_t \\
(1 - \phi L) y_t &= \varepsilon_t .
\end{aligned}
$$

Proposition 2.1 then allows us to write

$$
\begin{aligned}
y_t &= (1 - \phi L)^{-1} \varepsilon_t = \left(1 + \phi L + \phi^2 L^2 + \cdots\right) \varepsilon_t \\
&= \varepsilon_t + \phi\, \varepsilon_{t-1} + \phi^2\, \varepsilon_{t-2} + \cdots
\end{aligned}
\tag{2.9}
$$

If $|\phi| < 1$, the sequence $(\phi, \phi^2, \ldots)$ converges. It follows that the **stationarity condition** of an AR(1) is

$$|\phi| < 1.$$

Alternatively, this condition is equivalent to the solution of the equation in the lag operator.

$$1 - \phi L = 0$$

being greater than one in **absolute value**, given that if $|\phi| < 1$,

$$L = \frac{1}{\phi} > 1.$$

As an illustration, let's see how the autocorrelation function of an AR(1) behaves. First, let's multiply (2.8) by y_{t-k} while taking the expectations of both sides:

$$E(y_t y_{t-k}) - \phi E(y_{t-1} y_{t-k}) = E(y_{t-k} \varepsilon_t). \tag{2.10}$$

If $k > 0$, using (2.9), we obtain

$$
\begin{aligned}
E(y_{t-k}\varepsilon_t) &= E\left[(\varepsilon_{t-k} + \phi \varepsilon_{t-k-1} + \phi^2 \varepsilon_{t-k-2} + \cdots) \varepsilon_t\right] \\
&= E(\varepsilon_{t-k}\varepsilon_t) + \phi E(\varepsilon_{t-k-1}\varepsilon_t) + \phi^2 E(\varepsilon_{t-k-2}\varepsilon_t) + \cdots \\
&= 0.
\end{aligned}
$$

Therefore, still for $k > 0$, by introducing this last result into (2.10), we finally get

$$\gamma_k - \phi\, \gamma_{k-1} = 0. \tag{2.11}$$

Divide (2.11) by γ_0, which leads to the first-order difference equation (without constant term):

$$\rho_k = \phi\, \rho_{k-1}. \tag{2.12}$$

The solution to this last equation is easily obtained by successive iterations:

$$\begin{aligned}\rho_k &= \phi(\phi\,\rho_{k-2}) = \phi^2 \rho_{k-2}\\ &= \phi^2(\phi\,\rho_{k-3}) = \phi^3 \rho_{k-3}\\ &\vdots\\ &= \phi^{k-2}(\phi\rho_1) = \phi^{k-1}\rho_1\\ &= \phi^k \rho_0\\ &= \phi^k \quad (\rho_0 = \gamma_0/\gamma_0 = 1).\end{aligned} \tag{2.13}$$

In light of (2.13), given that $|\phi| < 1$, we will have the following proposition: *The autocorrelation function* (ACF) *of an* AR(1) *process decreases exponentially to 0 if* $\phi > 0$. *It also decreases to zero, but oscillating, if* $\phi < 0$.

This set of results generalises to the AR(p) process as defined by (2.7). First, this autoregressive model can be written using the lag operator:

$$(1 - \phi L - \phi^2 L^2 - \cdots - \phi^p L^p)y_t = \varepsilon_t,$$

from which we can extract the polynomial equation

$$1 - \phi L - \phi^2 L^2 - \cdots - \phi^p L^p = 0. \tag{2.14}$$

In the case of an AR(1) model, the **stationarity condition** was that the parameter ϕ, solution of the first-degree equation in L, is greater than one in absolute value. In light of (2.14), this condition implies that its p roots—some of which may be confused or complex—are **greater than one in modulus**. In modulus and not in absolute value, because, let's repeat, solutions can present themselves in the form of conjugate complexes. (It is also said, to express this same condition, that the roots of (2.14) are **outside the unit circle**.)

The autocorrelation function, the analogue of (2.12) for an AR(p) model, is given by the difference equation

$$\rho_k = \phi_1 \rho_{k-1} + \phi_2 \rho_{k-2} + \cdots + \phi_p \rho_{k-p}, \qquad k > 0. \tag{2.15}$$

The roots of the characteristic equation corresponding to (2.15)

$$m^p - \phi_1 m^{p-1} - \cdots - \phi_{p-1} m - \phi_p = 0$$

are either real or complex. In the first case, if these roots are also less than one in absolute value, then the solutions of (2.15) consist of damped exponentials. In the second case, if the conjugate complex roots are less than one in modulus, then these solutions are damped sinusoidal oscillations. Consequently, it follows that the FAC is here a combination of damped exponentials and also damped oscillations.

ARMA and MA Models

The ARMA models consist of two parts: an autoregressive part and a moving average part designated by the acronym MA, this being an abbreviation of "Moving Average". We will start with the study of this latter class of models.

Definition 2.4 We call a **moving average process of order q** a process $(y_t,\quad t \in \mathbb{Z})$ such that

$$y_t = \varepsilon_t + \theta_1 \varepsilon_{t-1} + \theta_2 \varepsilon_{t-2} + \cdots + \theta_q \varepsilon_{t-q}, \quad \forall t \in \mathbb{Z}, \tag{2.16}$$

where for all i, θ_i is a real number and (ε_t) is a white noise process with zero mean and constant variance σ^2.

We then note

$$y_t \sim \text{MA}(q).$$

The MA(1) process is written as follows

$$y_t = \varepsilon_t + \theta \varepsilon_{t-1}, \quad \varepsilon_t \sim bb(0, \sigma^2). \tag{2.17}$$

It is always stationary in view of the fact that (ε_t) is a stationary white noise. Using the lag operator, (2.17) becomes

$$\begin{aligned} y_t &= \varepsilon_t - \theta\, \varepsilon_{t-1} \\ &= (1 - \theta L)\varepsilon_t. \end{aligned}$$

If $|\theta| < 1$, so

$$(1 - \theta L)^{-1} y_t = (1 + \theta L + \theta^2 L^2 + \cdots) y_t = \varepsilon_t. \tag{2.18}$$

In other words, to obtain an autoregressive (infinite) representation of the process, the restriction $|\theta| < 1$ must be imposed. Such a restriction is known as the **invertibility condition** of the MA(1).

Let's see how to determine the corresponding autocorrelation function. The variance, γ_0 in this case, is given by

$$\begin{aligned}\gamma_0 = E\left[y_t^2\right] &= E\left[(\varepsilon_t - \theta\varepsilon_{t-1})(\varepsilon_t - \theta\varepsilon_{t-1})\right] \\ &= E(\varepsilon_t^2) + \theta^2 E(\varepsilon_{t-1}^2) \\ &= (1+\theta^2)\sigma^2.\end{aligned}$$

The first-order autocovariance is equal to

$$\begin{aligned}\gamma_1 = E(y_t y_{t-1}) &= E\left[(\varepsilon_t - \theta\varepsilon_{t-1})(\varepsilon_{t-1} - \theta\varepsilon_{t-2})\right] \\ &= -\theta E(\varepsilon_{t-1}^2) \\ &= -\theta\sigma^2.\end{aligned}$$

Finally, obviously, all autocovariances of order higher than one are zero. It follows that the autocorrelation function is determined by

$$\begin{aligned}\rho_0 &= 1 \\ \rho_1 &= \gamma_1/\gamma_0 = (-\theta\sigma^2)/(1+\theta^2)\sigma^2 = (-\theta)/(1+\theta^2) \\ \rho_k &= 0, \qquad \text{for all } k > 1.\end{aligned}$$

These results generalise without too much trouble to the process MA(q), which is always stationary. Using the lag operator, (2.17) is written as:

$$y_t = \varepsilon_t + \theta_1 L\varepsilon_t + \theta_2 L^2\varepsilon_t + \cdots + \theta_q L^q\varepsilon_t = (1 + \theta_1 L + \theta_2 L^2 + \cdots + \theta_q L^q)\,\varepsilon_t. \tag{2.19}$$

The **invertibility condition** requires that the roots of the polynomial equation associated with (2.19)

$$1 + \theta_1 L + \theta_2 L^2 + \cdots + \theta_q L^q = 0$$

are **outside the unit circle** (or, if you prefer, that they are greater than one in modulus or absolute value).

The variance of the MA(q) process is equal to

$$\begin{aligned}\gamma_0 = E(y_t^2) &= E(\varepsilon_t + \theta_1\varepsilon_{t-1} + \cdots + \theta_q\varepsilon_{t-q})^2 \\ &= \sigma^2 + \theta_1^2\sigma^2 + \cdots + \theta_q^2\sigma^2 \quad \text{(double products cancellation)} \\ &= (1 + \theta_1^2 + \cdots + \theta_q^2)\sigma^2. \end{aligned} \tag{2.20}$$

For $k = 1, 2, \ldots, q$, we have

$$\begin{aligned}\gamma_k = E(y_t y_{t-k}) &= E\left[(\varepsilon_t + \theta_1\varepsilon_{t-1} + \cdots + \theta_q\varepsilon_{t-q})(\varepsilon_{t-k} + \theta_1\varepsilon_{t-k-1} + \cdots + \theta_q\varepsilon_{t-k-q})\right] \\ &= E(\theta_k\varepsilon_{t-k}^2 + \theta_{k+1}\theta_1\varepsilon_{t-k-1}^2 + \cdots + \theta_q\theta_{q-k}\varepsilon_{t-q}^2) \\ &= (\theta_k + \theta_{k+1}\theta_1 + \cdots + \theta_q\theta_{q-k})\sigma^2.\end{aligned}$$

If $k > q$, there is no ε of the same date that intervenes in the calculation of γ_k and consequently, the expectations are null.

Ultimately, the corresponding autocovariances are:

$$\gamma_k = \begin{cases} (\theta_k + \theta_{k+1}\theta_1 + \cdots + \theta_q\theta_{q-k})\sigma^2 & k = 1, 2, \ldots, q \\ 0 & k > q. \end{cases} \tag{2.21}$$

It is then sufficient to divide (2.20) and (2.21) by γ_0 to obtain the autocorrelation function.

Example 2.2 Consider the MA(2) process

$$y_t = \varepsilon_t + \theta_1\varepsilon_{t-1} + \theta_2\varepsilon_{t-2}.$$

The autocorrelations are given by

$$\begin{aligned} \rho_0 &= 1 \\ \rho_1 &= (\theta_1 + \theta_2\theta_1)/(1 + \theta_1^2 + \theta_2^2) \\ \rho_2 &= \theta_2/(1 + \theta_1^2 + \theta_2^2) \\ \rho_3 &= \rho_4 = \cdots = 0. \end{aligned}$$

Autoregressive Moving Average Processes

The ARMA modelling generalises in a way the autoregressive and moving average models by combining them. Their main advantage over their two competitors is to be parsimonious in the number of parameters to estimate.

Definition 2.5 We call an **autoregressive moving average** process of order (p, q) a process $(y_t,\ \ t \in \mathbb{Z})$ such that.

$$y_t = \phi_1 y_{t-1} + \cdots + \phi_p y_{t-p} + \varepsilon_t + \theta_1\varepsilon_{t-1} + \cdots + \theta_q\varepsilon_{t-q}, \quad \forall t \in \mathbb{Z}, \tag{2.22}$$

where (ε_t) is a white noise process with zero mean and constant variance σ^2.

We then note

$$y_t \sim \text{ARMA}(p, q).$$

An ARMA process can still be written compactly using the lag operator. Indeed, if we define the two polynomials associated with (2.22) as follows:

$$\begin{cases} \phi(L) = 1 - \phi_1 L - \cdots - \phi_p L^p \\ \theta(L) = 1 + \theta_1 L + \cdots + \theta_q L^q, \end{cases} \tag{2.23}$$

it then follows

$$\phi(L)y_t = \theta(L)\varepsilon_t. \tag{2.24}$$

As the MA part of the model is always stationary, the stationarity of (2.24) depends entirely on that of its autoregressive part. It is therefore sufficient to use the results already obtained, so that the stationarity condition requires that the roots of the equation $\phi(L) = 0$ are **outside the unit circle**. Similarly, the invertibility condition of (2.24) is identical to that of the MA(q) process: the roots of the equation $\theta(L) = 0$ must be **outside the unit circle**.

As for the autocorrelation function of the ARMA(p, q) model, it combines the properties of the corresponding functions of the autoregressive and moving average parts of this model. To show this, let's consider the ARMA(1,1) process:

$$y_t - \phi y_{t-1} = \varepsilon_t + \theta \varepsilon_{t-1}. \tag{2.25}$$

Multiply (2.25) by y_{t-k} and then take the expectations of both sides: for all $k = 0, 1, 2, \ldots,$

$$\begin{aligned} \gamma_k &= E(y_t y_{t-k}) = E(\phi y_{t-1} y_{t-k} + \varepsilon_t y_{t-k} + \theta \varepsilon_{t-1} y_{t-k}) \\ &= \phi\, \gamma_{k-1} + E(\varepsilon_t y_{t-k}) + \theta E(\varepsilon_{t-1} y_{t-k}) \end{aligned} \tag{2.26}$$

The two expectations in (2.26) are zero for $k > 1$. If $k = 1$, the first one is null, while the second one is equal to

$$E(\varepsilon_{t-1} y_{t-1}) = E[\varepsilon_{t-1}(\phi y_{t-2} + \varepsilon_{t-1} + \theta\, \varepsilon_{t-2})] = \sigma^2. \tag{2.27}$$

Finally, when $k = 0$, the two expectations are given by

$$\begin{cases} E(\varepsilon_t y_t) = E[\varepsilon_t(\phi y_{t-1} + \varepsilon_t + \theta\, \varepsilon_{t-1})] = \sigma^2 \\ E(\varepsilon_{t-1} y_t) = E[\varepsilon_{t-1}(\phi y_{t-1} + \varepsilon_t + \theta\, \varepsilon_{t-1})] = \phi \sigma^2 + \theta \sigma^2. \end{cases} \tag{2.28}$$

Therefore, using (2.26)– (2.28), the autocovariance function is written as:

$$\begin{aligned} \gamma_0 &= \phi\, \gamma_1 + \sigma^2 + \theta\phi\, \sigma^2 + \theta^2 \sigma^2 \\ \gamma_1 &= \phi\, \gamma_0 + \theta\, \sigma^2 \\ \gamma_k &= \phi\, \gamma_{k-1}, \qquad k = 2, 3, \ldots. \end{aligned} \tag{2.29}$$

If we substitute γ_1 into the first of the previous equations and vice versa, we obtain:

$$\gamma_0 = \frac{1 + \theta^2 + 2\phi\theta}{1 - \phi^2}\sigma^2 \tag{2.30}$$

and then,

$$\gamma_1 = \frac{(\phi + \theta)(1 + \phi\theta)}{1 - \phi^2}\sigma^2. \tag{2.31}$$

Knowing that ρ_0 is always equal to one, it comes by dividing (2.31) and the third equation of (2.29) by γ_0:

$$\begin{cases} \rho_1 = \frac{(\phi+\theta)(1+\phi\theta)}{1+\theta^2+2\phi\theta} \\ \rho_k = \phi\rho_{k-1}, \quad k = 2, 3, \ldots. \end{cases} \tag{2.32}$$

It follows from (2.32) that for all $k > 1$, the behavior of the autocorrelation function is entirely governed by a first-order recurrence equation. If $\phi < 0$, the autocorrelations will show damped oscillations. On the other hand, if this parameter ϕ is positive, they will decrease exponentially.

Example 2.3 We consider the ARMA(1,1) process, where $\phi = 0,4$ and $\theta = 0,8$. The successive autocorrelations will be given by

$$\begin{aligned} &\rho_0 = 1, \ \rho_1 = 0,695 \\ &\rho_2 = 0,278, \ \rho_3 = 0,111, \ \rho_4 = 0,044, \quad \text{etc.} \end{aligned}$$

Of course, the autocorrelation function for the general ARMA(p, q) model presents a structure similar to the one just described: the autocorrelations up to order q depend on the autoregressive and moving average parameters, while those of a higher order result from a recurrence equation of order p determined exclusively by the parameter ϕ.

Modelling Volatility

It is well known that some time series, especially financial ones, have a variance that is not constant over time. This is why specific models of volatility variation have been developed.

Let's start by considering the autoregressive model of order one satisfying the usual stationarity condition

$$y_t = \phi y_{t-1} + \varepsilon_t, \quad |\phi| < 1, \tag{2.33}$$

while assuming that the disturbances admit a variance that evolves over time:

$$\varepsilon_t = v_t\sigma_t, \ \forall t, \quad v_t \sim N(0, 1)$$
$$\sigma_t^2 = E(\varepsilon_t^2 |\, I_{t-1} = \{\varepsilon_{t-1}\}) = \alpha_0 + \alpha_1\varepsilon_{t-1}^2, \quad \alpha_0 > 0, \ \ 0 < \alpha_1 < 1. \tag{2.34}$$

Definition 2.6 Let be $I_{t-1} = \{\varepsilon_{t-1}, \varepsilon_{t-2}, \ldots\}$ the set of information available at the period t-1 and $v_t \sim IIN(0, 1)$. Then ε_t is said to follow an ARCH(q) process if $\varepsilon_t = \sigma_t v_t$ and if

$$\sigma_t^2 = E(\varepsilon_t^2 |\, I_{t-1}) = \alpha_0 + \sum_{j=1}^{q} \alpha_j \varepsilon_{t-j}^2, \tag{2.35}$$

where $\alpha_i > 0, \ \ i = 0, 1, \ldots, q$ and $\sum\limits_{j=1}^{q} \alpha_j < 1$.

We then note

$$\varepsilon_t = \sigma_t v_t \sim \text{ARCH}(q).$$

Such a process, satisfying the stated conditions, is said to be an autoregressive process of order one with disturbances following an ARCH(1) process, that is, conditionally heteroskedastic autoregressive. The generalisation of (2.34) to the case where there are q lags taken into consideration is direct.

The **unconditional variance** of ε_t, denoted σ^2, is obtained by taking the expectation of (2.34) and then applying the law of iterated expectations

$$\sigma^2 = E\big[E(\varepsilon_t^2 | I_{t-1})\big] = \alpha_0 + \sigma^2 \sum_{j=1}^{q} \alpha_j$$
$$= \frac{\alpha_0}{1 - \sum\limits_{j=1}^{q} \alpha_j},$$

given that $\sum_{j=1}^{q} \alpha_j < 1$.

It is possible to model volatility more adequately by generalising the ARCH process, as Bollerslev (1986) did. We then obtain structures of the GARCH type (acronym for *Generalised Autoregressive Conditional Heteroskedastic models*). Thus, the GARCH(p, q) process will be written as:

$$\varepsilon_t = v_t\sigma_t, \quad v_t \sim IIN(0, 1)$$
$$\sigma_t^2 = E(\varepsilon_t^2 |\, I_{t-1}) = \omega + \sum_{j=1}^{q} \alpha_j \varepsilon_{t-j}^2 + \sum_{i=1}^{p} \beta_i \sigma_{t-i}^2, \tag{2.36}$$

where p refers to the number of conditional variance lags taken into account.

In practice, for the sake of parsimony, most of the time, a GARCH (1,1) process is used to model the heteroscedastic variance of a series.

2.3 Non-stationary Processes

Most economic series are not stationary, notably because they incorporate a trend, although it is necessary to distinguish between deterministic and stochastic trends. Nevertheless, we will first examine a non-stationary model widely used in finance or econometrics: the random walk.

The Random Walk Process

The random walk can, at first, be intuitively understood, using the example of the drunkard. He leaves the bar where he was to go to the bus stop located right in front of this bar. He therefore walks towards this stop, but in a totally random way given his high consumption of whisky. Thus, he takes steps, of identical length, to his left or right with the same probability. However, he will not reach the bus stop. He will inevitably deviate from it.

Definition 2.7 translates this particular experience on a theoretical level:

Definition 2.7 Let the process (y_t) be generated by

$$y_t = y_{t-1} + \varepsilon_t, \quad t = 1, 2, \ldots, \tag{2.37}$$

where ε_t is white noise with zero expectation and constant variance equal to σ^2. Then (y_t) is said to be a random walk.

If the generator of the process is given by

$$y_t = \alpha + y_{t-1} + \varepsilon_t, \quad t = 1, 2, \ldots,$$

where α is a constant, (y_t) is in this case a random walk with drift.

Here are the respective graphs of a random walk without drift and with drift. The first of them represents one of the possible empirical trajectories (based on 500 data) of the process

$$y_t = y_{t-1} + \varepsilon_t, \quad \varepsilon_t \sim GWN(0, 16),$$

where $GWN(0, 16)$ is the abbreviation for Gaussian or normal white noise, with zero mean and variance equal to 16 (Fig. 2.1).

As for the second graph, it provides a realisation of the random walk with drift (Fig. 2.2)

$$y_t = 2 + y_{t-1} + \varepsilon_t, \qquad \varepsilon_t \sim GWN(0, 9).$$

Note the great similarity of the representation obtained with the most common economic series: Gross Domestic Product (GDP), consumption, etc.

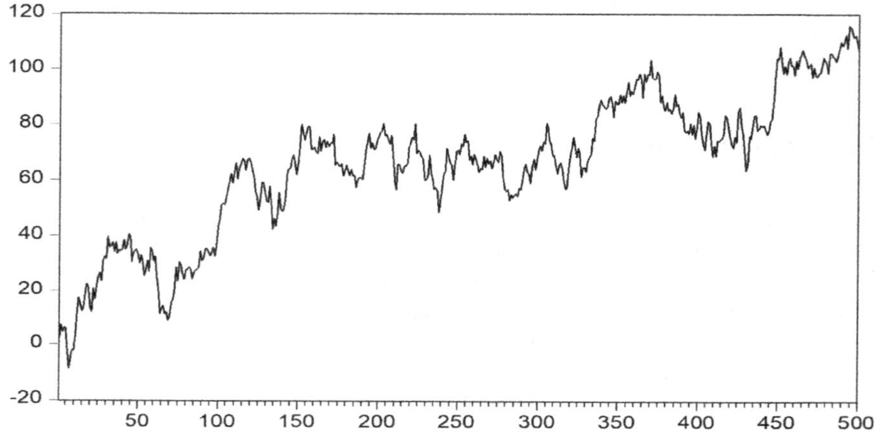

Fig. 2.1 A simple random walk

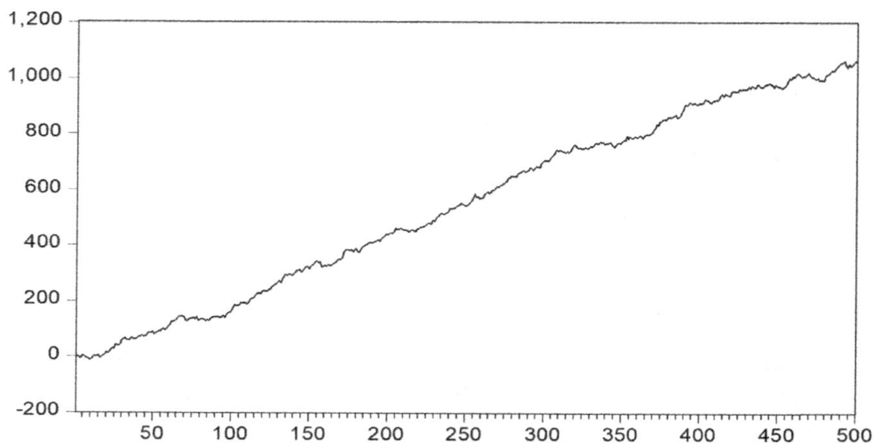

Fig. 2.2 A random walk with drift

A Non-stationary Process

Let's demonstrate that the random walk constitutes a weakly non-stationary process—see definition 2.2 of second-order stationarity. It is sufficient to show that it does not satisfy at least one of the three conditions stated in this definition.

By making repeated substitutions in (2.37), it comes

$$\begin{aligned} y_t &= y_{t-1} + \varepsilon_t = (y_{t-2} + \varepsilon_{t-1}) + \varepsilon_t \\ &= y_{t-2} + \varepsilon_t + \varepsilon_{t-1} \\ &\vdots \\ &= y_0 + \varepsilon_t + \varepsilon_{t-1} + \cdots + \varepsilon_1 \\ &= y_0 + \sum_{j=1}^{t} \varepsilon_j. \end{aligned} \tag{2.38}$$

With the help of (2.38), we can establish what the first two moments of the random walk are:

(i) $E(y_t) = E(y_0) + \sum_{j=1}^{t} E(\varepsilon_j) = y_0$;

(ii) $\operatorname{Var}(y_t) = E\big[(y_t - y_0)^2\big] = E\left[(\sum_{j=1}^{t} \varepsilon_j)^2\right] = \sum_{j=1}^{t} E(\varepsilon_j^2) = t\sigma^2$;

(iii) the covariance is given by

$$\begin{aligned} \operatorname{Cov}(y_t, y_{t-k}) &= E\big[(y_t - y_0)(y_{t-k} - y_0)\big] \\ &= E\big[(\varepsilon_t + \cdots + \varepsilon_{t-k} + \cdots)(\varepsilon_{t-k} + \varepsilon_{t-k-1} + \cdots)\big] \\ &= E\left(\sum_{j=1}^{t-k} \varepsilon_j^2\right) \\ &= (t-k)\sigma^2. \end{aligned}$$

Since both the variance and the covariance depend on time, the random walk process is clearly **non-stationary**.

To determine the autocorrelation of order k of the process, that is to say ρ_k, it will suffice to divide the autocovariance γ_{t-k} by the product of the standard deviations of y_t and of y_{t-k}. We then obtain

$$\rho_k = \frac{t-k}{[t(t-k)]^{1/2}} = \left[\frac{t-k}{t}\right]^{1/2}. \tag{2.39}$$

When k takes small values, the corresponding quotient in (2.39) is close to unity. However, as k increases, the values of the autocorrelations decrease, but slightly.

Hence the conclusion: *the* ***autocorrelation function*** *of a random walk process decreases slowly,* ***not exponentially***.

The delay operator L was introduced in Sect. 2.1 of this chapter. We will now use it to refine our understanding of the random walk. Indeed, (2.37) can be written

$$y_t - Ly_t = (1 - L)y_t = \varepsilon_t, \tag{2.37'}$$

where $1 - L$ constitutes a first-degree polynomial in L, which makes, by nullifying it, the equation

$$1 - L = 0.$$

The unique solution to this first-degree equation is obviously unity. In this measure, the random walk is said to be a **process with a unit root**.

Alternatively, let's now introduce the **difference operator** defined by

$$\Delta = 1 - L,$$

which allows us to write (2.37′) as follows:

$$\Delta y_t = (1 - L)y_t = \varepsilon_t.$$

Given that ε_t is a stationary white noise, it was therefore sufficient to differentiate the random walk once to make it stationary. In this measure, it is said to be **integrated of order one**, which is noted $I(1)$, while the stationary white noise process is **integrated of order zero**, or $I(0)$.

Let's generalise to the case of a process (x_t) integrated of order d, with $d \in \mathbb{N}^*$:

Definition 2.8 The process (x_t) is said to be integrated of order d, which is noted

$$x_t \sim I(d), \quad d = 1, 2, \ldots,$$

if it is stationary after having been differentiated d times, that is to say if

$$\Delta^d x_t \sim I(0).$$

Deterministic Trends and Non-stationary Series

Let the series y_t incorporating a constant, a linear deterministic trend and a disturbance u_t following an AR(1) process, with a white noise of zero mean and constant variance:

$$\begin{aligned} y_t &= \delta_0 + \delta_1 t + u_t, \\ u_t &= \phi\, u_{t-1} + \varepsilon_t, \qquad |\phi| < 1, \qquad \varepsilon_t \sim WN(0, \sigma^2). \end{aligned} \tag{2.40}$$

This series is non-stationary due to the presence of the linear trend. However, it is easy to make it stationary by removing this trend. Since

$$y_{t-1} = \delta_0 + \delta_1 (t-1) + u_{t-1},$$

it comes by subtracting this last expression from (1.39):

$$\Delta y_t = \delta_1 + \Delta u_t. \tag{2.41}$$

The series (2.40) can thus be described as **stationary around a trend** (abbreviated as TS).

Let's now assume that the disturbance u_t is generated by a random walk and therefore that $\rho = 1$. Knowing that

$$y_t = \delta_0 + \delta_1 t + u_{t-1} + \varepsilon_t = \delta_0 + \delta_1 t + (y_{t-1} - \delta_0 + \delta_1 (t-1)) + \varepsilon_t,$$

it follows that

$$\Delta y_t = \delta_1 + \varepsilon_t. \tag{2.42}$$

The series y_t has been made **stationary**, but this time **by differentiation** (abbreviated as DS).

In light of (2.41) and (2.42), one might think that the distinction between TS and DS series is hardly relevant. However, this is not the case if we refer to the nature of shocks or innovations in each case.

In (2.40), u_t measures the deviation between y_t and its affine trend (affine, because it incorporates the constant) during the same period. Let's now examine the effect of an innovation ε_t on the current and future deviations of the series from this affine trend. To do this, it will be enough to express u_t in terms of the innovations ε_t :

$$\begin{aligned} u_t &= \phi(\phi u_{t-2} + \varepsilon_{t-1}) + \varepsilon_t \\ &= \varepsilon_t + \phi \varepsilon_{t-1} + \phi^2 u_{t-2} \\ &= \varepsilon_t + \phi \varepsilon_{t-1} + \phi^2 \varepsilon_{t-2} + \phi^3 u_{t-3} \\ &\vdots \end{aligned}$$

$$= \sum_{j=0}^{+\infty} \phi^j \varepsilon_{t-j}. \tag{2.43}$$

From (2.43), we deduce that the effect of ε_t on u_t at the horizon $t+s$ is given by

$$\frac{\partial u_{t+s}}{\partial \varepsilon_t} = \phi^s. \tag{2.44}$$

It is then easy to distinguish the two cases considered:

- when $|\phi| < 1$, the higher s is, the closer ϕ^s gets to 0; consequently, the effect of the innovation on the deviations between the series and its affine trend tends towards zero when $s \to +\infty$. This effect is therefore **transient**.
- on the contrary, if we are dealing with a DS series, for which $\rho = 1$, then the effect of the innovation ε_t on this series has a **permanent** character.

2.4 Unit Root Tests

There are many unit root tests. All aim to determine whether a series is stationary or not. The oldest among them is the Dickey–Fuller (1979). This is the one we will therefore examine first.

Dickey–Fuller Test

Consider the autoregressive model of order 1

$$y_t = \phi y_{t-1} + \varepsilon_t, \quad y_0 = 0, \quad \varepsilon_t \sim IID(0, \sigma^2). \tag{2.45}$$

The least squares estimator of ϕ is then

$$\hat{\phi}_T = \frac{\sum_{t=1}^{T} y_{t-1} y_t}{\sum_{t=1}^{T} y_{t-1}^2}.$$

As we have already pointed out, it is enough to take $\phi = 1$ in the previous AR(1) to obtain a random walk. By subtracting y_{t-1} from both sides of (2.45), we get

$$\Delta y_t = (\phi - 1) y_{t-1} + \varepsilon_t. \tag{2.46}$$

The most direct method to verify the existence of a unit root in (2.46) is therefore to test

$$H_0 : \phi - 1 = 0 \quad \text{(non-stationarity)}$$

against the alternative hypothesis $\phi - 1 < 0$, that is, stationarity. (So, it is a one-sided left test.)

The least squares estimator in (2.46) is equal to

$$(\hat{\phi} - 1) = \frac{\sum_{t=1}^{T} y_{t-1} \Delta y_t}{\sum_{t=1}^{T} y_{t-1}^2} = \frac{\sum_{t=1}^{T} y_{t-1} \varepsilon_t}{\sum_{i=1}^{T} y_{t-1}^2}.$$

Under H_0, the test statistic is given by

$$z = T\left(\hat{\phi} - 1\right) = T\frac{\sum_{t=1}^{T} y_{t-1} \varepsilon_t}{\sum_{t=1}^{T} y_{t-1}^2} = \frac{T^{-1}\sum_{t=1}^{T} y_{t-1} \varepsilon_t}{T^{-2}\sum_{t=1}^{T} y_{t-1}^2}.$$

The problem now is to determine the limit law of $T\left(\hat{\rho} - 1\right)$ under the null hypothesis of an $I\,(1)$ series, that is non-stationary or unit root. Let's say right away that the statistic z does not follow a known standard distribution such as a normal, Chi-square, Student's, etc.

From this point of view, in order to progress, we will first need to introduce the concept of Brownian motion.

Definition 2.9 The continuous-time process $\{B(t),\ t \geq 0\}$ is said to be a standard Brownian motion if

(i) $B(0) = 0$;
(ii) $B(t)$ admits independent increases;
(iii) for all $s, t \geq 0$, $B(t + s) - B(s) \sim N(0, t)$;
(iv) every trajectory $B(t)$ is continuous.

In reality, this Brownian motion is the continuous-time generalisation of random walk. It can indeed be shown that it is obtained as the limit of such a random walk.

Here is the proposition that tells us what is the limit law of the statistic z of the Dickey–Fuller test:

Proposition 2.2 *Let the random walk without drift* $y_t = y_{t-1} + \varepsilon_t$, *where* $y_0 = 0$ and $\varepsilon_t \sim IIN(0, \sigma^2)$. *Under the* ***null hypothesis of non-stationarity,*** *the test statistic* z *converges in distribution towards a* ***Dickey–Fuller distribution*** *(denoted* DF_ρ*):*

$$z = T(\hat{\phi} - 1) = \frac{(1/T)\sum_{t=1}^{T} y_{t-1}\varepsilon_t}{(1/T^2)\sum_{t=1}^{T} y_{t-1}^2} \xrightarrow{D} DF_\rho = \frac{\frac{1}{2}\left[((B(1))^2 - 1\right]}{\int_0^1 [B(r)]^2 dr}.$$

The Dickey–Fuller distribution involves the square of a Brownian motion in the numerator and a functional of Brownian motions in the denominator. (As a reminder, a functional is a function that takes another function as an argument.) As we have already said, this asymptotic distribution is not standard and can only be obtained by stochastic simulation methods.

Alternatively, in (2.46), to test the null hypothesis of a unit root, one could apply an analogue of the test t, where the relevant statistic, denoted τ, would be defined by

$$\tau = (\hat{\phi} - 1)/\sigma_{\hat{\phi}},$$

with $\sigma^2_{\hat{\phi}} = s^2/\sum_{t=1}^{T} y^2_{t-1}$ and $s^2 = (T-1)^{-1}\sum_{t=1}^{T}(y_t - \phi y_{t-1})^2$.

Therefore, taking into account the previous elements, the alternative test statistic becomes

$$\tau = \frac{\hat{\phi} - 1}{\sigma_{\hat{\phi}}} = \frac{\sum_t y_{t-1}\varepsilon_t}{\sum_t y^2_{t-1}} \times \frac{\left(\sum_t y^2_{t-1}\right)^{1/2}}{(s^2)^{(1/2)}} = \frac{\sum_t y_{t-1}\varepsilon_t}{(s^2)^{(1/2)}(\sum_t y^2_{t-1})^{1/2}},$$

which leads, after multiplying top and bottom by T^{-1}, to the final expression

$$\tau = \frac{T^{-1}\sum\limits_t y_{t-1}\varepsilon_t}{(s^2)^{1/2}(T^{-2}\sum\limits_t y^2_{t-1})^{1/2}}.$$

Proposition 2.3 gives the asymptotic distribution of this alternative statistic.

Proposition 2.3 *Let the random walk without drift* $y_t = y_{t-1} + \varepsilon_t$, *where* $y_0 = 0$ and $\varepsilon_t \sim IIN(0, \sigma^2)$. *Under the* ***null hypothesis of non-stationarity,*** *the test statistic* τ *converges in distribution towards a* ***Dickey–Fuller distribution****, denoted* DF_τ :

$$\tau = \frac{T^{-1}\sum_{t=1}^{T} y_{t-1}\varepsilon_t}{(s^2)^{1/2}\left(T^{-2}\sum_{t=1}^{T} y^2_{t-1}\right)^{1/2}} \xrightarrow{D} DF_\tau = \frac{\frac{1}{2}\left[((B(1))^2 - 1\right]}{\left[\int_0^1 [B(r)]^2 dr\right]^{1/2}}.$$

This distribution is therefore the quotient of a Brownian motion and a functional of Brownian motions. It differs from that of z only by the presence of a square root in the denominator. It is also obtained by simulation.

Developments of the Dickey–Fuller Test

The DF test as just introduced is based on the consideration of a simple random walk, which led to the process defined by (2.45). However, more general forms are possible for Δy, notably by adding a **drift** and/or a **deterministic trend**.

The corresponding models are then

$$\begin{cases} \Delta y_t = \kappa + (\phi - 1)y_{t-1} + \varepsilon_t \\ \Delta y_t = \kappa + \gamma t + (\phi - 1)y_{t-1} + \varepsilon_t, \end{cases}$$

where κ is a drift and γt a deterministic trend. The associated test statistics will be denoted respectively z_c and z_{ct} (c for constant and t for the trend).

We can still rewrite the last of these models in a more compact way using vector notation:

$$\Delta y_t = \mathbf{s}'\boldsymbol{\delta} + (\phi - 1)y_{t-1} + \varepsilon_t, \tag{2.47}$$

where $\mathbf{s}' = (1, t)$ and $\boldsymbol{\delta} = (\kappa, \gamma)$.

Then comes the main result:

Proposition 2.4 *Under the null hypothesis of a unit root, hence of* ***non-stationarity****, the statistic z_{ct} follows a* ***Dickey–Fuller distribution of type*** *2:*

$$z_{ct} = T(\hat{\phi} - 1) \sim DF_{ct}.$$

This distribution differs from those seen previously, even though it also represents a combination of functionals of Brownian movements. For this reason, it has been indexed by *ct*. It is also obtained by simulation.

A second extension of the DF test involves allowing the presence of autocorrelation in the error term, which corresponds quite well to the profile of a typical economic series.

Suppose that the random variable is not white noise, but follows an AR(p) process. It is then a matter of performing a regression on an equation similar to (2.47), in which however we have incorporated p lagged terms, that is

$$\Delta y_t = \mathbf{s}'\boldsymbol{\delta} + \phi_0 y_{t-1} + \sum_{j=1}^{p} \beta_j \Delta y_{t-j} + \varepsilon_t. \tag{2.48}$$

It is the addition of these terms that justifies the name of the test **Augmented Dickey–Fuller** (abbreviated as ADF test, Augmented Dickey–Fuller test).

The two corresponding test statistics are then given by

$$\tau_{ADF} = \hat{\phi}_0 / \hat{\sigma}_{\hat{\phi}_0} \text{ et } z_{ADF} = T\hat{\phi}_0 / (1 - \sum_{j=1}^{p} \beta_j).$$

Under the null hypothesis of a unit root ($H_0 : \phi_0 = 1$) and assuming the appropriate choice of p, the asymptotic distributions of these statistics are, for the same choice of regressors $\mathbf{s}'$, identical to the Dickey–Fuller distributions obtained in the simple case:

$$\begin{cases} \tau_{ADF} \sim DF_{ct} \\ z_{ADF} \sim DF_{ct}. \end{cases}$$

From a practical point of view, two types of problems arise when implementing an ADF test:

1. Which components of the vector $\mathbf{s}'$ should be included in the regression? In this case, the choice is between a constant, a constant and a linear trend, or nothing at all. The consequence of an inappropriate selection is to reduce the power of the test. It is then best to refer to the profile of the series examined.
2. How many lags should be taken into consideration in the regression? The answer, not very helpful to be honest, is to introduce enough lags to eliminate the autocorrelation of the estimation residuals. However, the frequency of the data (annual, quarterly, monthly, etc.) provides a first indication of lag selection.

These are the practical difficulties that we will now face through the examination of the series of the gross domestic product (GDP) of France, excluding price increases. (Note that the data are expressed in millions of euros and cover the period from the first quarter of 1975 to the second quarter of 2021-base 2010.)

Example 2.4 As the data are quarterly, at least 5 lags should be taken into consideration. Similarly, given that the series shows regular growth, except, of course, in periods of recession, tests with constant or trend should be performed.
When a constant and two lags are incorporated into the regression, $\tau_c = -1,84$ and the p-value associated is equal to 0.6837. The ADF test, this time performed with four lags and incorporating both a constant and a trend, gives $\tau_{ct} = -2,315$ with a p-value of 0.425.
In both cases, we are led to accept the null hypothesis of a unit root in the GDP series. Moreover, when we perform the same tests using the first differences, we obtain a p-value less than 0.001. It follows that we then reject the null hypothesis of non-stationarity and we conclude that the real GDP is integrated of order one: real GDP $\sim I(1)$.

Other Tests of Stationarity

An interesting variant of the ADF test, due to Elliott et al. (1996), is the Dickey–Fuller using Generalised Least Squares (DF-GLS, for *Dickey–Fuller-Generalised Least Squares*). Compared to the ADF, the DF-GLS tests the simple hypothesis $H_0 : \phi - 1 = 0$, against the also simple alternative, $H_1 : \phi - 1 = \bar{a}$. This value $\bar{a}$ is chosen in such a way that

$$\begin{cases} \bar{a} = 1 - (7/T) & \textit{if } \ \mathbf{s}' = (1, 0) \\ \bar{a} = 1 - (13,5/T) & \textit{if } \ \mathbf{s}' = (1, \ t). \end{cases}$$

For the rest, the test procedure relies on the prior estimation of the constant and the trend. We thus apply the least squares to the data transformed according to

$$d(y_t|\,\overline{a}) = \begin{cases} y_t & if\ t = 1 \\ y_t - \overline{a}y_{t-1} & if\ t > 1. \end{cases}$$

Constant and trend are also transformed based on the same time scheme.

Then, once we have estimated the constant κ and the trend parameter γ, we subtract them from the original series in order to obtain

$$y_t^d = y_t - \mathbf{s}'\boldsymbol{\delta}(\overline{a}).$$

Finally, in the last step, we perform the following regression, which uses the transformed data y_t^d and not directly y_t :

$$\Delta y_t^d = (\phi - 1)y_{t-1}^d + \beta_1 \Delta y_{t-1}^d + \cdots + \beta_p \Delta y_{t-p}^d + \varepsilon_t.$$

Now we must also say a few words about another test that is associated with the names of Kwiatkowski et al. (1992) and which, for this reason, is called the KPSS test.

Unlike those developed previously, this one has as its null hypothesis the **stationarity** of y_t. The alternative hypothesis is obviously that y_t is a random walk, incorporating a unit root, therefore a series $I\,(1)$.

In short, the test procedure involves four steps:

1. We perform the auxiliary regression

$$y_t = \alpha + \delta t + \varepsilon_t.$$

2. The estimation of the parameters of this regression provides the residuals

$$\hat{\varepsilon}_t = y_t - \hat{\alpha} - \hat{\delta} t.$$

3. We then calculate the cumulative partial sums

$$S_t = \sum_{s=1}^{t} \hat{\varepsilon}_s, \qquad t = 1, .\ ., T.$$

4. We obtain *in fine* the test statistic

$$KPSS = \sum_{t=1}^{T} \frac{S_t^2}{\hat{\sigma}^2},$$

where $\widehat{\sigma}^2$ is an estimator of the variance of the residuals with correction for the error term.

This *KPSS* statistic asymptotically follows a law which is, like for the other statistics detailed previously, a functional of Brownian movements and which must, like them, be tabulated by simulation methods.

2.5 Monte Carlo Simulations

In principle, a stochastic simulation involves creating as many random samples as desired, using numerical methods implemented on a computer.

Introduction

Today, everyone agrees to establish the birth of stochastic simulations in 1949, when Metropolis and Ulam published the article "The Monte Carlo Method" in the *Journal of American Statistical Association*.

The reference to the famous casino should not be misleading. It essentially means that roulette is the perfect way to "produce" randomness.

The basic ingredient for performing such simulations is the generation of a set of random numbers or rather, **pseudo-random**. Indeed, a computer is a deterministic instrument incapable of truly imitating randomness, so the sequence of generated numbers reproduces the same sequence identically after a certain number of steps.

Example 2.5 Consider the sequence of numbers

$$04825102748957138132|\mathbf{048251}\ldots$$

At first glance, it appears to be random. However, we immediately notice that after a sequence of twenty digits, the same numbers are found, namely 048251, etc.

The whole problem is therefore to ensure that the repetition of the same sequence occurs after a very long period, for example after a million digits.

In this last case, the sequence of numbers obtained will resemble a random sample drawn from a uniform law on [0, 1]. We will then speak of a sequence of pseudo-random numbers.

Generation of Random Numbers

The algorithms used to generate pseudo-random sequences often rely on the same **multiplicative congruential** method. Let us recall from this point of view that if m is a positive integer, two relative integers x and y are said to be **congruent modulo** m if their difference is a multiple of m, which is noted

$$x \equiv y(\text{mod} m).$$

Thus, for example, 9 and 3 are congruent modulo 3, since their difference, namely 6, is a multiple of 3. Alternatively, two integers x and y are **congruent modulo** m if and only if their division by the positive integer m gives the same remainder. (In the chosen example, this remainder will obviously be equal to zero.)

A **congruential algorithm** generates a set of pseudo-random numbers $y_1, y_2, \ldots$ based on the following recurrence formula: for $n \geq 1$ and x_0 a fixed initial value,

$$\begin{cases} x_n \equiv ax_{n-1}(\text{mod} m) \\ y_n = x_n/m, \end{cases} \tag{2.49}$$

where a and m are given positive integers.

The initial number x_0 is called the **seed**, a the **multiplier** and m the **modulus** of the algorithm. For its part, $y_n = x_n/m$ is said to be a **pseudo-random number**, which constitutes an approximation of the value taken by a random variable distributed according to a uniform distribution on the interval [0, 1].

Given (2.49), the process of generating such numbers pseudo-random is organised according to the following iterative procedure:

(i) we start by dividing ax_{n-1} by m and we assign the remainder of this division to x_n;
(ii) in turn, x_n is divided by m, which makes the pseudo-random number y_n, necessarily less than 1;
(iii) we initiate the whole procedure from the seed x_0 and we repeat it N times, in order to obtain the desired pseudo-random sequence.

The ability of a congruential algorithm to produce a sequence without repetitive sequences entirely depends on the selection of the triplet (a, m, x_0), as the following example proves.

Example 2.6 Suppose we choose $a = 4, m = 8$ and $x_0 = 1$. We first obtain y_1 by initialising the algorithm, that is $x_1 = 4.1(mod 8) \Rightarrow x_1 = 4$ and $y_1 = 4/8 = 1/2$. Continuing the iterative procedure, we get the sequence: $1/2|3/4|/2|3/4|1/2|3/4|$ .... We therefore observe that with $m = 8$, the sequence $(1/2,\ 3/4)$ repeats identically at $n = 3, 5, 7, \ldots$.

In practice, to remedy this problem, we select a very high-value m that is a function of the computer used. Thus, for a 32-bit processor (this anagram is a contraction of the

English "binary digits", i.e. binary numbers, knowing that the numbers in question are 0 or 1), we take as the value of m an integer of the order of 2^{31}. In total, we then obtain a sequence of pseudo-random numbers, which only cycle after two or three billion numbers.

We will not go further into the theoretical considerations related to the construction of "good" random number generators. We will simply assert that such sequences of numbers are easily produced on any personal computer.

Generation of Normal Random Variables

Several methods exist to generate a random variable, including those of the inverse application and rejection. The first of these relies on a probabilistic proposition that establishes a link between the uniform distribution and the distribution function of the considered random variable.

Proposition 2.4 *Let* $U \sim U([0, 1)]$ *and F an absolutely continuous cumulative distribution function. Then the random variable X defined by*

$$X = F^{-1}(U)$$

admits F as its ***cumulative distribution function***.

It follows in any case from this proposition that the generation of a standard normal variable—to limit ourselves to this single case—goes through two main steps:

1. generate a set of pseudo-random numbers from a uniform distribution on [0,1];
2. use these pseudo-random numbers to generate via the inverse of the standardised normal distribution function, the values of this random variable.

Here is now the program (a script, as it is called in the terminology of GRETL) written to simulate both random numbers and the values of a standard normal random variable. (For more details on GRETL and its programming language, refer to Appendix A of this book.)

We will simply note that personal comments introduced in the script are preceded by the hashtag #, so they cannot be read by GRETL.

```
nulldata 1000          # creation of a series of 1000 "blank" data
set seed 3,213,789     # fixation of the seed for replication
genr u = uniform (0, 1)
genr z = normal (0, 1)
```

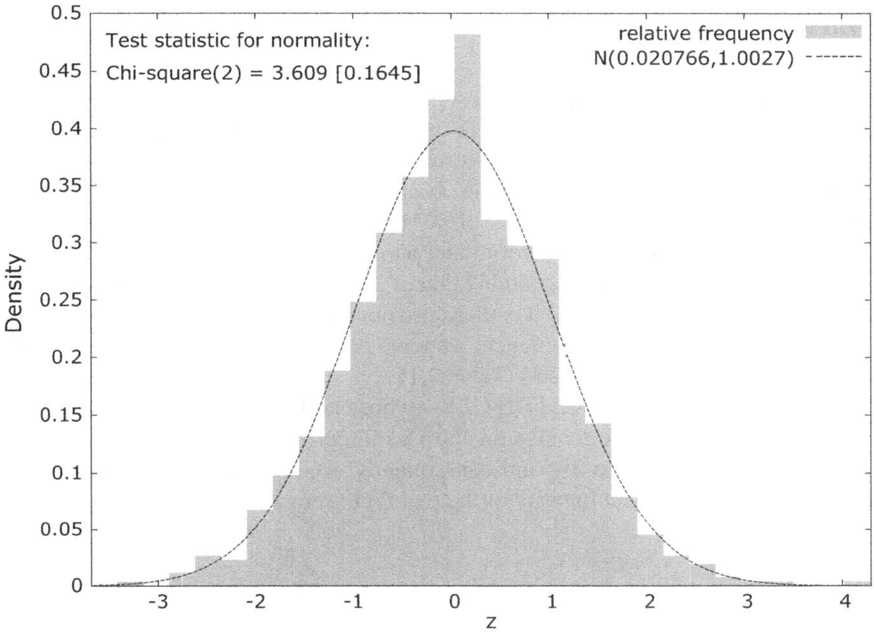

Fig. 2.3 Density simulation of a normal variable

The instructions in this script speak for themselves. First, there is the creation of a "white" series consisting of 1000 potential observations. Then comes the setting of the seed equal to the integer 3,213,789—it is obviously possible to choose other values. (The mention of the seed is made, for the sole purpose of more easily reproducing the simulations.) The last two steps consist of generating a pseudo-random sequence, as well as 1000 normally distributed observations.

To appreciate the value of the simulations thus carried out, Fig. 2.3 superimposes on the histogram of the frequencies of the observations the curve of the density of a centered reduced normal variable.

Upon examination of the figure and especially the p-value of the Doornik-Hansen normality test—equal to 0.16, it follows that we accept the null hypothesis of normality. (We will return in detail in a later chapter on this test.)

2.6 Applications

We will start by implementing the various stationarity tests that have been detailed in Sect. 2.4 before carrying out a simulation of white noise.

Unit Root Tests

We have already carried out an ADF test on the quarterly series of real French GDP in example 2.4. We will now complete the analysis of the non-stationarity of this series by considering the two other unit root tests.

Here is a graph of this series, whose observations, remember, span from the first quarter of 1975 to the second quarter of 2021-base 2010 (Fig. 2.4).

Clearly, the real GDP series incorporates a trend. We can also discern the presence of several recessions, including notably that of 2008–2009 and especially that due to the pandemic in 2020. A priori, it is therefore non-stationary, which is confirmed by the three tests used (For convenience, we have reproduced the results for the ADF test already given in example 2.4) (Table 2.1).

The first two tests—ADF and DF-GLS—whose null hypothesis is the presence of a unit root, lead to the same conclusion: the p-value corresponding to the two statistics is significantly higher than 5% and consequently, we are justified in accepting the null hypothesis of non-stationarity of the real GDP series.

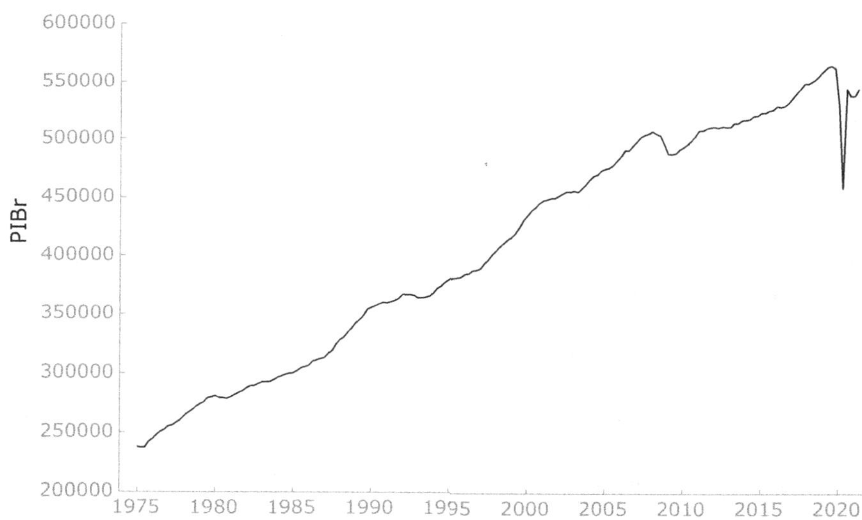

Fig. 2.4 Real French GDP

Table 2.1 Unit root tests (constant + trend)

	ADF*	DF-GLS**	KPSS
Test statistic	−2.315	−1.172	0.401
p-value	0.425	0.472	< 0.01

*2 lags ** 4 lags

Table 2.2 Unit root and first differences

	ADF*	DF-GLS**	KPSS
Test statistic	−8.148	−17.362	0.039
p-value	< 0.001	0.000	> 0.1

*3 lags ** 0 lags

The KPPS test has as its null hypothesis, let us recall, the stationarity of the series considered. Given a p-value of less than one percent, we are led to reject this null hypothesis and accept that of non-stationarity.

Let's now apply the same tests to real GDP, but this time expressed in first differences (Table 2.2).

Given their respective p-values, the ADF and DF-GLS tests lead to the acceptance of the alternative hypothesis of stationarity, while the KPSS test, with a p-value significantly higher than 10%, leads to the conclusion that the series differentiated once is stationary.

In light of these results, the real GDP series can be said to be integrated of order 1, that is

$$\text{GDPr} \sim I(1).$$

White Noise and Stationarity

We will start by simulating Gaussian white noise. This is the very example of a stationary process, as we have shown in example 2.1 of this chapter.

Here is the script used to generate it assuming that its mean is zero and its variance is equal to 4.

Nulldata 60	
Set seed 427,563	
Genr WN = normal (0.4)	# WN is white noise

Figure 2.5 provides a graphical representation of the temporal profile of this white noise process. (We have assumed that the simulated data were hourly data and that consequently, we were considering sixty hours.)

To see if this simulated series is indeed stationary, we apply the three unit root tests, already implemented in the previous section (Table 2.3).

The first two tests, whose null hypothesis is the non-stationarity of the white noise series, lead to the rejection of this hypothesis and the acceptance of the alternative, namely the stationarity of this simulated process, which is in agreement with the theory.

However, with a p-value much greater than 10%, it follows that the KPSS test leads us to accept the null hypothesis of stationarity of the simulated process $WN(0, 4)$.

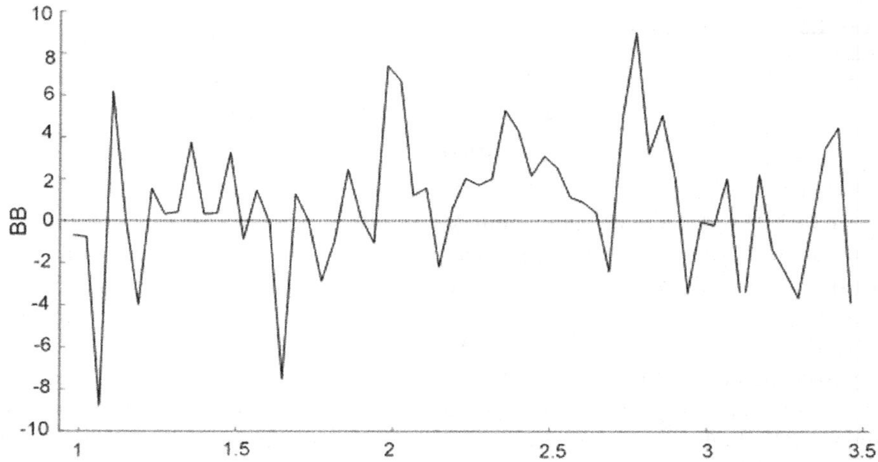

Fig. 2.5 A white noise process

Table 2.3 Stationarity tests

	ADF*	DF-GLS**	KPSS
Test statistic	−6.193	−17.362	0.039
p-value	< 0.0001	0.001	> 0.10

*0 lag ** 2 lags

2.7 Exercises

2.7.1 Let the AR(2) model be

$$y_t = 1.6y_{t-1} - 0.7y_{t-2} + \varepsilon_t, \quad \varepsilon_t \sim WN(0, \sigma^2).$$

Is this autoregressive process stationary?

2.7.2 Consider the elementary Keynesian model

$$\begin{cases} y_t = c_t + i_t \\ c_t = \alpha y_{t-1} + \varepsilon_t, & 0 < \alpha < 1 \\ i_t = \beta(c_{t-1} - c_{t-2}), & \alpha > 0, \end{cases}$$

where c_t is the consumption dependent on national income y, investment i_t is determined by the acceleration mechanism, while ε_t and ν_t are disturbances. Of course, the first equation is a pure identity.

We ask to write this system in the multivariate form of a vector autoregression of order 2.

2.7.3 Compute the mathematical expectation of y_t in the model

$$y_t = \alpha + \phi y_{t-1} + \varepsilon_t, \qquad \varepsilon_t \sim WN(0, \sigma^2),$$

where $|\phi| < 1$.

2.7.4 Consider the ARCH and GARCH volatility models. We then ask:

(i) why is it better to use a GARCH(1,1) model rather than a pure ARCH(q)?
(ii) why, in your opinion, are GARCH-type models frequently used to process the financial time series?

2.7.5 In Chapter 1—see "The Used Data" section—we use the series of US annual aggregate consumption and income for the period 1950–1985. We will now test the stationarity of the time series of level consumption, while applying the three tests that were developed for this purpose.

The following table presents the corresponding statistics and their associated p-values (Table 2.4).

Table 2.4 Unit root tests on c_t

	ADF*	DF-GLS*	KPSS
Test statistic	−2.01013	−1.25697	0.23768
p-value	0.5952	0.888	> 0.10

*With constant and trend

What conclusions do you draw from the analysis of the elements of this table regarding the stationarity or non-stationarity of the series c_t? In the case where c_t is non-stationary, i.e. I (1), what should be done to make it stationary?

2.7.6 A pair of (unloaded) dice is thrown continuously until all possible outcomes 2, 3, …, 12 occur at least once. Develop a simulation procedure that allows you to estimate the number of throws necessary for these results to occur.

Reference

Bollerslev, T. (1986), Generalized autoregressive conditional heteroscedasticity, *Journal of Econometrics, 31*, 307–327.

Dickey, D., & Fuller, W. (1979), Distribution of the estimators for autoregressive time series with a unit root, *Journal of the American Statistical Association, 74*, 427–431.

Elliot, G., Rothenberg, T.J., & Stock, J.H. (1996), Efficient tests for an autoregressive unit root, *Econometrica, 64*, 813–836.

Kwiatkowski, D.P., Phillips, P.C.B., Schmidt, P., & Shin, Y. (1992), Testing the null hypothesis of stationarity against the alternative of a unit root: How sure are we that economic time series have a unit root? *Econometrica, 54*, 159–178.

Chapter 3
Estimating the Model

3.1 Least Squares

The linear dynamic model has already been presented in the previous chapter. We will simply recall that it can be written as

$$\begin{aligned} y_t = \beta_0 + \beta_{10}x_{t1} + \beta_{11}x_{t-1,1} + \cdots + \beta_{1k_1}x_{t-k_1,1} + \beta_{20}x_{t1} + \beta_{22}x_{t-1,2} + \cdots + \beta_{2k_2}x_{t-k_2,2} \\ + \cdots + \beta_{m0}x_{tm} + \beta_{m1}x_{t-1,m} + \cdots + \beta_{mk_m}x_{t-k_m,m} + u_t, \quad t = 1, \ldots, T. \end{aligned}$$

By introducing a double summation, the above equation can still be written:

$$y_t = \beta_0 + \sum_{j=0}^{k_i} \sum_{i=1}^{m} \beta_{ij} x_{t-j,\,i} + u_t, \quad t = 1, \ldots, T. \tag{3.1}$$

If the number of parameters to be estimated is, in total, equal to K, the system of equations (3.1) is expressed in matrix form as

$$\boldsymbol{y} = \boldsymbol{X\beta} + \boldsymbol{u}, \tag{3.2}$$

where $\mathbf{y}$ and $\mathbf{u}$ are column vectors of dimension $T \times 1$, $\boldsymbol{\beta}$ a column vector with K components and $\mathbf{X}$ a matrix of dimension $T \times K$.

It should be emphasised that the model (3.2) is linear in the parameters $\beta_1, \ldots, \beta_K$ and not in the explanatory variables, whether lagged or not. Variables are also known as **regressors**.

Supplementary Information The online version contains supplementary material available at https://doi.org/10.1007/978-3-031-72910-2_3.

F. J. Bismans and O. Damette, *Dynamic Econometrics*,
https://doi.org/10.1007/978-3-031-72910-2_3

Model Assumptions and Their Interpretation

Let's start with the error term $u_t, \quad t = 1, \ldots, T,$, also known as the **disturbance** or **perturbation**. Let's consider it as a random variable and assume that, on average, the values of this disturbance cancel out, i.e. for all t, $E(u_t) = 0$. As for its variance, let's assume that it is equal to a strictly positive number σ^2, which, moreover, is constant over time. In other words, for all t,

$$\mathrm{Var}(u_t) = E\big[(u_t - E(u_t))^2\big] = E\big(u_t^2\big) = \sigma^2. \tag{3.3}$$

This assumption is known as the **homoskedasticity** of the error term. (It contrasts with heteroskedasticity, which means that the variance is no longer constant over time, but varies with the values of t.)

It is also essential to formulate a hypothesis as to the behaviour of the sequence (u_t). From this point of view, we will assume that there is no **autocorrelation of** the perturbations, so that

$$\mathrm{Cov}(u_i, u_j) = E\big[(u_i - E(x_i))\big(u_j - E(x_j)\big)\big] = E\big(u_i u_j\big) = 0. \tag{3.4}$$

Although it is not essential for applying least squares, we will also make the assumption that the disturbances are independently normally distributed, so that for all t:

$$u_t \sim N(0, \sigma^2).$$

Since the error term **u** is a vector, it is interesting to give these different hypotheses a matrix form. In this respect, the expectation of **u** is given by

$$E(\boldsymbol{u}) = E\begin{pmatrix} u_1 \\ \vdots \\ u_T \end{pmatrix} = \begin{pmatrix} E(u_1) \\ \vdots \\ E(u_T) \end{pmatrix} = \begin{pmatrix} 0 \\ \vdots \\ 0 \end{pmatrix} = \boldsymbol{0}.$$

The variance–covariance matrix of **u is** equal to

$$\begin{aligned} \mathrm{Var}(\boldsymbol{u}) &= E\big[(\boldsymbol{u} - E(\boldsymbol{u}))(\boldsymbol{u} - E(\boldsymbol{u}))'\big] = E(\boldsymbol{u}\boldsymbol{u}') \\ &= E\begin{pmatrix} u_1^2 & u_1u_2 & \cdots & u_1u_T \\ u_2u_1 & u_2^2 & \cdots & u_2u_T \\ \vdots & \vdots & \ddots & \vdots \\ u_Tu_1 & u_Tu_2 & \cdots & u_T^2 \end{pmatrix} = \begin{pmatrix} \sigma^2 & 0 & \cdots & 0 \\ 0 & \sigma^2 & \cdots & 0 \\ \vdots & \vdots & \ddots & \vdots \\ 0 & 0 & \cdots & \sigma^2 \end{pmatrix} = \sigma^2 \boldsymbol{I}_T, \end{aligned}$$

where $\boldsymbol{I}_T$ denotes the identity matrix of dimension $T \times T$ and the prime ($'$) represents the transposition operator.

The first hypothesis, which summarises all those put forward previously, can therefore be formulated as follows:

(H1)
u *follows a multivariate normal distribution with mean vector zero and variance–covariance matrix* $\sigma^2 \boldsymbol{I}_T$, *i.e.*

$$\boldsymbol{u} \sim MN(\boldsymbol{0}, \sigma^2 \boldsymbol{I}_T).$$

This assumption itself implies that the vector **y** is also random. In fact, since the latter is a function of the normally distributed errors **u**, it follows—at least if the product vector $\boldsymbol{X\beta}$ is composed of constants—that **y** follows a multivariate normal distribution of expectation

$$E(\boldsymbol{y}) = E(\boldsymbol{X\beta}) + E(\boldsymbol{u})$$

with a variance–covariance matrix equal to

$$\text{Var}(\boldsymbol{y}) = \text{Var}(\boldsymbol{X\beta}) + \text{Var}(\boldsymbol{u}) = \sigma^2 \boldsymbol{I}_T.$$

To sum up, we have

$$\boldsymbol{y} \sim MN(\boldsymbol{X\beta}, \sigma^2 \boldsymbol{I}_T).$$

Note the restriction in the previous sentence ("at least if the product vector $\boldsymbol{X\beta}$ is made up of constants"). This means that the regressors have all been considered to be **non-stochastic** or even **deterministic**.

However, while this assumption may sometimes be consistent with reality, it is not clear why the dependent variable should always be random, even though the regressors are not. Consequently, we can weaken the non-stochasticity hypothesis by assuming that the explanatory variables are random, but exogenous, i.e. that

$$E(\boldsymbol{u}|\boldsymbol{X}) = \boldsymbol{0}. \tag{3.5}$$

The interpretation of (3.5) is immediate: the expectation of the perturbation vector **u** is zero conditional on the entire matrix **X.** Logically, if we adopt condition (3.5), it also follows that

$$\text{Var}(\boldsymbol{u}|\boldsymbol{X}) = \sigma^2 \boldsymbol{I}_T. \tag{3.6}$$

Let us now introduce a second hypothesis which at first sight appears mathematical, but which in fact has a direct economic interpretation:

(H2)
the matrix **X** *of observations has rank* K, *where* $K < T$. *Alternatively,* **X** *is also said to have* ***full*** *rank.*

This condition on the rank of the matrix means that the K columns of **X** are linearly independent, which can be interpreted economically as the fact that all the explanatory variables count and that none of them can be "replaced" by a combination of the others. Assumption (H2) is then equivalent to the absence of **multicollinearity** between the regressors.

The condition of complete rank has several implications, including this one, which takes the form of a lemma, the use of which we shall soon see:

Lemma 3.1 *Let* **X** *be a matrix of dimension* $T \times K$, *where* $K < T$. *Then the product matrix* $\boldsymbol{X}'\boldsymbol{X}$ *is of rank* K and *therefore invertible.*

Least Squares Estimators

Let's turn first to the estimation of the vector $\boldsymbol{\beta}$ in (3.2). We immediately obtain a first theorem:

Theorem 3.1 Under *assumptions* (H1) and (H2), *the ordinary least squares* (OLS) *estimator of* $\boldsymbol{\beta}$ *is given by*

$$\hat{\boldsymbol{\beta}} = (\boldsymbol{X}'\boldsymbol{X})^{-1}\boldsymbol{X}'\boldsymbol{y}. \tag{3.7}$$

Moreover, this estimator is unique.

Demonstration

The OLS estimator, denoted by $\hat{\boldsymbol{\beta}}$, must minimise the sum of the squares of the disturbances, i.e.

$$\sum_{t=1}^{T} u_t^2 = \boldsymbol{u}'\boldsymbol{u} = (\boldsymbol{y} - \boldsymbol{X}\boldsymbol{\beta})'(\boldsymbol{y} - \boldsymbol{X}\boldsymbol{\beta}) = S(\boldsymbol{\beta}).$$

As we indicated in Chapter 1, it is sufficient to verify the first and second-order conditions for a minimum. We therefore first cancel the gradient of $S(\boldsymbol{\beta})$, i.e.

$$\frac{\partial}{\partial \boldsymbol{\beta}} S(\boldsymbol{\beta}) = \frac{\partial}{\partial \boldsymbol{\beta}}\left[(\boldsymbol{y} - \boldsymbol{X}\boldsymbol{\beta})'(\boldsymbol{y} - \boldsymbol{X}\boldsymbol{\beta})\right] = \frac{\partial}{\partial \boldsymbol{\beta}}\left[\boldsymbol{y}'\boldsymbol{y} - 2\boldsymbol{y}'\boldsymbol{X}\boldsymbol{\beta} + \boldsymbol{\beta}'\boldsymbol{X}'\boldsymbol{X}\boldsymbol{\beta}\right].$$

Using the two vector derivation formulae—see Appendix B—we get

$$-2X'y + 2X'X\beta = 0,$$

which leads to a system of T linear equations with K unknowns

$$X'X\beta = X'y.$$

Since, by Lemma 2.1, the matrix $X'X$ is of rank K and therefore invertible, the system in question—called the system of **normal equations**—has as its solution

$$\hat{\beta} = (X'X)^{-1}X'y.$$

The second-order conditions for a minimum of $S(\beta)$ require the Hessian matrix of the second partial derivatives of $S(\beta)$, evaluated at $\hat{\beta}$, to be positive definite. Now, since

$$\frac{\partial^2}{\partial\beta\partial\beta'}S(\beta) = \frac{\partial}{\partial\beta'}\left(-2X'y + 2X'X\beta\right) = 2X'X,$$

so all we need to do is show that $X'X$ is positive definite, i.e. for all $a \neq 0, a'X'Xa > 0$. If we then set $z = aX$, we get

$$a'X'Xa = z'z = \sum_{k=1}^{K} z_k^2.$$

Since $a \neq 0$ and $\mathbf{X}$ has rank K, i.e. the columns of $\mathbf{X}$ are linearly independent, the vector $\mathbf{Xa}$ is a linear combination of the columns of $\mathbf{X}$ and cannot be zero. The sum of squares $z'z$ is therefore strictly positive and $X'X$ is positive.

The uniqueness of the vector $\hat{\beta}$ is ensured by the fact that, by Lemma 3.1, $\text{rk}(X'X) = K$. Indeed, the system of normal equations $X'X\beta = X'y$—see Appendix B—then has a unique solution given by

$$\hat{\beta} = (X'X)^{-1}X'y.$$

It follows from this theorem that the ordinary least squares estimator of β is a **global minimizer** of $S(\beta)$. It should also be noted that, from a numerical point of view, $\hat{\beta}$ is easily obtained (by computer, of course!), given that all that is needed is to use matrix operations such as transposition, multiplication and inversion of matrices.

Let's now introduce the concepts of estimation residuals and adjusted values of the dependent variable.

Definition 3.1 The vector of OLS estimation **residuals** is given by

$$\hat{\boldsymbol{u}} = \boldsymbol{y} - \boldsymbol{X}'\boldsymbol{\beta}.$$

In addition, the **fitted values** of **y** are the vector $\hat{\boldsymbol{y}} = \boldsymbol{X}\hat{\boldsymbol{\beta}}$.

From this definition, we immediately derive the proposition that the vector of residuals is orthogonal to each of the regressors of the matrix **X,** i.e. mathematically:

Theorem 3.2 *The vector of residuals* $\hat{\boldsymbol{u}}$ *is* ***orthogonal to*** *each of the K columns of* **X**, i.e.

$$\boldsymbol{X}'\hat{\boldsymbol{u}} = \boldsymbol{0}.$$

Furthermore, if one of the regressors is a constant, $\sum u_t = 0. \sum u_t = 0.$

Demonstration

We immediately have

$$\boldsymbol{X}'\hat{\boldsymbol{u}} = \boldsymbol{X}'(\boldsymbol{y} - \boldsymbol{X}\hat{\boldsymbol{\beta}}) = \boldsymbol{X}'\boldsymbol{y} - \boldsymbol{X}'\boldsymbol{X}(\boldsymbol{X}'\boldsymbol{X})^{-1}\boldsymbol{X}'\boldsymbol{y} = \boldsymbol{0}.$$

Let us then assume, without loss of generality, that the first of the regressors is a constant and therefore that the first column of **X** is a vector of T units. It then follows from the previous equality that

$$\boldsymbol{x}_1'\hat{\boldsymbol{u}} = \sum_{t=1}^{T} \hat{u}_t = 0.$$

Measuring Statistical Fit

Let's start with a few writing conventions:

- *RSS* is the sum of the squares of the OLS estimation residuals, i.e. $RSS = \hat{\boldsymbol{u}}'\boldsymbol{u} = \sum (y_t - \hat{y}_t)^2$;
- *ESS* is the explained sum of squares from the regression of **y** on **x**, i.e. $ESS = \sum (\hat{y}_t - \overline{y})^2$,
- *TSS* is the total sum of squares, *i.e.* $TSS = \sum_t (y_t - \overline{y})^2$.

By virtue of Definition 3.1 and Theorem 3.2, we can write that

$$y'y = \hat{\beta}'X'X\hat{\beta} + (y - X\hat{\beta})'(y - X\hat{\beta}). \tag{3.8}$$

From (3.8), a proposition immediately follows, which we will consider as a lemma.

Lemma 3.2 *The sum of the squares of the least squares residuals is equal to the difference between the sum of the squares of the values of the dependent variable and the sum of the squares of the adjusted values of this same variable*:

$$\hat{u}'\hat{u} = y'y - \hat{y}'\hat{y}.$$

Demonstration

Since $\hat{u}'\hat{u} = (y - \hat{\beta}X)'(y - \hat{\beta}X) = y'y - 2\hat{\beta}'X'y + \hat{\beta}'X'X\hat{\beta}$ and, moreover, $X'X\hat{\beta} = X'y$, it comes

$$\begin{aligned}\hat{u}'\hat{u} &= y'y - 2\hat{\beta}'X'y + \hat{\beta}'X'X\hat{\beta} = y'y - 2\hat{\beta}'X'X\hat{\beta} + \hat{\beta}'X'X\hat{\beta} \\ &= y'y - \hat{\beta}'X'X\hat{\beta} \\ &= y'y - \hat{y}'\hat{y}.\end{aligned}$$

The lemma is thus proved.

Now we come to the essential theorem—already informally introduced in "Decomposition of the Variance of y" section, which tells us that the total sum of squares is equal to the addition of the explained sum of squares and the residuals sum of squares:

Theorem 3.3 *If one of the regressors is a constant, then*

$$TSS = ESS + RSS.$$

Demonstration

From Lemma 3.2, it follows that $y'y = \hat{y}'\hat{y} + \hat{u}'\hat{u}$. By subtracting $T\,\overline{y}^2$ from the two members of this last equality, we obtain

$$y'y - T\,\overline{y}^2 = \hat{y}'\hat{y} - T\,\overline{y}^2 + \hat{u}'\hat{u}.$$

As elsewhere,

$$\sum (y_t - \overline{y})^2 = \sum (y_t^2 - 2\overline{y}y_t + \overline{y}^2)$$
$$= \sum y_t^2 - 2T\overline{y}^2 + T\overline{y}^2$$
$$= \mathbf{y}'\mathbf{y} - T\overline{y}^2,$$

it follows that

$$TSS = \hat{\mathbf{y}}'\hat{\mathbf{y}} - T\overline{y}^2 + RSS.$$

All that remains is to demonstrate that $\hat{\mathbf{y}}'\hat{\mathbf{y}} - T\overline{y}^2$ is indeed equal to the explained sum of squares. The latter is itself equal to

$$ESS = \sum_{t=1}^{T} (\hat{y}_t - \overline{y})^2 = \hat{\mathbf{y}}'\hat{\mathbf{y}} - 2\overline{y}\sum_{t=1}^{T} \hat{y}_t + T\overline{y}^2.$$

Moreover, according to Lemma 2.2, if one of the regressors is a constant, then

$$\overline{y}\sum_{t=1}^{T} \hat{y}_t = \overline{y}\sum_{t=1}^{T} (y_t - \hat{u}_t) = \overline{y}\sum_{t=1}^{T} y_t = T\overline{y}^2.$$

Substituting this last value into the ESS sum gives us

$$ESS = \hat{\mathbf{y}}'\hat{\mathbf{y}} - 2T\overline{y}^2 + T\overline{y}^2 = \hat{\mathbf{y}}'\hat{\mathbf{y}} - T\overline{y}^2$$

and therefore $TSS = ESS + RSS$. □

The various sums defined in this way, together with the equality just demonstrated, will be used to introduce a measure of the value of the least squares adjustment, a measure we will call the squared coefficient of correlation (sometimes called the squared multiple correlation coefficient).

Definition 3.2 The **coefficient of correlation**, R^2, is the number equal to

$$R^2 = \frac{ESS}{TSS} = 1 - \frac{RSS}{TSS}. \tag{3.9}$$

When one of the regressors is a constant, given that $TSS = ESS + RSS$ and the sums SCR and SCT are both non-negative, it follows that

$$0 \le R^2 \le 1.$$

The interpretation of the coefficient R^2 as a tool for measuring the value of a statistical fit is straightforward:

- if the sum of the *RSS* residuals is small relative to the total sum, then this coefficient is close to unity and we can therefore conclude that the adjustment made is good;
- Conversely, if the residual sum represents a significant proportion of the total sum of squares, then R^2 approaches 0 and the regression performed is not very satisfactory.

However, we should be wary of being too enthusiastic about this measure. Firstly, it only makes sense if the theoretical model is linear. Secondly, the value of the coefficient R^2 automatically increases as regressors are added to the model. To remedy this problem, a more appropriate indicator has been proposed: the **corrected or adjusted coefficient of correlation**, $\overline{R}^2$.

The correction consists of weighting (3.9) by the number of regressors, more precisely by the ratio $(T-1)/(T-K)$, which must be greater than 1, when $K > 1$. We then have the definition:

Definition 3.3 The **adjusted coefficient** $\overline{R}^2$ is given by

$$\overline{R}^2 = 1 - \frac{(T-1)\,RSS}{(T-K)\,TSS}. \tag{3.10}$$

Distribution and Moments of $\hat{\boldsymbol{\beta}}$

The vector $\hat{\boldsymbol{\beta}}$ is now considered as an estimator of $\boldsymbol{\beta}$,. i.e. as a multivariate random variable that follows a specific probability distribution with equally specific moments.

Theorem 3.4 *The least squares estimator* $\hat{\boldsymbol{\beta}}$ *is* ***normally*** *distributed with mean vector equal to* $\boldsymbol{\beta}$ *and variance–covariance matrix* $\sigma^2(\boldsymbol{X}'\boldsymbol{X})^{-1}$:

$$\hat{\boldsymbol{\beta}} \sim MN\left(\boldsymbol{\beta},\ \sigma^2(\boldsymbol{X}'\boldsymbol{X})^{-1}\right). \tag{3.11}$$

Demonstration

This is done in three stages: (i) proving the normality of $\hat{\boldsymbol{\beta}}$; (ii) calculating its expectation and (iii) calculating its variance–covariance matrix.

(i) Since the matrix $(\boldsymbol{X}'\boldsymbol{X})^{-1}\boldsymbol{X}'$ is non-stochastic, i.e. made up of constants, $\hat{\boldsymbol{\beta}}$ is linear in $\mathbf{y}$, which is itself distributed according to a distribution $N(\boldsymbol{X\beta},\ \sigma^2\boldsymbol{I}_T)$. It follows that $\hat{\boldsymbol{\beta}}$ is normally distributed.

(ii) first, we have

$$\hat{\boldsymbol{\beta}} = (\boldsymbol{X}'\boldsymbol{X})^{-1}\boldsymbol{X}'\boldsymbol{y} = (\boldsymbol{X}'\boldsymbol{X})^{-1}\boldsymbol{X}'(\boldsymbol{X}\boldsymbol{\beta} + \boldsymbol{u}) = \boldsymbol{\beta} + (\boldsymbol{X}'\boldsymbol{X})^{-1}\boldsymbol{X}'\boldsymbol{u}. \tag{3.12}$$

If we take the expectation of this last expression, we get

$$\begin{aligned} E(\hat{\boldsymbol{\beta}}) &= \boldsymbol{\beta} + E\left[(\boldsymbol{X}'\boldsymbol{X})^{-1}\boldsymbol{X}'\boldsymbol{u}\right] \\ &= \boldsymbol{\beta} + (\boldsymbol{X}'\boldsymbol{X})^{-1}\boldsymbol{X}'E(\boldsymbol{u}) \quad (\boldsymbol{X} \text{ is non stochastic}) \\ &= \boldsymbol{\beta}.(E(\boldsymbol{u}) = \boldsymbol{0}) \end{aligned}$$

(iii) From (3.12), we derive that $\hat{\boldsymbol{\beta}} - \boldsymbol{\beta} = (\boldsymbol{X}'\boldsymbol{X})^{-1}\boldsymbol{X}'\boldsymbol{u}$. The variance–covariance matrix of $\hat{\boldsymbol{\beta}}$ is then

$$\begin{aligned} \operatorname{Var}(\hat{\boldsymbol{\beta}}) &= E\left[(\hat{\boldsymbol{\beta}} - E(\hat{\boldsymbol{\beta}}))(\hat{\boldsymbol{\beta}} - E(\hat{\boldsymbol{\beta}}))'\right] \\ &= E\left[(\hat{\boldsymbol{\beta}} - \boldsymbol{\beta})(\hat{\boldsymbol{\beta}} - \boldsymbol{\beta})'\right] \\ &= E\left[(\boldsymbol{X}'\boldsymbol{X})^{-1}\boldsymbol{X}'\boldsymbol{u}\boldsymbol{u}'\boldsymbol{X}(\boldsymbol{X}'\boldsymbol{X})^{-1}\right] \\ &= (\boldsymbol{X}'\boldsymbol{X})^{-1}\boldsymbol{X}'E(\boldsymbol{u}\boldsymbol{u}')\boldsymbol{X}(\boldsymbol{X}'\boldsymbol{X})^{-1}. \end{aligned}$$

As $E(\boldsymbol{u}\boldsymbol{u}') = \sigma^2 \boldsymbol{I}_T$, we conclude that

$$\operatorname{Var}(\hat{\boldsymbol{\beta}}) = \sigma^2 (\boldsymbol{X}'\boldsymbol{X})^{-1}.$$

Note that $E(\hat{\boldsymbol{\beta}}) = \boldsymbol{\beta}$ and therefore $\hat{\boldsymbol{\beta}}$ is an unbiased estimator of $\boldsymbol{\beta}$. We will return to this question later.

3.2 The Maximum Likelihood Method

Having presented the ordinary least squares estimation technique, we now turn to the maximum likelihood technique. We will begin, however, with a few general points about this method.

General Information

Let be the parameter vector $\boldsymbol{\theta} = (\theta_1, \ldots \theta_k) \in \mathbb{R}^k$ and $(X_1, X_2, \ldots, X_n)$ a n-vector of absolutely continuous, identically and independently distributed random variables, each with density function $f_{X_i}(\cdot\,;\,\boldsymbol{\theta})$, where $\boldsymbol{\theta} \in \Theta \subset \mathbb{R}^k$. The **joint density** of $\mathbf{X}$ is then given by

$$f_X(x_1, \ldots, x_n;\, \boldsymbol{\theta}) = f_{X_1}(x_1;\, \boldsymbol{\theta}) \cdot \ldots \cdot f_{X_n}(x_n;\, \boldsymbol{\theta}).$$

Now consider the sample $\boldsymbol{x} = (x_1, \ldots, x_n)$, realisation of the random vector $\mathbf{X}$. If these observations are considered as data and the joint density as a function of the parameter vector, then we have the following definition:

Definition 3.4 The **likelihood function** of the parameter vector $\boldsymbol{\theta}$, $L(\boldsymbol{\theta}\,; x_1, \ldots, x_n)$, is the application $L : \mathbb{R}^k \to \mathbb{R}$ such that

$$L(\boldsymbol{\theta}\,; x_1, \ldots, x_n) = f_X(x_1, \ldots, x_n\,; \boldsymbol{\theta}) = \prod_{i=1}^{n} f(x_i;\ \boldsymbol{\theta}), \tag{3.13}$$

where $f_X(x_1, \ldots, x_n;\ \boldsymbol{\theta})$ is the joint density of the vector X.

By taking the logarithms in (3.13), we obtain the **log-likelihood function** of $\boldsymbol{\theta}$, i.e.

$$\ln L(\boldsymbol{\theta}\,; x_1, \ldots, x_n) = \sum_{i=1}^{n} \ln f(x_i;\ \boldsymbol{\theta}). \tag{3.14}$$

The maximum likelihood principle requires us to select an estimator of $\boldsymbol{\theta}$, that maximises $L(\boldsymbol{\theta}\,; x_1, \ldots, x_n)$. We therefore have:

Definition 3.5 A **maximum likelihood estimator** of $\boldsymbol{\theta}$ is a vector $\hat{\boldsymbol{\theta}} \in \Theta$ such that for all $\tilde{\boldsymbol{\theta}} \in \Theta$,

$$L(\hat{\boldsymbol{\theta}}\,; x_1, \ldots, x_n) \geq L(\tilde{\boldsymbol{\theta}}\,; x_1, \ldots, x_n)$$

Alternatively, using (3.14), we can of course write that the estimator in question is such that

$$\ln L(\hat{\boldsymbol{\theta}}\,; x_1, \ldots, x_n) \geq \ln L(\tilde{\boldsymbol{\theta}}\,; x_1, \ldots, x_n).$$

Assuming that the likelihood function is continuously differentiable up to order two, the first-order conditions for a local maximum require the cancellation of the partial derivatives of this function, which gives the **likelihood equations:**

$$\frac{\partial}{\partial \theta_i} \ln L(\boldsymbol{\theta}, \boldsymbol{x}) = 0,\ i = 1, \ldots, k.$$

Any solution to the above system represents a stationary point of $\ln L(\boldsymbol{\theta}\,; \boldsymbol{x})$. For this point to be a maximant of the likelihood function, the second-order conditions must also be satisfied, in other words, the matrix of second partial derivatives of this likelihood function must **be negative** (see Appendix B, Definition 42).

From a practical point of view, we proceed as follows to determine a maximum likelihood estimator:

1. write down the likelihood function for the sample, then put it into logarithmic form;
2. cancel the vector of partial derivatives of the log-likelihood function, which gives the likelihood equations;
3. solve these likelihood equations and obtain the vector $\hat{\boldsymbol{\theta}}$;
4. verify that the second-order conditions are satisfied, so that $\hat{\boldsymbol{\theta}}$ is indeed a maximum likelihood estimator.

This last stage is often overlooked.

Example 3.1 Given a random sample of size n, where each observation x_j ($j = 1, \ldots, n$) is generated by an exponential density

$$f(x_j;\, \theta) = \theta\, e^{-\theta x_j},\ x_j > 0,\ \theta > 0.$$

The likelihood function is $L(\theta\,;\,\boldsymbol{x}) = \prod_j \theta\, e^{-\theta x_j}$ and the log-likelihood function is

$$\ln L(\theta\,;\,\boldsymbol{x}) = \sum_{j=1}^{n} (\ln\theta - \theta x_j) = n\ln\theta - \theta \sum_{j=1}^{n} x_j. \tag{3.15}$$

Let's cancel the first derivative of (3.15):

$$\frac{d\ln L(\theta\,;\,\boldsymbol{x})}{d\theta} = \frac{n}{\theta} - \sum_{j=1}^{n} x_j = 0,$$

to obtain

$$\hat{\theta} = \frac{n}{\sum\limits_{j=1}^{n} x_j}.$$

This last estimator is indeed the maximum likelihood estimator, since the second-order condition is satisfied:

$$\frac{d^2\ln L\big(\hat{\theta}, \boldsymbol{x}\big)}{d\theta^2} = -\ n/\hat{\theta}^2 < 0.$$

The Maximum Likelihood Estimator of $\boldsymbol{\beta}$

We again start with the simple dynamic linear model $\boldsymbol{y} = \boldsymbol{X\beta} + \boldsymbol{u}$, where $\boldsymbol{u} \sim MN(\boldsymbol{0}, \sigma^2 \boldsymbol{I}_T)$ and we seek to determine the maximum likelihood estimator (MLE) of $\boldsymbol{\beta}$. Alternatively, this model can also be expressed as

$$y_t = \boldsymbol{x'}_t \boldsymbol{\beta} + u_t,\ t = 1, \ldots, T, \tag{3.16}$$

where the line vector $\boldsymbol{x'}_t$ has K components, namely the unit and the various explanatory variables as well as the lags associated with each of them.

Designating the least squares estimator by $\hat{\boldsymbol{\beta}}_{LS}$ and the maximum likelihood estimator by $\hat{\boldsymbol{\beta}}_{ML}$, we will show that these two estimators coincide in the case of the linear model under consideration.

Theorem 3.5 *Let* $\boldsymbol{y} = \boldsymbol{X\beta} + \boldsymbol{u}$, *satisfying assumptions* (H1) *and* (H2). *Then*

$$\hat{\boldsymbol{\beta}}_{ML} = \hat{\boldsymbol{\beta}}_{LS}.$$

Demonstration

We will use the four-step procedure recommended in the previous section, starting by determining the likelihood function.

1. Bearing in mind that for all t, $y_t \sim N(\boldsymbol{x}'_t\boldsymbol{\beta}, \sigma^2)$, it follows that the density of y_t is given by

$$f(y_t;\ \boldsymbol{\beta}, \sigma^2) = \frac{1}{\sigma\sqrt{2\pi}} e^{-\frac{1}{2\sigma^2}(y_t - \boldsymbol{x'}_t\boldsymbol{\beta})^2}.$$

   Since the observations y_t are independently distributed, the likelihood function of the sample $\boldsymbol{y} = (y_1, \ldots, y_T)$ is:

$$L(\boldsymbol{\beta}, \sigma^2;\ \boldsymbol{y}) = \prod_{t=1}^{T} \frac{1}{\sigma\sqrt{2\pi}} e^{-\frac{1}{2\sigma^2}(y_t - \boldsymbol{x'}_t\boldsymbol{\beta})^2}$$
$$= \left(\sigma^2 2\pi\right)^{-T/2} e^{-\frac{1}{2\sigma^2}\sum_{t=1}^{T}(y_t - \boldsymbol{x'}_t\boldsymbol{\beta})^2}. \tag{3.17}$$

   The log-likelihood is obtained by taking the logarithm of (3.17), i.e.

$$\ln L(\boldsymbol{\beta}, \sigma^2;\ \boldsymbol{y}) = -\frac{T}{2}\ln 2\pi - \frac{T}{2}\ln\sigma^2 - \frac{1}{2\sigma^2}\sum_{t=1}^{T}\left(y_t - \boldsymbol{x'}_t\boldsymbol{\beta}\right)^2. \tag{3.18}$$

   This function can also be written as:

$$\ln L(\boldsymbol{\beta}, \sigma^2; \boldsymbol{y}) = -\frac{T}{2}\ln 2\pi - \frac{T}{2}\ln \sigma^2 - \frac{1}{2\sigma^2}(\boldsymbol{y} - \boldsymbol{X\beta})'(\boldsymbol{y} - \boldsymbol{X\beta}). \tag{3.19}$$

2. We cancel the matrix derivative of (3.19) with respect to the vector $\boldsymbol{\beta}$, which leads to:

$$\begin{aligned}\frac{\partial}{\partial \boldsymbol{\beta}} \ln L(\boldsymbol{\beta}, \sigma^2; \boldsymbol{y}) &= \frac{\partial}{\partial \boldsymbol{\beta}}\left[\left(-\frac{1}{2\sigma^2}(\boldsymbol{y}'\boldsymbol{y} - 2\boldsymbol{y}'\boldsymbol{X\beta} + \boldsymbol{\beta}'\boldsymbol{X}'\boldsymbol{X\beta})\right)\right] \\ &= -\frac{1}{2\sigma^2}\left(-2\boldsymbol{X}'\boldsymbol{y} + 2\boldsymbol{X}'\boldsymbol{X\beta}\right) \\ &= \mathbf{0}.\end{aligned}$$

3. By solving the system of likelihood equations $\boldsymbol{X}'\boldsymbol{y} - \boldsymbol{X}'\boldsymbol{X\beta} = \mathbf{0}$, using Lemma 3.1 we obtain:

$$\hat{\boldsymbol{\beta}}_{ML} = (\boldsymbol{X}'\boldsymbol{X})^{-1}\boldsymbol{X}'\boldsymbol{y}. \tag{3.20}$$

4. Since $\boldsymbol{X}'\boldsymbol{X}$ is positive (see Theorem 3.1), the matrix of second partial derivatives of the log-likelihood, i.e.

$$\frac{\partial^2 \ln L}{\partial \boldsymbol{\beta} \partial \boldsymbol{\beta}'} = -\frac{1}{\sigma^2}\boldsymbol{X}'\boldsymbol{X}$$

is negative, so that (3.20) represents the maximum likelihood estimator of $\boldsymbol{\beta}$ and is equal to the OLS estimator.

Estimation of the Variance σ^2

The log-likelihood function (3.19) contains σ^2 in the second and third terms of the right-hand side. Using the recommended procedure, we can therefore attempt to determine the maximum likelihood estimator of the variance of the perturbations.

Theorem 3.6 *Under assumptions* (H1) *and* (H2), *the maximum likelihood estimator of σ^2 is*

$$\hat{\sigma}^2_{MV} = (1/T)\,\hat{\boldsymbol{u}}'\hat{\boldsymbol{u}}, \tag{3.21}$$

where $\hat{\boldsymbol{u}} = \boldsymbol{y} - \boldsymbol{X}\hat{\boldsymbol{\beta}}$.

Demonstration

The proof strategy consists of replacing $\boldsymbol{\beta}$ by its estimator in (3.19). In this way, we obtain the **concentrated log-likelihood** function (the subscript "c" distinguishes the latter), which now depends only on the variance:

$$\ln L_c(\sigma^2 ; \boldsymbol{y}) = -\frac{T}{2} \ln 2\pi - \frac{T}{2} \ln \sigma^2 - \frac{1}{2\sigma^2}(\boldsymbol{y} - \boldsymbol{X}\hat{\boldsymbol{\beta}})'(\boldsymbol{y} - \boldsymbol{X}\hat{\boldsymbol{\beta}}).$$

Applying the first-order condition for a maximum first renders the likelihood equation

$$\frac{\partial \ln L_c(\sigma^2 ; \boldsymbol{y})}{\partial \sigma^2} = -\frac{T}{2\sigma^2} + \frac{1}{2\sigma^4}(\boldsymbol{y} - \boldsymbol{X}\hat{\boldsymbol{\beta}})'(\boldsymbol{y} - \boldsymbol{X}\hat{\boldsymbol{\beta}}) = 0,$$

which then leads to

$$\hat{\sigma}^2(\hat{\boldsymbol{\beta}}) = \frac{1}{T}(\boldsymbol{y} - \boldsymbol{X}\hat{\boldsymbol{\beta}})'(\boldsymbol{y} - \boldsymbol{X}\hat{\boldsymbol{\beta}}).$$

Since the least squares and maximum likelihood estimators of $\boldsymbol{\beta}$ are identical, we finally obtain

$$\hat{\sigma}^2 = (1/T)\, \hat{\boldsymbol{u}}'\hat{\boldsymbol{u}}.$$

The variance estimator is therefore equal to the average of the squares of the residuals. The problem with this estimator is that it is **biased**. Let's recall the definition of an unbiased estimator.

Definition 3.6 Let $\hat{\boldsymbol{\theta}}$ be a *n-estimator* of the vector $\boldsymbol{\theta}$.$\hat{\boldsymbol{\theta}}$ is said to be **unbiased** if $E(\hat{\boldsymbol{\theta}}) = \boldsymbol{\theta}$. Otherwise $\hat{\boldsymbol{\theta}}$ is **biased**.

We have said that the estimator (3.21) of the variance is biased. This is what we are going to demonstrate. First, let's introduce the matrix

$$\boldsymbol{M} = \boldsymbol{I}_T - \boldsymbol{X}'(\boldsymbol{X}'\boldsymbol{X})^{-1}\boldsymbol{X}.$$

It is easy to check that this matrix is symmetrical ($\boldsymbol{M} = \boldsymbol{M}'$) and that it is idempotent ($\boldsymbol{M}\boldsymbol{M} = \boldsymbol{M}$). In the same way, it follows:

$$\boldsymbol{M}\boldsymbol{X} = (\boldsymbol{I}_T - \boldsymbol{X}(\boldsymbol{X}'\boldsymbol{X})^{-1}\boldsymbol{X}')\boldsymbol{X} = \boldsymbol{X} - \boldsymbol{X} = \boldsymbol{0}. \tag{3.22}$$

We will also need the following lemma.

Lemma 3.3 *The vectors of the estimation residuals and the disturbances are linked by the relation*

$$\hat{\boldsymbol{u}} = \boldsymbol{M}\boldsymbol{u}.$$

Demonstration

Using (3.22), we immediately obtain

$$\hat{\boldsymbol{u}} = \boldsymbol{y} - \boldsymbol{X}\hat{\boldsymbol{\beta}} = \boldsymbol{y} - \boldsymbol{X}(\boldsymbol{X}'\boldsymbol{X})^{-1}\boldsymbol{X}'\boldsymbol{y} = \boldsymbol{M}\boldsymbol{y} = \boldsymbol{M}(\boldsymbol{X}\boldsymbol{\beta} + \boldsymbol{u}) = \boldsymbol{M}\boldsymbol{u}.$$

Finally, to show that the estimator (2.21) is biased, we need to use the trace operator of a matrix.

Definition 3.7 Let $\boldsymbol{A} = (a_{ij})_{n\times n}$. The trace of **A** is equal to

$$\mathrm{tr}(\boldsymbol{A}) = \sum_{i=1}^{n} a_{ii}.$$

Example 3.2 Consider the square matrix $\boldsymbol{A} = \begin{pmatrix} 2 & 1 \\ -2 & 3 \end{pmatrix}$.

Its trace is equal to $\mathrm{tr}(\boldsymbol{A}) = 2 + 3 = 5$,, i.e. the sum of the elements of its main diagonal.

Two properties of the trace will soon be useful to us (see Appendix B):

(i) the trace is a linear form so that $\forall \boldsymbol{A}, \boldsymbol{B} \in \mathbb{R}^{m\times n}$ and $\forall \alpha, \beta \in \mathbb{R}$,

$$\mathrm{tr}(\alpha\boldsymbol{A} + \beta\boldsymbol{B}) = \alpha\mathrm{tr}(\boldsymbol{A}) + \beta\mathrm{tr}(\boldsymbol{B}).$$

(ii) the trace of a product of two matrices does not depend on the order of multiplication of these matrices: $\forall \boldsymbol{A} \in \mathbb{R}^{m\times n}$ and $\forall \boldsymbol{B} \in \mathbb{R}^{n\times m}$,

$$\mathrm{tr}(\boldsymbol{AB}) = \boldsymbol{tr}(\boldsymbol{BA}).$$

Here is the theorem:

Theorem 3.7 *The estimator of the variance of the residuals has an estimation bias such that*

$$E(\hat{\sigma}^2) = \frac{T-K}{T}\sigma^2. \tag{3.23}$$

Demonstration

Using the properties of the matrix **M** and Lemma 3.3, we get

$$\hat{\boldsymbol{u}}'\hat{\boldsymbol{u}} = \boldsymbol{u}'\boldsymbol{M}'\boldsymbol{M}\boldsymbol{u} = \boldsymbol{u}'\boldsymbol{M}\boldsymbol{u}.$$

Furthermore, as $\boldsymbol{u}'\boldsymbol{M}\boldsymbol{u}$ is a quadratic form, and therefore a scalar, it is equal to its trace, so using the two properties of the trace, we obtain

$$\begin{aligned} E\left[\mathrm{tr}(\boldsymbol{u}'\boldsymbol{M}\boldsymbol{u})\right] &= \mathrm{tr}\left[E(\boldsymbol{u}'\boldsymbol{M}\boldsymbol{u})\right] = \mathrm{tr}\left[E(\boldsymbol{M}\boldsymbol{u}\boldsymbol{u}')\right] \\ &= \mathrm{tr}(\boldsymbol{M}\sigma^2\boldsymbol{I}_T) = \mathrm{tr}(\sigma^2\boldsymbol{M}) \\ &= \sigma^2\mathrm{tr}(\boldsymbol{M}). \end{aligned}$$

However, we also have

$$\begin{aligned} \mathrm{tr}(\boldsymbol{M}) &= \boldsymbol{tr}\left[\boldsymbol{I}_T - \boldsymbol{X}(\boldsymbol{X}'\boldsymbol{X})^{-1}\boldsymbol{X}'\right] = \mathrm{tr}(\boldsymbol{I}_T) - \mathrm{tr}\left[(\boldsymbol{X}'\boldsymbol{X})^{-1}\boldsymbol{X}'\boldsymbol{X}\right] \\ &= T - K. \end{aligned}$$

It follows from the two previous results that

$$\begin{aligned} E(\hat{\sigma}^2) &= E\left((1/T)\,\hat{\boldsymbol{u}}'\hat{\boldsymbol{u}}\right) = (1/T)E(\boldsymbol{u}'\boldsymbol{M}\boldsymbol{u}) = (1/T)E\left[\mathrm{tr}(\boldsymbol{u}'\boldsymbol{M}\boldsymbol{u})\right] \\ &= \frac{T-K}{T}\sigma^2. \end{aligned}$$

This proves the theorem.

Looking at (3.23), we see that to obtain a centred estimator of the variance, it is sufficient to divide the sum of the squares of the residuals by $(T - K)$ rather than by T. We therefore have.

Definition 3.8 The unbiased variance estimator, s^2, is given by

$$s^2 = \frac{1}{T-K}\hat{\boldsymbol{u}}'\hat{\boldsymbol{u}}. \tag{3.24}$$

3.3 Model Selection Criteria

In "Estimation by Maximum Likelihood" section, we introduced the coefficient of correlation (as well as its "adjusted" version) as an instrument for assessing the value of a statistical adjustment made by least squares. This coefficient can also be used to compare two or more econometric models. But there are also information criteria for comparison purposes.

The Akaike Criterion

When statistical adjustment is carried out by maximum likelihood, other specific measures of model comparison are available. The oldest of these is based on the **Akaike Information Criterion** (*AIC*).

The basic idea is quite simple: it involves a trade-off between maximising log-likelihood and adding additional regressors.

More precisely, we have the statistic

$$AIC = -2 \ln L(\hat{\boldsymbol{\beta}}, \hat{\sigma}^2) + 2K, \tag{3.25}$$

where $\ln L(\hat{\boldsymbol{\beta}}, \hat{\sigma}^2)$ is the maximum log-likelihood and K is the number of regressors.

It follows that the *AIC* criterion is therefore negatively related to the maximum likelihood value and positively related to the number of explanatory variables.

The decision rule associated with the criterion is therefore to select, from a set of models, the one which **minimises** the criterion and is therefore **the most parsimonious** in terms of the number of regressors used.

In the case of the linear model, the maximum log-likelihood is given by

$$\begin{aligned}\ln L(\hat{\boldsymbol{\beta}}, \hat{\sigma}^2) &= -\frac{T}{2}\ln 2\pi - \frac{T}{2}\ln\left[\frac{1}{T}(\boldsymbol{y} - \boldsymbol{X}\hat{\boldsymbol{\beta}})'(\boldsymbol{y} - \boldsymbol{X}\hat{\boldsymbol{\beta}})\right] - \frac{T}{2} \\ &= -\frac{T}{2}\ln 2\pi - \frac{T}{2}\ln\left[\frac{1}{T}RSS\right] - \frac{T}{2} \\ &= -\frac{T}{2}(1 + \ln 2\pi - \ln T) - \frac{T}{2}\ln RSS,\end{aligned}$$

where *RSS* is, of course, the sum of the squares of the estimated residuals.

Substituting the latter expression into (3.25) gives

$$\begin{aligned}AIC &= T(1 + \ln 2\pi - \ln T) + T\ln RSS + 2K \\ &= T(1 + \ln 2\pi) + T\ln\left(\frac{RSS}{T}\right) + 2K,\end{aligned}$$

which can be rewritten in the final form:

$$AIC = T\ln\left(\frac{RSS}{T}\right) + 2K + T(1 + \ln 2\pi). \tag{3.26}$$

The application of the Akaike criterion suffers from several shortcomings, the most important of which is that it sometimes leads to the selection of the least parsimonious model.

Other Selection Criteria

To overcome the limitations inherent in the Akaike criterion, two other criteria have been proposed:

- the Bayesian Information Criterion (*BIC*), due to Schwarz, is defined by

$$BIC = -2 \ln L(\hat{\boldsymbol{\beta}}, \hat{\sigma}^2) + K \ln T. \tag{3.27}$$

  Multiplying K by the logarithm of T means that the penalty associated with the introduction of new regressors increases with the size of the sample.
- the Hannan–Quinn information criterion (*HQC*) is given by

$$HQC = -2 \ln L(\hat{\boldsymbol{\beta}}, \hat{\sigma}^2) + 2K \ln(\ln T). \tag{3.28}$$

  The penalty calculation here is based on a logarithm of a logarithm (in short, an application of the law of the iterated logarithm)

Here is a reminder of the decision rule to be applied, whatever the criterion used:

> If $\mathcal{M}_1$ and $\mathcal{M}_2$ are two models estimated by maximum likelihood, we will select the one that minimises one or more of the distinguished criterion functions:
>
> $$\min\left(AIC_{\mathcal{M}_1}, CIA_{\mathcal{M}_2}\right), \min\left(BIC_{\mathcal{M}_1}, CIB_{\mathcal{M}_2}\right), \min\left(HQC_{\mathcal{M}_1}, HQC_{\mathcal{M}_2}\right).$$

3.4 Optimality Properties of Linear Estimators

We will first consider these properties in small samples—at finite distance, we say—before examining them asymptotically.

The Gauss–Markov Theorem

Optimality here is based on the fact that the least squares estimator has the smallest variance in the class of all the linear and unbiased estimators. The OLS estimator is then said to be the *best linear unbiased estimator* (BLUE).

Definition 3.9 Let $\mathcal{L}$ be the class of all linear unbiased estimators of $\boldsymbol{\beta}$. Then $\hat{\boldsymbol{\beta}}$ is the **best linear unbiased estimator** (BLUE) of $\boldsymbol{\beta}$ if for any other estimator $\tilde{\boldsymbol{\beta}} \in \mathcal{L}$,

$$\text{Var}(\tilde{\boldsymbol{\beta}}) - \text{Var}(\hat{\boldsymbol{\beta}}) \geq \mathbf{0},$$

where $\text{Var}(\tilde{\boldsymbol{\beta}}) - \text{Var}(\hat{\boldsymbol{\beta}})$ is a **positive semi-definite** matrix.

Before stating the theorem and in order to prove it, we will rely on the most general assumptions concerning the potential stochasticity of the explanatory variables. We will therefore adopt assumptions (3.5) and (3.6), i.e.

$$E(\boldsymbol{u}|\boldsymbol{X}) = \mathbf{0}$$
$$\text{Var}(\boldsymbol{u}|\boldsymbol{X}) = \sigma^2 \boldsymbol{I}_T.$$

Within this general framework, let us introduce the class of linear estimators $\tilde{\boldsymbol{\beta}}_L = \boldsymbol{Q}\boldsymbol{y}$, defined on the set of matrices $\mathbf{Q}$ of dimension $K \times T$ such that the following assumptions are satisfied:

(i) $E(\boldsymbol{u}|\ \boldsymbol{X}, \boldsymbol{Q}) = \mathbf{0}$,

(ii) $\text{Var}(\boldsymbol{u}|\ \boldsymbol{X}, \boldsymbol{Q}) = E(\boldsymbol{u}\boldsymbol{u}'|\ \boldsymbol{X}, \boldsymbol{Q}) = \sigma^2 \boldsymbol{I}_T$,

(iii) $\boldsymbol{Q}\boldsymbol{X} = \boldsymbol{I}_K$.

Then, knowing that the model used is $\boldsymbol{y} = \boldsymbol{X}\boldsymbol{\beta} + \boldsymbol{u}$,

$$E\big(\tilde{\boldsymbol{\beta}}_L|\ \boldsymbol{X}, \boldsymbol{Q}\big) = E(\boldsymbol{Q}(\boldsymbol{X}\boldsymbol{\beta} + \boldsymbol{u})|\ \boldsymbol{X}, \boldsymbol{Q}) = E(\boldsymbol{Q}\boldsymbol{X}\boldsymbol{\beta}|\ \boldsymbol{X}, \boldsymbol{Q}) + E(\boldsymbol{Q}\boldsymbol{u}|\ \boldsymbol{X}, \boldsymbol{Q}). \quad (3.29)$$

Taking the unconditional expectations from (3.29), we get

$$E\big(\tilde{\boldsymbol{\beta}}_L\big) = \boldsymbol{\beta} + E\big[\boldsymbol{Q}E(\boldsymbol{Q}\boldsymbol{u}|\ \boldsymbol{X}, \boldsymbol{Q})\big] = \boldsymbol{\beta}, \quad (3.30)$$

from which we deduce that all $\tilde{\boldsymbol{\beta}}_L$ are linear and unbiased. The class $\tilde{\boldsymbol{\beta}}_L$ is therefore made up of linear and unbiased estimators of $\boldsymbol{\beta}$.

Given (3.30), the conditional variance of $\tilde{\boldsymbol{\beta}}_L$ is given by

$$\text{Var}(\tilde{\boldsymbol{\beta}}_L|\ \boldsymbol{X}, \boldsymbol{Q}) = E\big[(\tilde{\boldsymbol{\beta}}_L - \boldsymbol{\beta})(\tilde{\boldsymbol{\beta}}_L - \boldsymbol{\beta})'|\ \boldsymbol{X}, \boldsymbol{Q}\big].$$

As $\tilde{\boldsymbol{\beta}}_L = \boldsymbol{Q}(\boldsymbol{X}\boldsymbol{\beta} + \boldsymbol{u}) = \boldsymbol{\beta} + \boldsymbol{Q}\boldsymbol{u}$, this variance can be written as

$$\text{Var}(\tilde{\boldsymbol{\beta}}_L|\ \boldsymbol{X}, \boldsymbol{Q}) = E(\boldsymbol{Q}\boldsymbol{u}\boldsymbol{u}'\boldsymbol{Q}'|\ \boldsymbol{X}, \boldsymbol{Q}) = \sigma^2 \boldsymbol{Q}\boldsymbol{Q}'. \quad (3.31)$$

Here is the announced theorem:

Theorem 3.8 *Let be the class of linear unbiased estimators $\tilde{\boldsymbol{\beta}}_L = \boldsymbol{Qy}$ and $\hat{\boldsymbol{\beta}}$ the least squares estimator. Then the matrix*

$$\operatorname{Var}(\tilde{\boldsymbol{\beta}}_L) - \operatorname{Var}(\hat{\boldsymbol{\beta}})$$

is a ***positive semidefinite*** *matrix for all* **Q** *satisfying assumptions (i)–(iii).*

Demonstration

Let $\boldsymbol{D} = \boldsymbol{Q} - (\boldsymbol{X}'\boldsymbol{X})^{-1}\boldsymbol{X}'$. It follows that

$$\boldsymbol{DX} = \boldsymbol{QX} - (\boldsymbol{X}'\boldsymbol{X})^{-1}\boldsymbol{X}'\boldsymbol{X} = \boldsymbol{0}.$$

We have also

$$\begin{aligned}\boldsymbol{QQ}' &= \left((\boldsymbol{X}'\boldsymbol{X})^{-1}\boldsymbol{X}' + \boldsymbol{D}\right)\left((\boldsymbol{X}'\boldsymbol{X})^{-1}\boldsymbol{X}' + \boldsymbol{D}\right)' \\ &= (\boldsymbol{X}'\boldsymbol{X})^{-1}\boldsymbol{X}'\boldsymbol{X}(\boldsymbol{X}'\boldsymbol{X})^{-1} + \boldsymbol{DX}(\boldsymbol{X}'\boldsymbol{X})^{-1} + (\boldsymbol{X}'\boldsymbol{X})^{-1}\boldsymbol{X}'\boldsymbol{D}' + \boldsymbol{DD}' \\ &= (\boldsymbol{X}'\boldsymbol{X})^{-1} + \boldsymbol{DD}'.\end{aligned}$$

From the previous equality and (3.31), we get

$$\operatorname{Var}(\tilde{\boldsymbol{\beta}}_L \mid \boldsymbol{X}, \boldsymbol{Q}) = \sigma^2\left[(\boldsymbol{X}'\boldsymbol{X})^{-1} + \boldsymbol{DD}'\right] = \operatorname{Var}(\hat{\boldsymbol{\beta}} \mid \boldsymbol{X}, \boldsymbol{Q}) + \sigma^2\boldsymbol{DD}' \tag{3.32}$$

Taking the unconditional expectations from (3.32), we then obtain

$$\operatorname{Var}(\tilde{\boldsymbol{\beta}}_L) = \operatorname{Var}(\hat{\boldsymbol{\beta}}) + \sigma^2 E(\boldsymbol{DD}').$$

Finally, given that the matrix $\boldsymbol{DD}'$ is positive semidefinite (see Appendix B.8, 44), it follows that

$$\operatorname{Var}(\tilde{\boldsymbol{\beta}}_L) - \operatorname{Var}(\hat{\boldsymbol{\beta}}) \geq \boldsymbol{0}.$$

Convergence in Probability

We will now look at the asymptotic properties of OLS estimators, i.e. the properties obtained when the sample size tends towards infinity. In particular, we want to show that $\hat{\boldsymbol{\beta}}$ is **convergent in probability**. Let us recall the general definition of such a form of convergence.

Definition 3.10 A sequence (X_n) of random variables **converges in probability** to X if for all $\varepsilon > 0$,

$$\lim_{n \to +\infty} P(|X_n - X| > \varepsilon) = 0.$$

We then note

$$\operatorname{plim}_{n \to +\infty} X_n = X.$$

Our objective is to demonstrate that

$$\operatorname{plim}_{T \to +\infty} \hat{\boldsymbol{\beta}}_T = \boldsymbol{\beta}.$$

We begin by substituting the hypothesis (H_2) of the initial model with this other hypothesis:

(H_2') The observation matrix **X is** a non-stochastic matrix of rank $K < T$ such that

$$\boldsymbol{L} = \operatorname{plim}_{T \to +\infty} \left(\frac{1}{T} \boldsymbol{X}'\boldsymbol{X} \right)$$

exists and is **non-singular**.

The justification for this assumption is simple: we expect the matrix $\boldsymbol{X}'\boldsymbol{X}$, each element of which is a scalar product of two vectors of dimension T, to tend towards infinity when T itself tends towards infinity. To avoid this matrix having no limit in probability, we simply divide each of its elements by T.

To show that the least squares estimator is convergent, we need the following lemma:

Lemma 3.4 *Under assumption* (H_2'), $\operatorname{plim}_{T \to +\infty} \left(\frac{1}{T} \boldsymbol{X}'\boldsymbol{u} \right) = \boldsymbol{0}$.

Demonstration

We obviously have $E(\boldsymbol{X}'\boldsymbol{u}/T) = \boldsymbol{0}$ for all T. The variance of $\boldsymbol{X}'\boldsymbol{u}/T$ is therefore equal to

$$\operatorname{Var}(\boldsymbol{X}'\boldsymbol{u}/T) = E\big[(\boldsymbol{X}'\boldsymbol{u}/T)(\boldsymbol{X}'\boldsymbol{u}/T)'\big] = (\sigma^2/T)(\boldsymbol{X}'\boldsymbol{X}/T),$$

so that its asymptotic variance is

$$\underset{T\to+\infty}{\text{plim}} E\left[\left(\frac{X'u}{T}\right)\left(\frac{X'u}{T}\right)'\right] = \underset{T\to+\infty}{\text{plim}} \frac{\sigma^2}{T}\cdot L = 0.$$

Since the asymptotic expectation and variance of $X'u/T$ are zero, it follows that

$$\underset{T\to+\infty}{\text{plim}} \frac{X'u}{T} = 0.$$

We can now state and prove the essential theorem of this section.

Theorem 3.9 *Under the assumption* (H$_2$'), *the least squares estimator of* $\boldsymbol{\beta}$ *is* ***convergent***.

Demonstration

Since

$$\hat{\boldsymbol{\beta}} = (X'X)^{-1}X'(X\boldsymbol{\beta}+u) = \boldsymbol{\beta} + (X'X)^{-1}X'u = \boldsymbol{\beta} + \left(\frac{X'X}{T}\right)^{-1}\left(\frac{X'u}{T}\right),$$

Given Lemma 3.4,

$$\underset{T\to+\infty}{\text{plim}}\ \hat{\boldsymbol{\beta}} = \boldsymbol{\beta} + \underset{T\to+\infty}{\text{plim}}\left[\left(\frac{X'X}{T}\right)^{-1}\left(\frac{X'u}{T}\right)\right] = \boldsymbol{\beta} + L^{-1}\underset{T\to+\infty}{\text{plim}}\left(\frac{X'u}{T}\right) = \boldsymbol{\beta}.$$

We could demonstrate in a similar way—which we will not do—that the variance estimator σ^2 is also convergent in probability, i.e. that it tends asymptotically towards the true value of the parameter.

Convergence and Asymptotic Efficiency

Having established that the $\hat{\boldsymbol{\beta}}$ least squares estimator is convergent, we will now demonstrate that it has another optimality property in large samples: it is said to be efficient in the sense that, asymptotically, its variance is minimal in the class of all centred estimators of $\boldsymbol{\beta}$.

Such an approach would imply, as was done for the Gauss–Markov theorem, that we compare the variance of the OLS or maximum likelihood estimator with that of any other estimator of a particular class of estimators of $\boldsymbol{\beta}$. Fortunately, there is another, simpler approach, based on consideration of the Fréchet-Rao-Cramer lower bound.

Definition 3.11 The Fréchet-Rao-Cramèr **lower bound** (LB) is the **inverse of** the **Fisher information matrix** defined by

$$\boldsymbol{I}(\boldsymbol{\beta}) = -E\big(\partial^2 \ln L(\boldsymbol{\beta}\,;\boldsymbol{y})/\partial\boldsymbol{\beta}\,\partial\boldsymbol{\beta}'\big).$$

Theorem 3.10 *The Fréchet-Rao-Cramér lower bound for the* $\boldsymbol{\beta}$ *estimator is given by* $\sigma^2(\boldsymbol{X}'\boldsymbol{X})^{-1}$.

Demonstration

Let's start by calculating the partial second derivative of the log-likelihood function:

$$\frac{\partial^2 \ln L(\boldsymbol{\beta})}{\partial\boldsymbol{\beta}\partial\boldsymbol{\beta}'} = \frac{\partial}{\partial\boldsymbol{\beta}}\left[-\frac{1}{\sigma^2}(-\boldsymbol{X}'\boldsymbol{y} + \boldsymbol{X}'\boldsymbol{X}\boldsymbol{\beta})\right] = -\frac{1}{\sigma^2}(\boldsymbol{X}'\boldsymbol{X}).$$

The expectation of this second derivative with a negative sign gives the Fisher information matrix

$$\boldsymbol{I}(\boldsymbol{\beta}) = -E\left(\frac{\partial^2 \ln L(\boldsymbol{\beta})}{\partial\boldsymbol{\beta}\partial\boldsymbol{\beta}'}\right) = \frac{1}{\sigma^2}(\boldsymbol{X}'\boldsymbol{X}), \tag{3.33}$$

from which we can immediately deduce that the bound we are looking for is equal to the inverse of this matrix, i.e. $\sigma^2(\boldsymbol{X}'\boldsymbol{X})^{-1}$.

Generally speaking, an estimator is said to be **efficient** if it achieves the lower bound, i.e. if it has the minimum variance in a given class of estimators.

However, what is of particular interest to us here is the notion of **asymptotic efficiency**. From this point of view, we define asymptotic BIFRAC as follows.

Definition 3.12 The **asymptotic Fréchet-Rao-Cramér lower bound** for the linear model under consideration is the matrix $K \times K$:

$$LB = \frac{\sigma^2}{T}\text{plim}_{T\to+\infty}\,(\boldsymbol{X}'\boldsymbol{X}/T)^{-1}. \tag{3.34}$$

Using this definition, we demonstrate that the asymptotic variance of $\hat{\boldsymbol{\beta}}$ reaches the lower bound defined in (3.34) and that, consequently, the estimator itself is asymptotically efficient, i.e. when the size of the reference sample—T *in* this case—tends towards infinity.

Theorem 3.11 $\hat{\boldsymbol{\beta}}$ *is an* ***asymptotically efficient*** *estimator of* $\boldsymbol{\beta}$.

Demonstration

We have proved (see Theorem 3.3) that $\hat{\boldsymbol{\beta}} \sim N\left(\boldsymbol{\beta},\ \sigma^2(\boldsymbol{X}'\boldsymbol{X})^{-1}\right)$. Therefore, whatever the sample size T,

$$\sqrt{T}(\hat{\boldsymbol{\beta}} - \boldsymbol{\beta}) \sim N\left(0,\ \sigma^2(\boldsymbol{X}'\boldsymbol{X}/T)^{-1}\right).$$

It immediately follows that the asymptotic distribution of the above distribution is

$$\sqrt{T}(\hat{\boldsymbol{\beta}} - \boldsymbol{\beta}) \sim N\left(0,\ \sigma^2 \text{plim}_{T\to+\infty}(\boldsymbol{X}'\boldsymbol{X}/T)^{-1}\right).$$

From this last result, it follows that the asymptotic distribution of $\hat{\boldsymbol{\beta}}$ is given by

$$\hat{\boldsymbol{\beta}} \sim AN\left(\boldsymbol{\beta},\ (\sigma^2/T)\ \text{plim}_{T\to+\infty}(\boldsymbol{X}'\boldsymbol{X}/T)^{-1}\right)$$

and that, as a result, the asymptotic variance of ($\hat{\boldsymbol{\beta}}$,AVar($\hat{\boldsymbol{\beta}}$)) is equal to

$$\text{AVar}(\hat{\boldsymbol{\beta}}) = (\sigma^2/T)\ \text{plim}_{T\to+\infty}(\boldsymbol{X}'\boldsymbol{X}/T)^{-1}. \tag{3.35}$$

Since the latter variance is equal to the lower bound (3.34), we conclude that the estimator $\hat{\boldsymbol{\beta}}$ is asymptotically efficient. □

Knowing by the assumption ($H_{2'}$) that $\boldsymbol{L} = \text{plim}_{T\to+\infty}\left(\boldsymbol{X}'\boldsymbol{X}/T\right)$, we can rewrite (3.35) and the corresponding asymptotic bound in the condensed form

$$\frac{1}{T}\text{plim}_{T\to+\infty}\,\sigma^2\left(\frac{\boldsymbol{X}'\boldsymbol{X}}{T}\right)^{-1} = \frac{\sigma^2}{T}\boldsymbol{L}^{-1}.$$

3.5 The General Linear Dynamic Model

So far, we have been content to consider the simple dynamic model

$$y_t = \beta_0 + \sum_{j=0}^{k_i}\sum_{i=1}^{m}\beta_{ij}x_{t-j,i} + u_t, \quad t = 1, \ldots, T. \tag{3.1}$$

The time has come to add an autoregressive part to this model, in which the dependent variable y_t is itself a function of its own time-shifted values.

To the initial model (3.1), we add p shifts of the dependent variable, which leads to the formalised expression

$$y_t = \beta_0 + \sum_{r=1}^{p} \alpha_r y_{t-r} + \sum_{j=0}^{k_i} \sum_{i=1}^{m} \beta_{ij} x_{t-j,\,i} + u_t, \quad t = 1, \ldots, T. \tag{3.36}$$

Such a model is known as an **AutoRegressive Distributed Lags (*ARDL*) model**. However, estimating it raises a number of new problems, which we shall now examine. Let's start by examining the issue of fitting a purely autoregressive model to a sample of observations.

Estimating an AR(1)

Either the process

$$y_t = \phi y_{t-1} + \varepsilon_t, \qquad t = 1, \ldots, T,$$

where $\varepsilon_t \sim IIN(0, \sigma^2)$. We will assume that this process is stationary (see "Autoregressive Processes" section), i.e. that $|\phi| < 1$.

We now need to estimate the pair (ϕ, σ^2), in the knowledge that the observations y_t are obviously dependent. We will therefore consider the conditional distribution $y_t|\, y_{t-1},$, which can easily be shown to be normal, with mean and variance equal to ϕy_{t-1} and σ^2 :

$$y_t|\, y_{t-1} \sim N(\phi y_{t-1}, \sigma^2).$$

The joint density of the T observations can then be factored as a product of the conditional densities by the marginal density

$$f(y_T, y_{T-1}, \ldots, y_1) = \left[\prod\nolimits_{t=2}^{T} f_t(y_t|\, y_{t-1}, \ldots, y_2)\right] f_1(y_1),$$

so that the likelihood function can be written as

$$\begin{aligned} L(\phi, \sigma^2;\; y_T, \ldots, y_1) &= \left[\prod\nolimits_{t=2}^{T} f_t(y_t|\, y_{t-1}, \ldots, y_2)\right] f_1(y_1) \\ &= \left[\prod\nolimits_{t=2}^{T} \frac{1}{\sigma\sqrt{2\pi}}\, e^{-\,(1/2\sigma^2)\,(y_t - \phi y_{t-1})^2}\right] f_1(y_1). \end{aligned} \tag{3.37}$$

It remains to determine the distribution of $f_1(y_1).f_1(y_1)$.

From this point of view, using relation (2.8) from the second chapter, we can be sure that for any t,

$$y_t = \varepsilon_t + \phi\, \varepsilon_{t-1} + \phi^2 \varepsilon_{t-2} + \cdots,$$

which leads us to

$$\begin{cases} E(y_t) = \sum_{j=0}^{+\infty} \phi^j E(\varepsilon_{t-j}) = 0 \\ \mathrm{Var}(y_t) = \sum_{j=0}^{+\infty} \phi^{2j} \mathrm{Var}(\varepsilon_{t-j}) = \sigma^2/(1-\phi^2). \end{cases}$$

We deduce that$y_1 \sim N(0, \sigma^2/(1-\phi^2))$ and therefore its likelihood in (3.37) is given by

$$f_1(y_1) = \frac{\sqrt{1-\phi^2}}{\sigma\sqrt{2\pi}} e^{-\frac{1-\phi^2}{2\sigma^2} y_1^2}. \tag{3.38}$$

Given (3.37) and (3.38), the log-likelihood of the AR(1) process under consideration is written as:

$$\begin{aligned} \ln L(\phi, \sigma^2) = -\frac{T}{2}\ln 2\pi - \frac{T}{2}\ln\sigma^2 - \frac{1}{2\sigma^2}\sum_{t=2}^{T}(y_t - \phi y_{t-1})^2 \\ + \frac{1}{2}\ln(1-\phi^2) - \frac{1}{2}\sigma^{-2}(1-\phi^2)y_1^2. \end{aligned} \tag{3.39}$$

The maximum likelihood estimators of (ϕ, σ^2) in (3.39) are then obtained by maximising the corresponding log-likelihood with respect to each of the two parameters.

The problem is that by calculating and cancelling the first partial derivatives of (3.39) with respect to ϕ and σ^2, we obtain a system of two non-linear equations.

To solve such a system, we need to use a log-likelihood maximisation algorithm or iterative procedure to determine the estimators $\hat{\phi}$ and $\hat{\sigma}^2$. (We will come back to this technique for maximising a non-linear function later.)

This particular optimisation procedure can of course be generalised to more complex models of the form AR(p). In this case, we simply have a system of $(p+1)$ non-linear equations to solve.

In any case, the estimators thus obtained will enjoy all the properties characteristic of maximum likelihood estimators, in particular those of convergence (in probability), normality and asymptotic efficiency.

Estimation of an ARDL Model

Consider the ARDL model

$$y_t = \sum_{r=1}^{p} \gamma_r y_{t-r} + z'_t \delta + u_t, \quad t = 1, \ldots, T, \tag{3.40}$$

where y_t represents the T observations on the dependent variable, γ_r is the parameter associated with this dependent variable lagged by r periods, z'_t is a row vector made up of the K non-stochastic explanatory variables, $\boldsymbol{\delta}$ is the column vector of the parameters associated with these K variables and finally, u_t is a normally distributed disturbance.

We can rewrite (3.40) in matrix form

$$\boldsymbol{y} = \boldsymbol{X\beta} + \boldsymbol{u}, \tag{3.41}$$

with $\mathbf{X}$ the matrix composed of T row vectors $\boldsymbol{x}'_t = (y_{t-1}, \ldots, y_{t-p}, z'_t)$ and $\boldsymbol{\beta}$ the column vector $(\gamma_1, \ldots, \gamma_p, \boldsymbol{\delta})'$.

We know, by hypothesis, that u_t is independent of $\boldsymbol{x}_t$. However, not all observations on $\mathbf{u}$ are independent of all observations on $\mathbf{X}$. For example, $\boldsymbol{x}_t$ includes the variable y_{t-1}, which is obviously not independent of u_{t-1}.

Therefore, even if for all t, we have

$$E(u_t|\, \boldsymbol{x}'_t) = E(u_t) = 0,$$

the fact remains that overall,

$$E(\boldsymbol{u}|\, \boldsymbol{X}) \neq \boldsymbol{0}.$$

The least squares estimator $\hat{\boldsymbol{\beta}} = (\boldsymbol{X}'\boldsymbol{X})^{-1}\boldsymbol{X}'\boldsymbol{y}$ in (3.41) is therefore biased, since

$$E(\hat{\boldsymbol{\beta}}) = \boldsymbol{\beta} + E((\boldsymbol{X}'\boldsymbol{X})^{-1}\boldsymbol{X}'\boldsymbol{u}),$$

the latter expectation in the right-hand side being non-zero.

As a result, the distribution of $\hat{\boldsymbol{\beta}}$ in a small sample, is not identifiable and we must therefore turn to an asymptotic approach. To do this, we will adopt the following assumptions:

Assumptions

1. $u_t \sim IIN(0, \sigma^2)$.
2. the elements z_{tk}, $t = 1, 2, \ldots, +\infty$, $k = 1, \ldots, K$, of $\mathbf{X}$ are uniformly bounded in absolute value.
3. the roots of the recurrence equation $\lambda^p - \sum_{r=1}^{p} \alpha_r \lambda^{p-r} = 0$ are all less than unity in modulus (stationarity assumption).
4. the matrix $\mathbf{X}$ is such that

$$\boldsymbol{L} = \text{plim}_{\,T\to+\infty} \left((1/T)\, \boldsymbol{X}'\, \boldsymbol{X}\right)$$

exists and is non-singular.

5. for any integer θ,

$$\lim_{T \to +\infty} (1/(T - \theta)) \sum_{t=1}^{T-\theta} z'_t z_{t+\theta}$$

exists and is **non-singular** for $\theta = 0$.

Here is a theorem based on these assumptions, which assures us that the OLS estimator is *CAN*, i.e. convergent and asymptotically normal.

Theorem 3.12 *The least squares estimator $\hat{\boldsymbol{\beta}} = (\boldsymbol{X}'\boldsymbol{X})^{-1}\boldsymbol{X}'\boldsymbol{y}$ of model (3.37) is* ***convergent*** *and the* ***asymptotic distribution*** *of $\sqrt{T}(\hat{\boldsymbol{\beta}} - \boldsymbol{\beta})$ is given by $N(\mathbf{0}, \sigma^2 \boldsymbol{L}^{-1})$.*

Demonstration

The demonstration will take place in two stages.

(i) As

$$\hat{\boldsymbol{\beta}} = \boldsymbol{\beta} + \left(\frac{\boldsymbol{X}'\boldsymbol{X}}{T}\right)^{-1} \frac{\boldsymbol{X}'\boldsymbol{u}}{T}, \tag{3.42}$$

it follows that

$$\text{plim}_{T \to +\infty} \hat{\boldsymbol{\beta}} = \boldsymbol{\beta} + \boldsymbol{L}^{-1} \text{plim}_{T \to +\infty} \frac{\boldsymbol{X}'\boldsymbol{u}}{T}.$$

Since the last term in the left-hand side is zero by Lemma 2.4, we end up with

$$\text{plim}_{T \to +\infty} \hat{\boldsymbol{\beta}} = \boldsymbol{\beta}.$$

(ii) The demonstration of asymptotic normality will only be sketched out, given that it involves a central limit theorem allowing a limited dependence between the sequence of random variables under consideration. Using this theorem, we arrive at the proposition that

$$\frac{\boldsymbol{X}'\boldsymbol{u}}{\sqrt{T}}$$

converges in distribution to $N(\mathbf{0}, \sigma^2 \boldsymbol{L})$.

Given this, using (3.42) we get

$$\sqrt{T}(\hat{\boldsymbol{\beta}} - \boldsymbol{\beta}) = \left(\frac{\boldsymbol{X}'\boldsymbol{X}}{T}\right)^{-1} \frac{\boldsymbol{X}'\boldsymbol{u}}{\sqrt{T}}.$$

Since $X'u/\sqrt{T}$ converges to $N(\mathbf{0}, \sigma^2 L)$ and

$$\text{plim}_{T \to +\infty} \left(\frac{X'X}{T} \right)^{-1} = L^{-1},$$

it follows that $\sqrt{T}(\hat{\beta} - \beta)$ converges in distribution to

$$N(\mathbf{0}, \sigma^2 L^{-1} L L^{-1}) = N(\mathbf{0}, \sigma^2 L^{-1}).$$

To finish with this theorem, remember that it only applies to relationships between stationary series.

3.6 Applications

We will limit ourselves to two applications. The first will use the simulated series of an AR (1) process to illustrate the estimation procedure using ordinary least squares and the selection of the appropriate model. The second will attempt to fit an autoregressive model to the French Gross Domestic Product (GDP) series, which we have already used in the previous chapter.

Estimation of an AR (1) Model

The simulated model includes 500 data points from 1st January 2020 to 14 May of the following year. It has the following characteristics:

$$\begin{cases} y_t = 0.6 y_{t-1} + \varepsilon_t, \quad \varepsilon_t \sim N(0, 4) \\ y_1 = 0. \end{cases}$$

Here is the graph for this series (Fig. 3.1):

This simulated series has been constructed in such a way that it is stationary, since the coefficient of y_{t-1} has been set equal to 0.6. It is clear from the graph that it is indeed stationary.

This being the case, it is easy to check the stationarity of the chronicle by applying the usual unit root tests (Table 3.1).

The augmented Dickey–Fuller test is associated with an extremely low p-value, so that we are led to reject the null hypothesis of non-stationarity and accept that of the absence of a unit root. The KPSS test confirms this result, since it leads us to accept the null hypothesis of stationarity.

We will now use conditional maximum **likelihood** to estimate two first-order autoregressive models, one without a constant and the other with a constant:

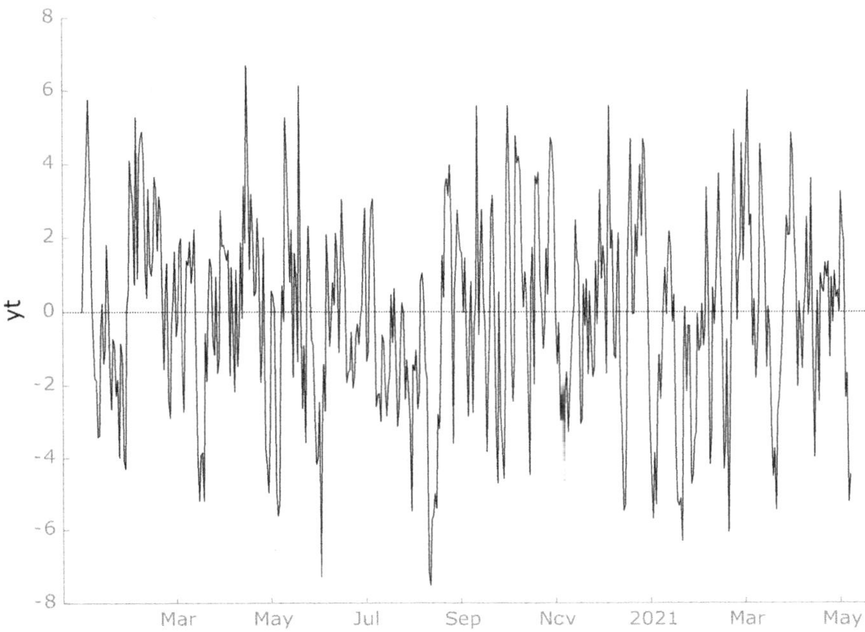

Fig. 3.1 Simulated series of a AR(1) process

Table 3.1 Stationarity tests

	ADF*	KPSS
Test statistics	− 10.47	0.085
p-value	< 0.0001	> 0.10

*0 lag

$$\begin{cases} y_t = 0.636 y_{t-1} \\ y_t = -0.043 + 0.636 y_{t-1}. \end{cases}$$

We have to choose between these two models, even though we know that only the first, by construction, is "true". Table 3.2 provides the appropriate selection tools.

The value of the coefficient of correlation for the model without constant is (very) slightly higher than that observed for the second model, so that to select M1.

Table 3.2 Model selection

	No constant (M1)	Constant (M2)
R^2	0.403	0.402
Akaike	2129.0	2130.8
Bayesian	2133.2	2139.2
Hannan–Quinn	2130.7	2134.1

Examination of the three information criteria leads to a similar conclusion. In fact, all three are minimal for the M1 model, so that the choice of this model is obvious. (Note that the coefficient of the constant is not significantly different from zero; we will come back to this in the next chapter.)

Autoregressive GDP Model

We have already presented the French real GDP series (1st quarter 1975–2nd quarter 2021) in Chapter 2 of this book and have shown that it is integrated of order 1. Consequently, to fit an autoregressive model to the data, we will use the first differences of GDP, which will allow us to respect the stationarity hypothesis. Figure 3.2 provides a graphical representation of this differentiated series.

The graph clearly shows no trend in the first differences. However, the impact of the recession caused by the pandemic is clearly visible.

For the rest, it is a matter of estimating the model

$$\Delta PIB_t = m_0 + \sum_r \phi_r \Delta PIB_{t-r} + \varepsilon_t, \quad \varepsilon_t \sim N(0, \sigma^2),$$

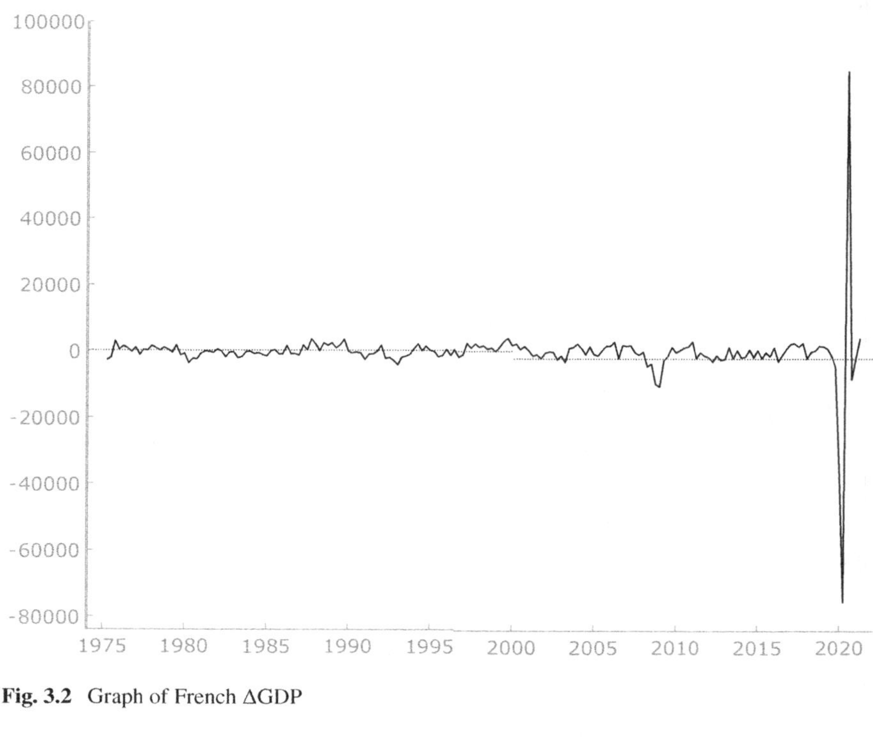

Fig. 3.2 Graph of French ΔGDP

where the number r of shifts must be determined and the constant m_0 omitted if necessary.

Examination of the autocorrelation function or correlogram—see "Some Preliminary Results" and "Autoregressive Processes" sections—offers an initial approach to assessing the number of lags r to be taken into account. In the present case, this examination of the function would lead to a maximum of two lags being retained.

Here are the results of the estimates made using the conditional maximum likelihood method, which is equivalent to the application of least squares to a large sample, as we have seen:

$$\begin{cases} \text{M1}: \ \Delta PIB_t = -0.226\ \Delta PIB_{t-1} \\ \text{M2}: \ \Delta PIB_t = 2114.4 - 0.269\ \Delta PIB_{t-1} \\ \text{M3}: \ \Delta PIB_t = 1659.8 - 0.33\ \Delta PIB_{t-1} - 0.22\ \Delta PIB_{t-2}. \end{cases}$$

Table 3.3 details the values of the different information criteria and the coefficient of correlation that will be used to select one of these three models.

Table 3.3 Selection criteria

	M1	M2	M3
R^2	0.072	0.07	0.112
AIC	3863.6	3854.8	3869.9
BIC	3866.8	3861.3	3882.8
HQC	3864.9	3857.4	3875.1

The first observation concerns the coefficient of correlation, which favours the third model, but which is above all very low whatever the model chosen. Generally speaking, it should not be given undue importance.

On the other hand, an examination of the information criteria leads to the same conclusion: M2 is the model that should be selected, since the values taken by the criteria associated with M2 are minimal. It follows that it is the AR (1) model with constant that best fits the French real (excluding inflation) GDP series.

The selection approach used in these applications is only a first step, which needs to be completed by implementing a battery of hypothesis tests. This is the subject of the next chapter.

3.7 Exercises

3.7.1 Consider the following Table 3.4 which gives the results of the estimation of two autoregressive models for French real GDP for the period 1975Q2–2019Q4.

Table 3.4 Autoregressive models

	Model 1: AR(1) ($T = 178$)		Model 2: AR(2) ($T = 177$)	
	Coefficient	t-ratio	Coefficient	t-ratio
Constant	877.85	5.1	732.05	3.97
$GDP_{t\text{-}1}$	0.518	7.96	0.428	3.65
$GDP_{t\text{-}2}$	–	–	0.171	2.27
	$R^2 = 0.260449$		$R^2 = 0.274516$	

1. Interpret the results of each of these regressions.
2. Calculate the adjusted correlation coefficients for the two models using the evidence provided by the table.

3.7.2 Knowing that the residual sum of squares for the previous model 1 is equal to

$$RSS = 10.638,$$

calculate the variance s^2 using formula (3.24).

3.7.3 Consider a random sample $(X_1, \ldots, X_T)$ drawn from a population distributed according to a normal distribution of known mean μ and unknown variance σ^2. You are asked to determine the maximum likelihood estimator of σ^2.

3.7.4 Let be $(X_1, \ldots, X_T)$ a set of IID random variables from a normally distributed population with mean μ and variance 1.

We ask:

1. write the likelihood function knowing that the unknown parameter is μ ;
2. establish the corresponding log-likelihood function;
3. determine the maximum likelihood estimator of μ ;
4. check that the estimator obtained in the previous step does indeed maximise the log-likelihood.

3.7.5 Let us return to the two autoregressive models whose estimates appear in Table 3.5. This table presents the values of the information criteria, as well as the corresponding correlation coefficients.

You are asked to select one of the two models and to explain the reasons for this choice.

Table 3.5 Choosing a model

	Model 1	Model 2
AIC	3144.16	3124.26
BIC	3150.53	3133.79
HQC	3146.74	3128.13
R^2	0.2646	0.2828

Chapter 4
Testing the Model

4.1 A General Test Structure

We begin by examining how to deal with the problem of imposing a set of restrictions on the parameters of the linear model. Note that the approach adopted is entirely based on maximum likelihood.

Test on Constrained Estimators

Imposing m linear constraints on the vector $\boldsymbol{\beta}$ in the model $\boldsymbol{y} = \boldsymbol{X}\boldsymbol{\beta} + \boldsymbol{u}$ amounts to giving a matrix **R**, of dimension $m \times K$, such that

$$\boldsymbol{R}\boldsymbol{\beta} = \boldsymbol{r},$$

where **r** is a vector $m \times 1$.

Example 4.1 Suppose we introduce the constraints $\beta_1 = \beta_2 = 0$ on the vector $\boldsymbol{\beta}$. The matrix **R** then reduces to a matrix of dimension $2 \times K$ and we have

$$\begin{pmatrix} 1\,0\,0 \cdots 0 \\ 0\,1\,0 \cdots 0 \end{pmatrix} \begin{pmatrix} \beta_1 \\ \beta_2 \\ \vdots \\ \beta_K \end{pmatrix} = \begin{pmatrix} 0 \\ 0 \end{pmatrix}.$$

Supplementary Information The online version contains supplementary material available at https://doi.org/10.1007/978-3-031-72910-2_4.

F. J. Bismans and O. Damette, *Dynamic Econometrics*,
https://doi.org/10.1007/978-3-031-72910-2_4

The question now arises as to how the maximum likelihood estimator—which is also the least squares estimator—is affected by the imposition of this system of linear restrictions. The following proposition provides an answer to this question.

Proposition 4.1 *The estimator of* $\boldsymbol{\beta}$ *which maximises* $L(\boldsymbol{\beta}, \sigma^2; \boldsymbol{y})$ *under the constraint* $\boldsymbol{R\beta} = \boldsymbol{r}$ *is the vector* $\hat{\boldsymbol{\beta}}_0$ *defined by*

$$\hat{\boldsymbol{\beta}}_0 = \hat{\boldsymbol{\beta}} + (\boldsymbol{X}'\boldsymbol{X})^{-1}\boldsymbol{R}'\left[\boldsymbol{R}(\boldsymbol{X}'\boldsymbol{X})^{-1}\boldsymbol{R}'\right]^{-1}(\boldsymbol{r} - \boldsymbol{R}\hat{\boldsymbol{\beta}}), \tag{4.1}$$

where $\hat{\boldsymbol{\beta}} = (\boldsymbol{X}'\boldsymbol{X})^{-1}\boldsymbol{X}'\boldsymbol{y}$ *is the maximum likelihood estimator evaluated over the entire parameter space.*

From the point of view of test theory, the null hypothesis now takes the form

$$H_0: \quad \boldsymbol{R\beta} = \boldsymbol{r}$$

against the alternative $H_1: \quad \boldsymbol{R\beta} \neq \boldsymbol{r}$.

The **likelihood ratio** statistic is obtained by taking the ratio between the maximum of the constrained likelihood and the maximum of the free likelihood. It is therefore written as

$$\lambda(\boldsymbol{y}) = L(\hat{\boldsymbol{\beta}}_0)/L(\hat{\boldsymbol{\beta}}), \tag{4.2}$$

where $\hat{\boldsymbol{\beta}}_0$ is the constrained estimator and $\hat{\boldsymbol{\beta}}$ is the free estimator. (Note that this statistic is necessarily between 0 and 1, since the denominator is represented by the free, unconstrained estimator).

A **likelihood ratio test** leads to rejection of the null hypothesis if and only if the statistic λ is greater than a critical value k which is determined by the significance level α chosen for the test, i.e. if and only if

$$\lambda(\boldsymbol{y}) > k_\alpha.$$

The likelihood constrained by H_0 is written as:

$$\begin{aligned} L(\hat{\boldsymbol{\beta}}_0, \hat{\sigma}_0^2) &= \Pi_{t=1}^{T}\left(1/(\hat{\sigma}_0\sqrt{2\pi})\right)e^{-(1/2\hat{\sigma}_0^2)\,(y_t - \boldsymbol{x}'_t\hat{\boldsymbol{\beta}}_0)^2} \\ &= \left(2\pi\hat{\sigma}_0^2\right)^{-\frac{T}{2}} e^{-\frac{1}{2\hat{\sigma}_0^2}(\boldsymbol{y}-\boldsymbol{X}\hat{\boldsymbol{\beta}}_0)'\,(\boldsymbol{y}-\boldsymbol{X}\hat{\boldsymbol{\beta}}_0)}. \end{aligned} \tag{4.3}$$

Since the (unconstrained) estimator of the variance is by Eq. (3.21) equal to

$$\hat{\sigma}_{MV}^2 = (1/T)\hat{\boldsymbol{u}}'\hat{\boldsymbol{u}} = (1/T)(\boldsymbol{y} - \boldsymbol{X}\hat{\boldsymbol{\beta}})'(\boldsymbol{y} - \boldsymbol{X}\hat{\boldsymbol{\beta}}),$$

it follows that (4.3) becomes

$$L(\hat{\boldsymbol{\beta}}_0, \hat{\sigma}_0^2) = (2\pi)^{-T/2}\left[\frac{1}{T}\left(\boldsymbol{y} - \boldsymbol{X}\hat{\boldsymbol{\beta}}_0\right)'\left(\boldsymbol{y} - \boldsymbol{X}\hat{\boldsymbol{\beta}}_0\right)\right]^{-T/2} e^{-T/2}. \tag{4.4}$$

The free likelihood is given by

$$L(\hat{\boldsymbol{\beta}}, \hat{\sigma}^2) = (2\pi)^{-T/2}\left[\frac{1}{T}(\boldsymbol{y} - \boldsymbol{X}\hat{\boldsymbol{\beta}})'(\boldsymbol{y} - \boldsymbol{X}\hat{\boldsymbol{\beta}})\right]^{-T/2} e^{-T/2}. \tag{4.5}$$

Substituting (4.4) and (4.5) into (4.3) gives a first expression for the likelihood ratio statistic:

$$\lambda(\boldsymbol{y}) = \frac{L(\hat{\boldsymbol{\beta}}_0, \hat{\sigma}_0^2)}{L(\hat{\boldsymbol{\beta}}, \hat{\sigma}^2)} = \left[\frac{\left(\boldsymbol{y} - \boldsymbol{X}\hat{\boldsymbol{\beta}}_0\right)'\left(\boldsymbol{y} - \boldsymbol{X}\hat{\boldsymbol{\beta}}_0\right)}{\left(\boldsymbol{y} - \boldsymbol{X}\hat{\boldsymbol{\beta}}\right)'\left(\boldsymbol{y} - \boldsymbol{X}\hat{\boldsymbol{\beta}}\right)}\right]^{-T/2}. \tag{4.6}$$

A first expression, we said, because the numerator of (4.6), after a few rather laborious transformations, can be written as

$$\left(\boldsymbol{y} - \boldsymbol{X}\hat{\boldsymbol{\beta}}_0\right)'\left(\boldsymbol{y} - \boldsymbol{X}\hat{\boldsymbol{\beta}}_0\right) = \left(\boldsymbol{y} - \boldsymbol{X}\hat{\boldsymbol{\beta}}\right)'\left(\boldsymbol{y} - \boldsymbol{X}\hat{\boldsymbol{\beta}}\right) + (\hat{\boldsymbol{\beta}} - \hat{\boldsymbol{\beta}}_0)'\boldsymbol{X}'\boldsymbol{X}(\hat{\boldsymbol{\beta}} - \hat{\boldsymbol{\beta}}_0). \tag{4.7}$$

Given that the constrained estimator is determined by Eq. (4.1) of the first proposition of this chapter, the immediately preceding equation undergoes a final transformation, i.e.

$$\begin{aligned}
&(\hat{\boldsymbol{\beta}} - \hat{\boldsymbol{\beta}}_0)'\boldsymbol{X}'\boldsymbol{X}(\hat{\boldsymbol{\beta}} - \hat{\boldsymbol{\beta}}_0) \\
&= \left(\boldsymbol{r} - \boldsymbol{R}\hat{\boldsymbol{\beta}}\right)'\left[\boldsymbol{R}(\boldsymbol{X}'\boldsymbol{X})^{-1}\boldsymbol{R}'\right]^{-1}\boldsymbol{R}(\boldsymbol{X}'\boldsymbol{X})^{-1}\boldsymbol{I}_K\boldsymbol{R}'\left[\boldsymbol{R}(\boldsymbol{X}'\boldsymbol{X})^{-1}\boldsymbol{R}'\right]^{-1}\left(\boldsymbol{r} - \boldsymbol{R}\hat{\boldsymbol{\beta}}\right) \\
&= \left(\boldsymbol{r} - \boldsymbol{R}\hat{\boldsymbol{\beta}}\right)'\left[\boldsymbol{R}(\boldsymbol{X}'\boldsymbol{X})^{-1}\boldsymbol{R}'\right]^{-1}\left(\boldsymbol{r} - \boldsymbol{R}\hat{\boldsymbol{\beta}}\right).
\end{aligned} \tag{4.8}$$

Therefore, using (4.7) and (4.8), the statistic (4.6) becomes

$$\lambda(\boldsymbol{y}) = \left[1 + \frac{(\boldsymbol{r} - \boldsymbol{R}\hat{\boldsymbol{\beta}})'\left[\boldsymbol{R}(\boldsymbol{X}'\boldsymbol{X})^{-1}\boldsymbol{R}'\right]^{-1}(\boldsymbol{r} - \boldsymbol{R}\hat{\boldsymbol{\beta}})}{(\boldsymbol{y} - \boldsymbol{X}\hat{\boldsymbol{\beta}})'(\boldsymbol{y} - \boldsymbol{X}\hat{\boldsymbol{\beta}})}\right]^{-T/2}.$$

We will therefore reject the null hypothesis if

$$\frac{(\boldsymbol{r} - \boldsymbol{R}\hat{\boldsymbol{\beta}})'\left[\boldsymbol{R}(\boldsymbol{X}'\boldsymbol{X})^{-1}\boldsymbol{R}'\right]^{-1}(\boldsymbol{r} - \boldsymbol{R}\hat{\boldsymbol{\beta}})}{(\boldsymbol{y} - \boldsymbol{X}\hat{\boldsymbol{\beta}})'(\boldsymbol{y} - \boldsymbol{X}\hat{\boldsymbol{\beta}})} > k_1^{-T/2} - 1 = k_2. \tag{4.9}$$

To determine the value of the constant, we need to know the probability distribution of this statistic λ, which, in conjunction with the level α, will then provide k_α.

Distribution of the Likelihood Ratio Statistic

Dividing the numerator and denominator of (4.9) by m and $(T - K)$, respectively gives

$$\frac{(\boldsymbol{r}-\boldsymbol{R}\hat{\boldsymbol{\beta}})'\left[\boldsymbol{R}(\boldsymbol{X}'\boldsymbol{X})^{-1}\boldsymbol{R}'\right]^{-1}(\boldsymbol{r}-\boldsymbol{R}\hat{\boldsymbol{\beta}})/m}{(\boldsymbol{y}-\boldsymbol{X}\hat{\boldsymbol{\beta}})'(\boldsymbol{y}-\boldsymbol{X}\hat{\boldsymbol{\beta}})/(T-K)} > \frac{T-K}{m}k_2 = k. \tag{4.10}$$

We will show that the statistic $\lambda(\boldsymbol{y})$ in (4.10) follows a Fisher-Snedecor distribution with degrees of freedom m and $(T - K)$. To do this, however, we will need several results of a probabilistic nature. (For the definition of the F distribution and the corresponding demonstrations, see Bismans, 2016, Chapter 4).

Review

1. If X_1 and X_2 are two **independent** random variables such that $X_1 \sim \chi^2(m_1)$ and $X_2 \sim \chi^2(m_2)$, then $Y = (X_1/m_1)/(X_2/m_2) \sim F(m_1, m_2)$.
2. If **A** is a **symmetrical** and **idempotent** matrix $T \times T$ and if, in addition, $\boldsymbol{x} \sim MN(\boldsymbol{0}, \sigma^2\boldsymbol{I}_T)$, then $(1/\sigma^2)\boldsymbol{x}'\boldsymbol{A}\boldsymbol{x} \sim \chi^2(\text{tr}(\boldsymbol{A}))$.
3. Let $\boldsymbol{x} \sim MN(\boldsymbol{0}, \sigma^2\boldsymbol{I}_T)$ and **A**, **B** be two **symmetrical** matrices. Then $Q_1 = (1/\sigma^2)\boldsymbol{x}'\boldsymbol{A}\boldsymbol{x}$ and $Q_2 = (1/\sigma^2)\boldsymbol{x}'\boldsymbol{B}\boldsymbol{x}$ are **independent** if and only if $\boldsymbol{AB} = \boldsymbol{0}$.

Here is the theorem.

Theorem 4.1 *Under the null hypothesis* $H_0 : \boldsymbol{R\beta} = \boldsymbol{r}$, *the statistic*

$$\frac{(\boldsymbol{r}-\boldsymbol{R}\hat{\boldsymbol{\beta}})'\left[\boldsymbol{R}(\boldsymbol{X}'\boldsymbol{X})^{-1}\boldsymbol{R}'\right]^{-1}(\boldsymbol{r}-\boldsymbol{R}\hat{\boldsymbol{\beta}})/m}{(\boldsymbol{y}-\boldsymbol{X}\hat{\boldsymbol{\beta}})'(\boldsymbol{y}-\boldsymbol{X}\hat{\boldsymbol{\beta}})/(T-K)} \tag{4.11}$$

follows a ***Fisher–Snedecor*** *distribution with degrees of freedom* m *and* $(T - K)$.

Demonstration

The proof will proceed in three stages: proving that (i) the numerator of (4.11) follows a Chi-square distribution; (ii) its denominator is also distributed according to a Chi-square distribution; (iii) the numerator and denominator are independent.

(i) As $\boldsymbol{r} = \boldsymbol{R\beta}$, we have

$$\begin{aligned} r - R\hat{\beta} &= R\beta - R\hat{\beta} = R\beta - R(X'X)^{-1}X'y = R\beta - R(X'X)^{-1}X'(X\beta + u) \\ &= -R(X'X)^{-1}X'u. \end{aligned}$$

It follows that the numerator of (4.10) is a **u-quadratic** form written as

$$u'X(X'X)^{-1}R'\left[R(X'X)^{-1}R'\right]^{-1}R(X'X)^{-1}X'u = u'Qu.$$

Note that the matrix **Q**, of dimension $T \times T$, is symmetric—this is immediate—and idempotent, i.e. $QQ = Q$ (check it!). Furthermore, using a property of the trace of a matrix (see Appendix B.17 and "The Gauss–Markov Theorem" section), we obtain

$$\begin{aligned} \mathrm{tr}(Q) &= \mathrm{tr}\left(\left[R(X'X)^{-1}R'\right]^{-1}R(X'X)^{-1}X'X(X'X)^{-1}R'\right) \\ &= \mathrm{tr}(I_m) \\ &= m. \end{aligned}$$

It then follows from Review 2 that $(1/\sigma^2)\, u'Qu \sim \chi^2(m)$.

(ii) The denominator of (4.11) is equal to $\hat{u}'\hat{u}/(T-K)$. Now, we have proved (Lemma 3.3, "The Gauss–Markov Theorem" section) that

$$\hat{u} = (I_T - X(X'X)^{-1}X')\,u = Mu,$$

where **M** is a symmetric idempotent matrix. Moreover, using the same property of the trace, we have:

$$\mathrm{tr}(M) = \mathrm{tr}(I_T - X(X'X)^{-1}X') = \mathrm{tr}(I_t) - \mathrm{tr}\left((X'X)^{-1}X'X\right) = T - K.$$

From then on, using these results and Review 2:

$$(1/\sigma^2)\hat{u}'\hat{u} = (1/\sigma^2)u'Mu \sim \chi^2(T-K).$$

(iii) All that remains is to show that the product MQ is zero. As $MX = (I_T - X(X'X)^{-1}X')X = \mathbf{0}$, we obtain

$$MQ = MX(X'X)^{-1}R\left[R(X'X)^{-1}R'\right]^{-1}R(X'X)^{-1}X' = \mathbf{0}.$$

It is possible to give another expression for the likelihood ratio statistic, which is intuitively meaningful. Indeed, this statistic is also equal to

$$\frac{(\hat{u}'_C\hat{u}_C - \hat{u}'_F\hat{u}_F)/m}{\hat{u}'_F\hat{u}_F/(T-K)} \sim F(m, T-K), \tag{4.12}$$

where $\hat{\boldsymbol{u}}'_C\hat{\boldsymbol{u}}_C$ is the sum of the squares of the constrained residuals and $\hat{\boldsymbol{u}}'_F\hat{\boldsymbol{u}}_F$ is the sum of the squares of the residuals of the free model.

4.2 Significance of the Estimated Coefficients

We will first look at how to test each of the coefficients in the linear model before discussing the significance of the linear relationship as a whole.

A Single Linear Restriction

We wish to test the null hypothesis that the coefficient $\beta_k, \quad k = 1, \ldots, K$, takes a given value $\overline{\beta}_k$. In matrix form, this restriction is written as:

$$\boldsymbol{R\beta} = (0, \ldots, 0, 1, 0, \ldots, 0)\begin{pmatrix} \beta_1 \\ \vdots \\ \beta_k \\ \vdots \\ \beta_K \end{pmatrix} = r = (\overline{\beta}_k).$$

In this case, the numerator of (4.11) must be modified as follows:

$$\begin{cases} \boldsymbol{R\hat{\beta}} - \boldsymbol{r} = (\hat{\beta}_k - \overline{\beta}_k) \\ \left[\boldsymbol{R}(\boldsymbol{X}'\boldsymbol{X})^{-1}\boldsymbol{R}'\right]^{-1} = \left[(\boldsymbol{X}'\boldsymbol{X})^{-1}_{kk}\right]^{-1}, \end{cases}$$

where $(\boldsymbol{X}'\boldsymbol{X})^{-1}_{kk}$ is the element at the intersection of the kth row and the kth column of $(\boldsymbol{X}'\boldsymbol{X})^{-1}$.

The likelihood ratio statistic can then be written as

$$\begin{aligned} \lambda(\boldsymbol{y}) &= \frac{(\hat{\beta}_k - \overline{\beta}_k)\left[(\boldsymbol{X}'\boldsymbol{X})^{-1}_{kk}\right]^{-1}(\hat{\beta}_k - \overline{\beta}_k)/1}{(\boldsymbol{y} - \boldsymbol{X\hat{\beta}})'(\boldsymbol{y} - \boldsymbol{X\hat{\beta}})/(T-K)} \\ &= \frac{(\hat{\beta}_k - \overline{\beta}_k)^2}{(\boldsymbol{X}'\boldsymbol{X})^{-1}_{kk}\, \hat{\boldsymbol{u}}'\hat{\boldsymbol{u}}/(T-K)} \quad \sim \quad F(1, T-K). \end{aligned} \tag{4.13}$$

This statistic is the ratio of two independent random variables following a Chi-square distribution with 1 and $(T - K)$ degrees of freedom respectively. We will show how this statistic can be transformed so that it follows a Student distribution with $(T - K)$ degrees of freedom. Let us recall the general definition of a Student distribution.

Definition 4.1 If $X \sim N(0,1)$ and $Y \sim \chi^2(m)$, then

$$X/\sqrt{Y/m} \sim t\,(m).$$

Here is the theorem:

Theorem 4.2

$$\frac{\hat{\beta}_k - \overline{\beta}_k}{\sqrt{\sigma^2_{\hat{\beta}_k}}} \sim t(T-K), \tag{4.14}$$

where $\sigma^2_{\hat{\beta}_k} = s^2(\boldsymbol{X}'\boldsymbol{X})^{-1}_{kk}$ *is the* ***standard error*** *of the* estimator $\hat{\beta}_k$.

Demonstration

To prove this theorem, we'll need a quick reminder of probability:

Review
The random variable $Y = \sum_{j=1}^{\nu} X_j^2$, where $\forall j,\ X_j \sim N(0,1)$, follows a **Chi-square distribution** with ν degrees of freedom

(i) In (4.13), the numerator is Chi-square distributed with one degree of freedom. Therefore, using review, the square root of this numerator follows a centred reduced normal distribution:

$$\hat{\beta}_k - \overline{\beta}_k \sim N(0,1). \tag{4.15}$$

(ii) Again in (4.13), as $\lambda(\boldsymbol{y}) \sim F(1, T-K)$, we get

$$(\boldsymbol{X}'\boldsymbol{X})^{-1}_{kk}\ \hat{\boldsymbol{u}}'\hat{\boldsymbol{u}}/(T-K) \sim \chi^2(T-K).$$

However, knowing from (3.24) that $s^2 = (1/(T-K))\hat{\boldsymbol{u}}'\hat{\boldsymbol{u}}$, we have

$$\begin{aligned}(\boldsymbol{X}'\boldsymbol{X})^{-1}_{kk}\ \hat{\boldsymbol{u}}'\hat{\boldsymbol{u}} &= (\boldsymbol{X}'\boldsymbol{X})^{-1}_{kk}(T-K)\,s^2\\ &= (T-K)\,\sigma^2_{\hat{\beta}_k} \sim \chi^2(T-K).\end{aligned}$$

Taking the square root of the latter expression, the denominator of (4.13) can be written as

$$\sqrt{\frac{(T-K)\sigma^2_{\hat{\beta}_k}}{T-K}} = \sigma^2_{\hat{\beta}_k}. \tag{4.16}$$

(iii) Given (4.15) and (4.16), we finally have

$$\frac{\hat{\beta}_k - \overline{\beta}_k}{\sqrt{\sigma^2_{\hat{\beta}_k}}} \sim t(T-K).$$

From a practical point of view, in econometrics, we are testing the economic significance of a regression coefficient; in other words, we want to know if the coefficient is not zero. Consequently, we take $\overline{\beta}_k = 0$ and the null and alternative hypotheses are then

$$H_0 : \beta_k = 0$$
$$H_1 : \beta_k \neq 0$$

Under H_0, the test statistic is noted $t_{\hat{\beta}_k}$ and follows a Student distribution with $(T-K)$ degrees of freedom:

$$t_{\hat{\beta}_k} = \frac{\hat{\beta}_k}{\sigma_{\hat{\beta}_k}} \sim t(T-K). \tag{4.17}$$

Example 4.2 In the previous chapter, "Autoregressive GDP Model" section, we fitted an AR(2) process to the French real GDP series. In addition, here are the t-tests on each of the estimated parameters:

$$\underset{(4.21)}{\Delta PIB_t = 1659.8} - \underset{(-4.55)}{0.33}\ \Delta PIB_{t-1} - \underset{(-3.03)}{0.22}\ \Delta PIB_{t-2}.$$

The values of the t-statistic associated with the three coefficients are shown in brackets. The values of Student's t distribution corresponding to 95% and 99% of the distribution are 1.645 and 2.326 respectively, with $T - K = 182$.
Since the values of t are well above 2.326, we reject the null hypothesis and conclude that the three coefficients are significantly different from zero. Alternatively, since all the p-values are less than one percent, we accept the alternative hypothesis that all three coefficients are significant.

Existence of an Explanatory Relationship

In the linear model, when the first regressor is a constant, it is usual to test for the existence of an explanatory relationship between y and $x_2, \ldots, x_K$,, i.e. to test for

$$H_0 : \beta_2 = \cdots = \beta_K = 0.$$

The most immediate solution from this point of view is to use the modified Eq. (4.12), bearing in mind that here *m is* equal to $(K - 1)$, i.e.

$$\frac{(\hat{\mathbf{u}}'_C\hat{\mathbf{u}}_C - \hat{\mathbf{u}}'_F\hat{\mathbf{u}}_F)/(K-1)}{\hat{\mathbf{u}}'_F\hat{\mathbf{u}}_F/(T-K)} \sim F(K-1, T-K). \tag{4.18}$$

In fact, this statistic is closely related to the coefficient R^2. Remember that *RSS* is the sum of the squares of the estimated residuals, *ESS* is the explained sum of the squares of the adjusted dependent variable from the regression and *TSS* is the total sum of the squares.

In this case, if the only regressor is a constant and if we denote by RSS_C the sum of the squares of the constrained residuals and by RSS_F the sum of the squares of the free residuals, then

$$RSS_C = TSS$$

and more,

$$RSS_C - RSS_F = TSS - RSS_F = ESS.$$

Hence (4.18) can be written as

$$\frac{ESS/(K-1)}{RSS/(T-K)} \sim F(K-1, T-k). \tag{4.19}$$

Alternatively, since

$$\begin{cases} R^2 = 1 - (RSS/TSS) \Leftrightarrow 1 - R^2 = (RSS/TSS) \\ R^2 = ESS/TSS, \end{cases}$$

it follows that

$$\begin{cases} RSS = (1 - R^2)TSS \\ ESS = R^2 TSS. \end{cases}$$

Consequently, (4.19) becomes given the previous identities

$$\frac{R^2 TSS/(K-1)}{(1-R^2)TSS/(T-K)},$$

so that

$$\frac{R^2/(K-1)}{1-R^2/(T-K)} \sim F(K-1, T-K). \tag{4.20}$$

Implementing the F-test therefore presents no difficulty whatsoever. Simply use the coefficient R^2 from the regression and apply formula (4.20). All that remains is to see whether the value obtained is greater than the chosen fractile (associated with 90%, 95%, 99%, etc., or the corresponding p-value), so as to reject the null hypothesis that the estimated coefficients are not significant.

Example 4.3 In Chapter three, "Autoregressive GDP Model" section, and in the immediately preceding section, we fitted an AR(2) process to the French real GDP series and indicated how the t-tests on the estimated coefficients would look.

$$\underset{}{\Delta PIB_t} = \underset{(4.21)}{1659.8} - \underset{(4.55)}{0.33}\ \Delta PIB_{t-1} - \underset{(3.03)}{0.22}\ \Delta PIB_{t-2}.$$

We now need to test the overall nullity of the coefficients of the two regressors $\Delta\ PIB_{t-1}$ and $\Delta\ PIB_{t-2}$. Given that $R^2 = 0.12$, with $K - 1 = 2$ and $T - K = 183$, the application of formula (4.20) gives a value for the likelihood ratio statistic equal to 12.

Since the critical value corresponding to 95% of the F distribution takes on a value of 3, it follows that we can reject the null hypothesis and accept the alternative hypothesis of the significance of the explanatory relationship as a whole.

4.3 Asymptotic Tests

We have already encountered the likelihood ratio test on several occasions. We will now introduce its asymptotic version, while pointing out that there are also two other tests applicable to large samples: the Wald and Lagrange multiplier tests.

An Intuitive Approach

We saw in (4.2) that the likelihood ratio statistic λ was given by the ratio between the maximum of the likelihood function constrained by H_0 and the maximum of the free, unconstrained likelihood.

We can, however, give an alternative presentation, insofar as we will no longer be working with the statistic itself, but with its transformation $-2 \ln \lambda$. We will therefore have

$$\begin{aligned} LR = -2 \ln \lambda &= -2\big[\ln L(\theta_0) - \ln L(\hat{\theta})\big] \\ &= 2\big[\ln L(\hat{\theta}) - \ln L(\theta_0)\big], \end{aligned} \tag{4.21}$$

where $L(\theta_0)$ is, of course, the constrained maximum and $L(\hat{\theta})$ the unconstrained maximum.

As before, we will accept the null hypothesis if the *LR* statistic is below a critical value determined by its asymptotic distribution—we will see that the latter is a Chi-square—and we will reject it if *LR* is above this critical value.

Suppose that the likelihood function $\ln L(\theta)$ has a quadratic form and is strictly concave. Let us then plot its maxima, constrained and free, in Fig. 4.1 (the consequence of strict concavity is that each of these maxima is unique).

Referring to (4.21), it can be seen from this graph that $LR/2$ is equal to the difference, i.e. the distance, between the values of $\ln L(\theta)$ in $\hat{\theta}$ and θ_0. Moreover, this distance is a function of the difference $\hat{\theta} - \theta_0$, **and of the curvature of** the log-likelihood function, defined by the absolute value of the second derivative of $\ln L(\theta)$ evaluated in $\hat{\theta}$:

$$C(\hat{\theta}) = \left| \frac{d^2 \ln L(\theta)}{d\theta^2} \right|.$$

Intuitively, the null hypothesis $H_0 : \theta = \theta_0$ will be rejected if the log-likelihoods $\ln L(\hat{\theta})$ and $\ln L(\theta_0)$ are too far apart, i.e. if $LR/2$ is too high.

The **Wald test**, which we are about to introduce, is based on the curvature $C(\hat{\theta})$ of the likelihood functions used. More precisely, this test no longer takes into account the difference between the likelihoods, but rather the square of the distance between $\hat{\theta}$ and θ_0, weighted by the degree of curvature of the likelihood function.

As a result, the Wald statistic is defined by

$$W = (\hat{\theta} - \theta_0)^2 C(\hat{\theta}). \tag{4.22}$$

However, we prefer this other form to (4.22)

$$W = (\hat{\theta} - \theta_0)^2 I(\hat{\theta}), \tag{4.23}$$

where $I(\hat{\theta})$ denotes the Fisher information quantity evaluated in $\hat{\theta}$ and defined—see for its multivariate version, Definition 3.11, "Convergence and Asymptotic Efficiency" section—by

Fig. 4.1 Likelihood ratio test

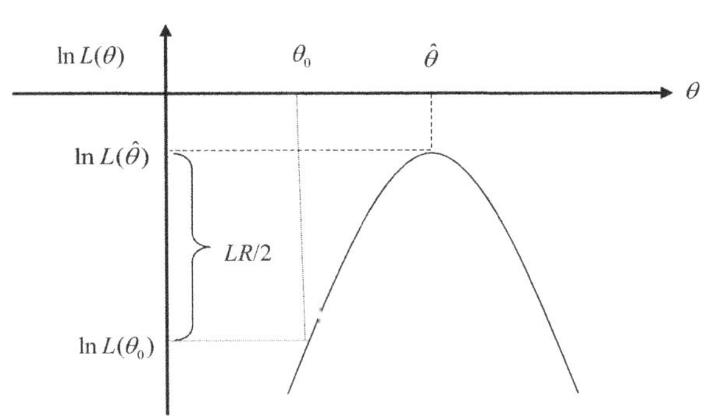

$$I(\theta) = -E\left(\frac{d^2 \ln L(\theta)}{d\theta^2}\right).$$

In other words, the weighting of the square $(\hat{\theta} - \theta_0)^2$ in (4.23) is achieved by taking into account the **mean curvature of** the likelihood.

The last asymptotic test considered, that of the **Lagrange multiplier**, is based on the concept of **score**, defined as follows:

Definition 4.2 Let the likelihood function $L(\boldsymbol{\theta}\ ; \boldsymbol{x})$ be assumed once to be continuously derivable on the open Θ. The **gradient** of this likelihood function is called the **score**, denoted $s(\boldsymbol{\theta})$,:

$$s(\boldsymbol{\theta}) = \text{grad}L(\boldsymbol{\theta}\ ; \boldsymbol{x}) = (\partial L(\boldsymbol{\theta}\ ; \boldsymbol{x})/\partial\theta_1, \ldots, \partial L(\boldsymbol{\theta}\ ; \boldsymbol{x})/\partial\theta_k)'.$$

In this case, the score will be represented by a scalar, given that we are interested, at this stage, only in the parameter θ.

For the rest, the idea behind the test is simple: if the null hypothesis $H_0 : \theta = \theta_0$ is true, then the constrained estimator θ_0 is close to the free estimator $\hat{\theta}$; since the latter maximises the likelihood, it satisfies the first-order condition $s(\hat{\theta}) = 0$ and, as a result, the score evaluated at θ_0 should be close to 0. (We will, however, work with the square of the score later on, so as to obtain a Lagrange multiplier statistic that is always positive).

However, as Fig. 4.2 shows, it may be that two different sets of data lead to identical score values, even though one of these sets makes the constrained estimator θ_0 closer to one or other of the maxima $\hat{\theta}_1$ and $\hat{\theta}_2$.

On this graph, the score $s(\theta_0)$ is the slope of the tangent at the intersection of likelihoods 1 and 2. They therefore have the same score, while θ_0 is closer to θ_1 than to θ_2. The quantity $[s(\theta_0)]^2$ must therefore be weighted by the **inverse** of $\ln L(\theta)$'s degree of curvature in θ_0 or, what is equivalent, by the **inverse** of Fisher's quantity of information in θ_0.

Then we have the Lagrange multiplier statistic:

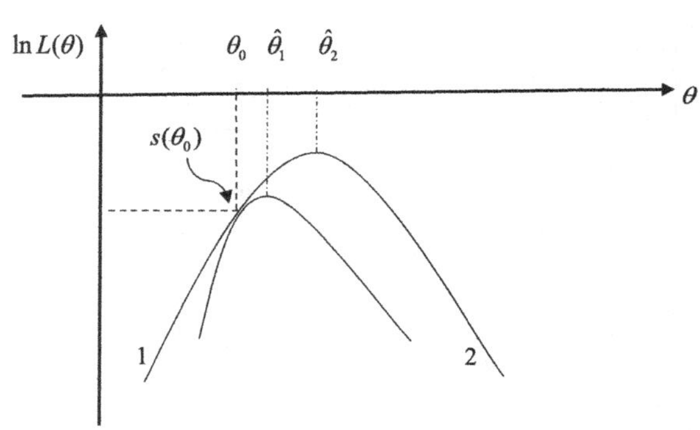

Fig. 4.2 Lagrange multiplier test

$$LM = [s(\theta_0)]^2 I(\theta_0)^{-1}. \quad (4.24)$$

The interpretation of (4.24) is straightforward: the higher the *LM* statistic, so the further θ_0 is from $\hat{\theta}$ and the greater the incentive to reject the null hypothesis $\theta = \theta_0.\theta = \theta_0$.

A final clarification on this asymptotic test: it is referred to, indifferently, as the score or the Lagrange multiplier, simply because we obtain the same statistic using the score or the Lagrangian.

$$\mathcal{L}(\theta, \nu) = \ln L(\theta) - \nu(\theta - \theta_0).$$

Asymptotic Distribution of the Three Tests

Let us generalise the above by considering the vector $\boldsymbol{\theta}$ with k components and the log-likelihood function $\ln L(\boldsymbol{\theta})$. Our first task is now to determine the multidimensional equivalents of the statistics (4.21), (4.23) and (4.24).

As far as the likelihood ratio statistic is concerned, we must necessarily take into account the fact that there are now m restrictions on the parameters. Therefore, the equivalent of (4.21) is

$$LR = 2\left(\ln L(\hat{\boldsymbol{\theta}}) - \ln L(\hat{\boldsymbol{\theta}}_0)\right). \quad (4.25)$$

The formulation of the equivalent of the Wald statistic is more complicated. It is necessary to introduce the column vector $\boldsymbol{r}(\boldsymbol{\theta}) = \boldsymbol{0}$ of the m restrictions imposed by H_0 on the k-vector $\boldsymbol{\theta}$, with $m < k$, as well as the $(m \times k)$—matrix $\mathbf{R}$ of the first partial derivatives evaluated in $\hat{\boldsymbol{\theta}}.\hat{\boldsymbol{\theta}}$.

The Wald statistic is then a quadratic form in $\boldsymbol{r}(\boldsymbol{\theta})$ while the Fisher information matrix takes into account the $\mathbf{R}$ matrix, so that

$$W = \boldsymbol{r}(\hat{\boldsymbol{\theta}})'\left[\boldsymbol{R}\boldsymbol{I}(\hat{\boldsymbol{\theta}})^{-1}\boldsymbol{R}'\right]^{-1}\boldsymbol{r}(\hat{\boldsymbol{\theta}}). \quad (4.26)$$

Finally, the analogue of the univariate Lagrange multiplier statistic (4.24) is given by

$$LM = \boldsymbol{s}(\boldsymbol{\theta}_0)'\boldsymbol{I}(\boldsymbol{\theta}_0)^{-1}\boldsymbol{s}(\boldsymbol{\theta}_0), \quad (4.27)$$

where $\boldsymbol{s}(\boldsymbol{\theta}_0)$ and $\boldsymbol{I}(\boldsymbol{\theta}_0)$ denote respectively the score vector and the Fisher information matrix, both evaluated in the estimator $\boldsymbol{\theta}_0$ constrained by the m restrictions.

We should also point out that there is an inequality that allows us to establish an order of comparison between the three statistics:

$$W \geq LR \geq LM \, .$$

There is now a final step to be taken, which is to determine the asymptotic distribution of the three statistics. From this point of view, they all follow the same distribution, which is a Chi-square distribution with an appropriate number of degrees of freedom.

There is a rather cumbersome theorem on the subject to state and prove—see Bismans, 2016, pp. 551–555. For this reason, we will simply give it in the form of a highly simplified and unproved proposition.

Proposition 4.2 *Given a set of regularity conditions, the likelihood ratio, Wald and Lagrange multiplier statistics all* ***asymptotically*** *follow the same* ***Chi-square distribution*** *with m degrees of freedom at* H_0*,.*

However, there is still an essentially practical question to be answered: which of these three tests should be used? The answer is, as it should be, nuanced and must take account of the following factors:

- by definition, the likelihood ratio test requires estimation of the free model, i.e. $\hat{\boldsymbol{\theta}}$, as well as the model constrained by the null hypothesis, i.e. $\boldsymbol{\theta}_0$; as a result, determining the *LR* statistic can be a very complex calculation.
- In contrast, the Wald statistic requires only the estimation of the free model, i.e. $\hat{\boldsymbol{\theta}}$, which greatly simplifies the calculations, especially when the estimation of $\boldsymbol{\theta}_0$ is complicated.
- the Lagrange multiplier statistic is based entirely on the sole consideration of the constrained model, i.e. on the estimation of $\boldsymbol{\theta}_0$; it follows that it is advantageous to use this statistic when the estimation of the constrained model is straightforward.

All in all, taking into account the advantages and disadvantages of each of these three asymptotic tests, the approach to be adopted in selecting one of these tests will be essentially pragmatic. First and foremost, it will seek to determine which statistic is easiest to estimate.

4.4 Validation of Model Assumptions: Autocorrelation

The linear model studied is based on a set of conditions summarised in hypothesis (H1):

$$\boldsymbol{u} | X \sim N(\boldsymbol{0}, \sigma^2 \boldsymbol{I}_T).$$

This synthetic hypothesis assumes, in particular, that there is no autocorrelation in the error term **u**. It is to this question that we now turn.

Presence of Autocorrelated Disturbances

The absence of autocorrelation in the model $y_t = \mathbf{x}'_t\boldsymbol{\beta} + u_t$, implies that

$$\mathrm{Cov}(u_t, u_{t-j}) = E(u_t, u_{t-j}) = 0, \quad j \neq 0.$$

This is a bold assumption, given that economic series often exhibit a greater or lesser degree of autocorrelation.

We can therefore assume that the disturbances follow an autoregressive process of order 1:

$$u_t = \rho\, u_{t-1} + v_t, \quad |\rho| < 1, \quad v_t \sim N(0, \sigma^2).$$

More generally, we will consider that the disturbance term u_t follows an autoregressive process of order p:

$$u_t = \rho_1 u_{t-1} + \rho_2 u_{t-2} + \cdots + \rho_p u_{t-p} + v_t. \tag{4.28}$$

An autocorrelation test would lead us to formulate the null hypothesis as follows:

$$H_0 \;:\; \rho_1 = \rho_2 = \cdots = \rho_p = 0,$$

i.e. the absence of autocorrelation versus the alternative of p-order autocorrelation.

The model constrained by H_0 then takes the simple form

$$y_t = \mathbf{x}'_t\boldsymbol{\beta} + v_t = \mathbf{x}'_t\boldsymbol{\beta} + u_t, \quad t = 1, \ldots, T, \tag{4.29}$$

while the free version is written as

$$y_t = \mathbf{x}'_t\boldsymbol{\beta} + \rho_1 u_{t-1} + \rho_2 u_{t-2} + \cdots + \rho_p u_{t-p} + v_t, \quad t = 1, \ldots, T. \tag{4.30}$$

Using these two models, we could construct a likelihood ratio test. However, implementing such an approach raises a major problem: in (3.30), the disturbance term $\mathbf{u}$ is not observable, so it would have to be estimated.

On the other hand, estimating the parameters of the constrained model as represented by (4.29) poses no problem, since it simply requires the application of ordinary least squares.

For all these reasons, we will abandon the idea of constructing a likelihood ratio test and prefer to use a Lagrange multiplier test, which only involves estimating the simple model constrained by the null hypothesis.

Lagrange Multiplier Test

The *LM* statistic—see Eq. (4.27)—takes the general form

$$LM = \boldsymbol{s}(\boldsymbol{\theta}_0)'\boldsymbol{I}(\boldsymbol{\theta}_0)^{-1}\boldsymbol{s}(\boldsymbol{\theta}_0),$$

where $\boldsymbol{s}(\boldsymbol{\theta}_0)$ and $\boldsymbol{I}(\boldsymbol{\theta}_0)$ stand for the score vector and the Fisher information matrix respectively. Let's take a look at how the two are presented in our case.

The likelihood function of the sample $(y_1, \ldots, y_T)$ for the constrained model is written as

$$L(\boldsymbol{\beta}, \sigma^2; \boldsymbol{y}) = \prod_{t=1}^{T} (2\pi\sigma^2)^{-1/2} e^{-(1/2\sigma^2)(y_t - \boldsymbol{x}'_t\boldsymbol{\beta})^2},$$

from which we derive the log-likelihood

$$\ln L(\boldsymbol{\beta}, \sigma^2) = -\frac{T}{2}\ln 2\pi - \frac{T}{2}\ln\sigma^2 - \frac{1}{2\sigma^2}(\boldsymbol{y} - \boldsymbol{X\beta})'(\boldsymbol{y} - \boldsymbol{X\beta}). \tag{4.31}$$

The score vector is obtained by calculating the first partial derivatives of (4.31) with respect to $\boldsymbol{\beta}$ and σ^2 :

$$\boldsymbol{s}(\boldsymbol{\beta}, \sigma^2) = \begin{pmatrix} \partial \ln L/\partial\boldsymbol{\beta} \\ \partial \ln L/\partial\sigma^2 \end{pmatrix} = \begin{pmatrix} (1/\sigma^2)\boldsymbol{X}'(\boldsymbol{y} - \boldsymbol{X\beta}) \\ (-T/2\sigma^2) + (1/2\sigma^4)(\boldsymbol{y} - \boldsymbol{X\beta})'(\boldsymbol{y} - \boldsymbol{X\beta}) \end{pmatrix}.$$

As $\boldsymbol{y} - \boldsymbol{X\beta} = \boldsymbol{u}$, the score row vector can finally be given the form

$$\boldsymbol{s}'(\boldsymbol{\beta}, \sigma^2) = \left(\frac{1}{\sigma^2}\boldsymbol{X}'\boldsymbol{u}, \quad -\frac{T}{2\sigma^2} + \frac{1}{2\sigma^4}\boldsymbol{u}'\boldsymbol{u}\right). \tag{4.32}$$

The matrix of second partial derivatives of the log-likelihood function is given by

$$\begin{pmatrix} \frac{\partial^2 \ln L}{\partial\boldsymbol{\beta}\partial\boldsymbol{\beta}'} & \frac{\partial^2 \ln L}{\partial\boldsymbol{\beta}\partial\sigma^2} \\ \frac{\partial^2 \ln L}{\partial\sigma^2\partial\boldsymbol{\beta}} & \frac{\partial^2 \ln L}{\partial(\sigma^2)^2} \end{pmatrix} = \begin{pmatrix} -\frac{1}{\sigma^2}(\boldsymbol{X}'\boldsymbol{X}) & -\frac{1}{\sigma^4}\boldsymbol{X}'(\boldsymbol{y} - \boldsymbol{X\beta}) \\ -\frac{1}{\sigma^4}\boldsymbol{X}'(\boldsymbol{y} - \boldsymbol{X\beta}) & \frac{T}{2\sigma^4} - \frac{1}{\sigma^6}(\boldsymbol{y} - \boldsymbol{X\beta})'(\boldsymbol{y} - \boldsymbol{X\beta}) \end{pmatrix}.$$

Given that $\boldsymbol{y} - \boldsymbol{X\beta} = \boldsymbol{u}$ and $E(\boldsymbol{u}) = \boldsymbol{0}$, the **Fisher information matrix** is equal to

$$\boldsymbol{I}(\boldsymbol{\beta}, \sigma^2) = -E\left(\frac{\partial^2 \ln L}{\partial(\boldsymbol{\beta}, \sigma^2)\partial(\boldsymbol{\beta}, \sigma^2)'}\right) = \begin{pmatrix} \frac{1}{\sigma^2}\boldsymbol{X}'\boldsymbol{X} & \boldsymbol{0} \\ \boldsymbol{0} & \frac{\boldsymbol{u}'\boldsymbol{u}}{\sigma^6} - \frac{T}{2\sigma^4} \end{pmatrix}. \tag{4.33}$$

The score and information matrix must obviously be evaluated as constrained estimators $(\tilde{\boldsymbol{\beta}}, \tilde{\sigma}^2)$, where

$$\begin{cases} \tilde{\boldsymbol{\beta}} = (\boldsymbol{X}'\boldsymbol{X})^{-1}\boldsymbol{X}'\boldsymbol{y} \\ \tilde{\sigma}^2 = (1/T)\tilde{\boldsymbol{u}}'\tilde{\boldsymbol{u}}. \end{cases}$$

Armed with these various tools, we can now state the relevant theorem for testing autocorrelation.

Theorem 4.3 *Let* $\boldsymbol{y} = \boldsymbol{X}\boldsymbol{\beta} + \boldsymbol{u}$, *be the model where for any* $t = 1, \ldots, T$, u_t *follows an autoregressive process of order* p

$$u_t = \rho_1 u_{t-1} + \rho_2 u_{t-2} + \cdots + \rho_p u_{t-p} + v_t, \quad \boldsymbol{v} \sim MN(\boldsymbol{0}, \sigma^2 \boldsymbol{I}_t).$$

So

(i) *the* ***test statistic*** *for the null hypothesis* $H_0 : \rho_1 = \rho_2 = \cdots = \rho_p = 0$ *is given by*

$$ML = T\,\frac{\tilde{\boldsymbol{u}}'\boldsymbol{X}(\boldsymbol{X}'\boldsymbol{X})^{-1}\boldsymbol{X}'\tilde{\boldsymbol{u}}}{\tilde{\boldsymbol{u}}'\tilde{\boldsymbol{u}}}, \tag{4.34}$$

where $\tilde{\boldsymbol{u}}$ *is the vector of regression residuals, constrained by* $H_0, y_t = \boldsymbol{x}'_t\boldsymbol{\beta} + u_t, \quad t = 1, \ldots, T.$

(ii) *the LM statistic is equal to* $T \cdot R^2$, *where* R^2 *is the coefficient associated to the* ***auxiliary regression***

$$\tilde{u}_t = \boldsymbol{x}'_t\boldsymbol{\gamma} + \rho_1\tilde{u}_{t-1} + \rho_2\tilde{u}_{t-2} \cdots + \rho_p\tilde{u}_{t-p} + v_t, \quad t = 1, \ldots, T. \tag{4.35}$$

Furthermore, asymptotically, under H_0,

$$T \cdot R^2 \sim \chi^2(p).$$

Demonstration

(i) The Lagrange multiplier statistic, evaluated in $(\tilde{\boldsymbol{\beta}}, \tilde{\sigma}^2)$, is, given (4.32) and (4.33), equal to

$$\begin{aligned} ML &= \boldsymbol{s}'(\tilde{\boldsymbol{\beta}}, \tilde{\sigma}^2)\boldsymbol{I}^{-1}(\tilde{\boldsymbol{\beta}}, \tilde{\sigma}^2)\,\boldsymbol{s}(\tilde{\boldsymbol{\beta}}, \tilde{\sigma}^2) \\ &= \begin{pmatrix} \frac{1}{\tilde{\sigma}^2}\tilde{\boldsymbol{u}}'\boldsymbol{X} & \boldsymbol{0} \end{pmatrix} \begin{pmatrix} \tilde{\sigma}^2(\boldsymbol{X}'\boldsymbol{X})^{-1} & \boldsymbol{0} \\ \boldsymbol{0} & \frac{2\tilde{\sigma}^4}{T} \end{pmatrix} \begin{pmatrix} \frac{1}{\tilde{\sigma}^2}\boldsymbol{X}'\tilde{\boldsymbol{u}} \\ \boldsymbol{0} \end{pmatrix} \\ &= \frac{\tilde{\boldsymbol{u}}'\,\boldsymbol{X}(\boldsymbol{X}'\boldsymbol{X})^{-1}\boldsymbol{X}'\tilde{\boldsymbol{u}}}{\tilde{\sigma}^2} \\ &= T\,\frac{\tilde{\boldsymbol{u}}'\,\boldsymbol{X}(\boldsymbol{X}'\boldsymbol{X})^{-1}\boldsymbol{X}'\tilde{\boldsymbol{u}}}{\tilde{\boldsymbol{u}}'\tilde{\boldsymbol{u}}}. \end{aligned}$$

(ii) In the constrained model $\tilde{\boldsymbol{u}} = \boldsymbol{y} - \boldsymbol{X}\tilde{\boldsymbol{\beta}}$, the vector $\tilde{\boldsymbol{u}}$ is orthogonal (= independent) of each column of **X**, because

$$\boldsymbol{X}'\tilde{\boldsymbol{u}} = \boldsymbol{X}'(\boldsymbol{y} - \boldsymbol{X}\tilde{\boldsymbol{\beta}}) = \boldsymbol{X}'\boldsymbol{y} - \boldsymbol{X}'\boldsymbol{X}(\boldsymbol{X}'\boldsymbol{X})^{-1}\boldsymbol{X}'\boldsymbol{y} = \boldsymbol{0}.$$

Given this last equality, it follows that the sum of the residuals is also zero. In fact, since there is a constant term, i.e. a column of units (e.g. the first), in **X**, we have:

$$\boldsymbol{x}'_1\tilde{\boldsymbol{u}} = (\boldsymbol{1}, \ldots, \boldsymbol{1})\begin{pmatrix} \tilde{u}_1 \\ \vdots \\ \tilde{u}_T \end{pmatrix} = \sum\nolimits_{t=1}^{T} \tilde{u}_t = 0.$$

Consequently, the least squares estimator of $\boldsymbol{\gamma}$ in the auxiliary regression (4.35) is given by

$$\hat{\boldsymbol{\gamma}} = (\boldsymbol{X}'\boldsymbol{X})^{-1}\boldsymbol{X}'\tilde{\boldsymbol{u}}.$$

The total sum of squares is

$$TSS = \tilde{\boldsymbol{u}}'\tilde{\boldsymbol{u}}$$

and the adjusted one,

$$\begin{aligned} ESS &= (\boldsymbol{X}\hat{\boldsymbol{\gamma}})'(\boldsymbol{X}\hat{\boldsymbol{\gamma}}) \\ &= \hat{\boldsymbol{\gamma}}'\boldsymbol{X}'\boldsymbol{X}\hat{\boldsymbol{\gamma}} \\ &= \tilde{\boldsymbol{u}}'\boldsymbol{X}(\boldsymbol{X}'\boldsymbol{X})^{-1}\boldsymbol{X}'\boldsymbol{X}(\boldsymbol{X}'\boldsymbol{X})^{-1}\boldsymbol{X}'\tilde{\boldsymbol{u}} \\ &= \tilde{\boldsymbol{u}}'\boldsymbol{X}(\boldsymbol{X}'\boldsymbol{X})^{-1}\boldsymbol{X}'\tilde{\boldsymbol{u}}. \end{aligned}$$

The square of the correlation coefficient is then:

$$R^2 = \frac{ESS}{TSS} = \frac{\tilde{\boldsymbol{u}}'\boldsymbol{X}(\boldsymbol{X}'\boldsymbol{X})^{-1}\boldsymbol{X}'\tilde{\boldsymbol{u}}}{\tilde{\boldsymbol{u}}'\tilde{\boldsymbol{u}}}.$$

It follows from (4.34) that the LM statistic is equal to $T.\ R^2$ and then, by virtue of Proposition 4.2, asymptotically follows a Chi-square distribution with p degrees of freedom:

$LM = T.\ R^2 \overset{As}{\sim} \chi^2(p)$. ◆

Practical Test Procedure

Based on Theorem 4.2, we can summarise the procedure for detecting the presence of autocorrelation in the estimated residuals as follows.

Testing Autocorrelation of Order *p*

1. The residuals $\tilde{u}_t$ are obtained by estimating the model constrained by the null hypothesis of no autocorrelation, i.e. $y_t = \boldsymbol{x}'_t\boldsymbol{\beta} + u_t$.
2. The auxiliary regression is then estimated
$$\tilde{u}_t = \boldsymbol{x}'_t\boldsymbol{\gamma} + \rho_1\tilde{u}_{t-1} + \rho_2\tilde{u}_{t-2}\cdots + \rho_p\tilde{u}_{t-p} + v_t,$$
which gives the coefficient R^2.
3. We form the statistic $T.\ R^2$.
4. The null hypothesis is rejected if $LM > k_\alpha$, where k_α is the fractile of order α of a variable X distributed according to a $\chi^2(p)$, fractile such as $P(X > k_\alpha) = \alpha$. (Alternatively, if the p-value is less than α.)

Here is an illustration of the test procedure, using French GDP data that has already been used on several occasions.

Example 4.4 In Example 4.2, we obtained a relationship between GDP in primary differences and its past, of the form

$$\underset{(4.21)}{\Delta PIB_t = 1659.8} - \underset{(4.55)}{0.33}\ \Delta PIB_{t-1} - \underset{(3.03)}{0.22}\ \Delta PIB_{t-2}.$$

A priori, we would expect the residuals in this equation not to be autocorrelated. Let's go through the steps described above to check this.
Using the residuals from this regression, we estimate the auxiliary equation with $p = 4$, which leads to $R^2 = 0.018$ and the statistic $LM = 181 \times 0.018 = 3.258$. Since the value of a Chi-square with 4 degrees of freedom and $\alpha = 0.05$ is equal to 9.49, we accept the null hypothesis that there is no autocorrelation in the residuals.

A few more details on this test procedure might prove useful.

1. When performing auxiliary regression, it is advisable to replace lost observations $\tilde{u}_{t-1}, \ldots, \tilde{u}_{t-p}$ with zeros rather than treating them as missing values.
2. The null hypothesis $H_0 : \rho_1 = \rho_2 = \cdots = \rho_p = 0$ can also be tested by assessing the overall significance of the coefficients in the auxiliary regression (4.35). We then obtain a statistic, denoted LMF, such as

$$LMF \sim F(p, T - p).$$

The null hypothesis of no autocorrelation is rejected when $LMF > k_\alpha, k_\alpha$ being the critical value relating to α for the F distribution, or, alternatively, when the p-value associated with LMF is less than the significance level $\alpha.\alpha$.

Example 4.5 Again, on the basis of the relationship in the previous example, the statistic $LMF = 0.299 \sim F(4.170)$ is associated with a p-value of 0.88, much higher than $\alpha = 0.05$ or 0.01, so that, as with the LM statistic, we accept the null hypothesis of no autocorrelation.

3. In general, the two statistics LM and LMF give probabilities very close to each other and can therefore be used interchangeably. However, as the distribution of the former is asymptotic, it is preferable to use the latter when the number of observations is small.

4.5 Validation of Model Assumptions: Homoskedasticity

There are several homoskedasticity tests. We will limit ourselves here to presenting two of them: the first, which is associated with the names of Breusch and Pagan, and the second, known as the ARCH (*AutoRegressive Conditional Heteroskedasticity*) test.

Breusch–Pagan Test

Consider the linear model

$$y_t = \boldsymbol{x}'_t\boldsymbol{\beta} + u_t, \quad t = 1, \ldots, T, \tag{4.36}$$

where the perturbations u_t are independently and normally distributed with zero expectation, but a variance temporally determined by the equation

$$\sigma_t^2 = h(\alpha_1 + \alpha_2 z_{t2} + \cdots + \alpha_p z_{tp}) = h(z'_t\boldsymbol{\alpha}). \tag{4.37}$$

The vector z'_t, of dimension p, includes a constant and the variables—a subset of $\boldsymbol{x}'_t$—which affect the evolution of the variance. The form of the function $h(\cdot)$ is not specified, but it is sufficiently general to leave room for several possibilities, including linear and quadratic heteroskedasticity:

$$h(z'_t\boldsymbol{\alpha}) = (z'_t\boldsymbol{\alpha})^d, \ d = 1, 2.$$

The null hypothesis is that of the homoskedasticity of disturbances

$$H_0 \ : \ \alpha_2 = \cdots = \alpha_p = 0$$

and the alternative, the heteroskedasticity of these same disturbances.

Under the null hypothesis, the constrained model is equal to (4.36), but with $u_t \sim N(0, \sigma^2)$. It is obviously much simpler to estimate than the free model. Consequently, a Lagrange multiplier test is particularly recommended in this case.

To cut to the chase, the Breusch–Pagan statistic is obtained by estimating the model constrained by H_0, which provides the residuals $\tilde{u}_t$, and then using these residuals to estimate the auxiliary regression.

$$\tilde{u}_t^2 = \alpha_1 + \alpha_2 z_{t2} + \cdots + \alpha_p z_{tp} + v_t. \tag{4.38}$$

Finally, the BP statistic can be written as

$$BP = \frac{ESS}{TSS/T} = T \cdot R^2, \tag{4.39}$$

where ESS and TSS are respectively the adjusted and total sums of squares in the regression (4.38), while R^2 is the squared correlation coefficient of this auxiliary regression. This statistic, of course, asymptotically follows a Chi-square distribution with $(p-1)$ degrees of freedom.

$$BP = T \cdot R^2 \sim \chi^2(p-1).$$

As with autocorrelation, we can also check the overall significance of the explanatory variables for heteroskedasticity in (4.38) by testing $H_0 : \alpha_2 = \cdots = \alpha_p = 0$, so that we obtain the statistic F_{hom} such that

$$F_{\text{hom}} \sim F(p-1, T-p).$$

Testing homoskedasticity

1. The residuals $\tilde{u}_t$ are obtained by estimating the model $y_t = \mathbf{x'}_t\boldsymbol{\beta} + u_t$, where the variance is constrained by the null hypothesis $H_0 : \alpha_2 = \cdots = \alpha_p = 0$.
2. The auxiliary regression $\tilde{u}_t^2 = \alpha_1 + \alpha_2 z_{t2} + \cdots + \alpha_2 z_{tp} + v_t$, is estimated, giving the coefficient of determination R^2.
3. We then form the statistic $BP = T.\ R^2$.
4. The null hypothesis is rejected if $BP > k_\alpha$, where k_α is the critical value of order α of a variable X distributed according to a distribution $\chi^2(p-1)$, fractile such as $P(X > k_\alpha) = \alpha$. (Alternatively, if the p-value is less than α.)

Alternatively, if we use the F_{hom} statistic, this point 4 is replaced by the following: we reject the null hypothesis when $F_{\text{hom}} > k_\alpha$ or, alternatively, when the p-value associated with k_α is less than the chosen significance threshold $\alpha.\alpha$.

ARCH Test

Consider again the linear model (4.36) $y_t = \boldsymbol{x'}_t\boldsymbol{\beta} + u_t$, in which the variance is not constant, but depends instead on its own past, for example the q periods prior to t. The perturbations u_t consequently follow an ARCH(q) process, the definition of which we gave in Chapter 2 (see Definition 2.6).

Under the assumption of conditional normality of the disturbances, the model in question is written as follows

$$\begin{aligned} &y_t|\, I_{t-1} \sim N(\boldsymbol{x'}_t\boldsymbol{\beta}, \sigma_t^2) \\ &\sigma_t^2 = \alpha_0 + \sum\nolimits_{j=1}^{q} \alpha_j u_{t-j}^2 \\ &u_t = y_t - \boldsymbol{x'}_t\boldsymbol{\beta}. \end{aligned} \tag{4.40}$$

Testing the homoskedasticity of disturbances means testing the null hypothesis in (4.40) $H_0 : \ \alpha_1 = \alpha_2 = \cdots = \alpha_q = 0$. The model so constrained is given by

$$y_t = \boldsymbol{x'}_t\boldsymbol{\beta} + u_t, \quad u_t \sim N(0, \alpha_0), \quad t = 1, \ldots, T. \tag{4.41}$$

Moreover, using the formalism developed in (4.37), we obtain the expression for the variance

$$\sigma_t^2 = h(\boldsymbol{z'}_t\boldsymbol{\alpha}) = h(\alpha_0 + \alpha_1 u_{t-1}^2 + \cdots + \alpha_q u_{t-q}^2). \tag{4.42}$$

If we compare (4.41) and (4.42) with (4.36) and (4.37), we see that the first set has exactly the same structure as the second. The Lagrange multiplier statistic is equal to $T \cdot R^2$, where R^2 is the coefficient associated to the auxiliary regression

$$\tilde{u}_t^2 = \alpha_0 + \alpha_1 \tilde{u}_{t-1}^2 + \cdots + \alpha_q \tilde{u}_{t-q}^2 + \varepsilon_t \tag{4.43}$$

Therefore, we have

$$\xi_{\text{ARCH}} = T \cdot R^2 \overset{As}{\sim} \chi^2(q).$$

The procedure for performing an ARCH homoskedasticity test can therefore be summarised as follows:

ARCH test

1. The residuals $\tilde{u}_t$ are obtained by estimating the model $y_t = \boldsymbol{x'}_t\boldsymbol{\beta} + u_t$, where the variance is constrained by the null hypothesis $H_0 : \alpha_2 = \cdots = \alpha_p = 0$.
2. The auxiliary regression $\tilde{u}_t^2 = \alpha_0 + \alpha_1\tilde{u}_{t-1}^2 + \cdots + \alpha_q\tilde{u}_{t-q}^2 + \varepsilon_t$, is estimated, giving the coefficient of determination R^2.
3. We then form the statistic $\xi_{\text{ARCH}} = T.\ R^2$.
4. The null hypothesis is rejected if $\xi_{\text{ARCH}} > k_\alpha$, where k_α is the fractile of order α of a variable X distributed according to a $\chi^2(q)$, fractile such as $P(X > k_\alpha) = \alpha$. (Alternatively, H_0 is rejected if the p-value associated with ξ_{ARCH} is less than α.)

As with the Breusch–Pagan test, it is also possible to test the overall significance of the coefficients in the auxiliary regression. We then obtain the statistic F_{ARCH} as follows

$$F_{\text{ARCH}} \sim F(q, T - q).$$

The corresponding decision rule is as follows: we reject the null hypothesis of homoskedasticity if $F_{\text{ARCH}} > k_\alpha$, with k_α the fractile of order α of a random variable X distributed according to a Fisher-Snedecor distribution, or alternatively, if the p-value associated with F_{ARCH} is less than $\alpha.\alpha$.

Example 4.6 As in the two previous examples, we will use the results of the regression between French real GDP, expressed in first differences, and its past.
The ξ_{ARCH} statistic with $q = 4$ is equal to 46.68. It is associated with a p-value of less than 0.0001, so the null hypothesis of homoskedasticity is rejected.
The conclusion is identical when an F-test is performed, since the statistic $F_{\text{ARCH}} = 9.8 \sim F(4.176)$ is associated with a p-value of less than 0.0001.

4.6 Validation of Hypotheses: Normality

The assumption of normality of the disturbance term is not essential for the estimation of the linear model by least squares—as we have seen. On the other hand, it is central to the application of the maximum likelihood estimation method.

Bera–Jarque Statistics

Bera and Jarque have proposed a test based on a Lagrange multiplier statistic asymptotically distributed according to a Chi-square distribution with two degrees of freedom. However, the procedure for determining the precise form of this statistic is very complex and involves the Pearson family of distributions. To avoid such complexities, we will adopt a more indirect, but also more intuitive, approach.

First, we need a reminder about the centred moments of order 3 and order 4 of a normal random variable, moments known respectively as the **skewness** and **kurtosis** coefficients of this distribution.

Reminder

If $X \sim N(\mu, \sigma^2)$, then the *central moments* of order 3 and 4 of X are given by

$$m_3 = E\Big[(X - E(X))^3\Big] = 0$$

$$m_4 = E\Big[(X - E(X))^4\Big] = 3\sigma^4.$$

Consider again the linear model

$$y_t = \boldsymbol{x}'_t\boldsymbol{\beta} + u_t, \quad t = 1, \ldots, T, \quad u_t \sim IIN(0, \sigma^2).$$

We wish to test the null hypothesis of normality, i.e. more precisely that $E(u_t^3) = 0$ and $E(u_t^4) = 3\sigma^4$ or alternatively,

$$\begin{cases} E(u^3)/\sigma^3 = 0 \\ E(u^4)/\sigma^4 = 3. \end{cases} \tag{4.44}$$

If $\hat{u}_t$ and $\hat{\sigma}^2 = (1/T)\sum \hat{u}_t^2$ denote the estimation residuals and their sample variance respectively, then the empirical correspondents of (4.44) are given by

$$\begin{cases} \tau_3 = (1/T)\sum_{t=1}^{T} (\hat{u}_t^3/\hat{\sigma}^3) \\ \tau_4 = (1/T)\sum_{t=1}^{T} (\hat{u}_t^4/\hat{\sigma}^4). \end{cases} \tag{4.45}$$

It can be shown that under the double null hypothesis, the asymptotic distribution of τ_3 is

$$\sqrt{T}\tau_3 \overset{As}{\sim} N(0, 6) \tag{4.46}$$

and that of τ_4

$$\sqrt{T}(\tau_4 - 3) \overset{As}{\sim} N(0, 24). \tag{4.47}$$

Finally, let's define the **Bera–Jarque statistic** as follows:

$$BJ = \frac{T}{6}\tau_3^2 + \frac{T}{24}(\tau_4 - 3)^2. \tag{4.48}$$

Testing Normality

It remains to show that the statistic (4.48) asymptotically follows a Chi-square distribution with two degrees of freedom, a proposition that we will record in the form of a theorem.

Theorem 4.4 *Under the null hypothesis of normality of the estimation residuals, the **Bera–Jarque statistic** is **asymptotically** distributed according to a Chi-square distribution with two degrees of freedom*

$$BJ \overset{As}{\sim} \chi^2(2).$$

Demonstration

The demonstration will take place in three stages.

(i) Given (4.46) and (4.47), we have

$$\frac{\sqrt{T}}{\sqrt{6}}\tau_3 \overset{As}{\sim} N(0, 1) \text{ and } \frac{\sqrt{T}}{\sqrt{24}}(\tau_4 - 3) \overset{As}{\sim} N(0, 1).$$

(ii) Since τ_3 and τ_4 each represents a normal random sum multiplied by a calibration factor, it is sufficient for these two sums to be independent for their covariance to be zero. Furthermore, since the terms of each of these sums are by hypothesis independent, it is sufficient, in the last instance, to check that their covariances are equal for all t. Finally, since $T \to +\infty, (\hat{u}_t/\hat{\sigma}) \to u_t$, we can limit ourselves to proving that

$$E\big(u_t^3(u_t^4 - 3)\big) = E(u_t^7) - 3E(u_t^3) = 0.$$

Whatever t is, the previous covariance is equal to zero, given that all odd-order moments of a normal random variable are zero.

(iii) In the review (see "A Single Linear Restriction" section), we stated the proposition that a sum of squares of independent standard normal random variables is distributed according to a Chi-square distribution with as many degrees of freedom as there are terms in the sum. From this proposition, it immediately follows that

$$BJ = \frac{T}{6}\tau_3^2 + \frac{T}{24}(\tau_4 - 3)^2 \overset{As}{\sim} \chi^2(2).$$

At finite distances, i.e. in small samples, the statistics τ_1 and τ_2 are not really independent; moreover, the second of them tends very slowly towards a normal distribution.

This is why, as proposed by Doornik and Hansen, the *BJ* statistic must be corrected for the bias introduced by the smallness of the sample. In this way, if z_1 and z_2 denote the transformed degree of asymmetry and kurtosis respectively, then the new normality test statistic **approximately** follows a Chi-square distribution with two degrees of freedom:

$$BJ_{\text{mod}} = z_1^2 + z_2^2 \approx \chi^2(2).$$

Example 4.7 We will again use the AR(2) model to estimate French real GDP expressed in terms of first differences. The Bera–Jarque statistic is equal to 158.05; it is associated with a *p*-value of 0.00000. The null hypothesis of the normality of the disturbance is therefore rejected.

4.7 Dealing with Heteroskedasticity and Autocorrelation

We will first consider the White (1980) estimator, which aims to "correct" the heteroskedasticity of the variance–covariance matrix of **u**, before looking at another, more general, convergent estimator of this matrix, which corrects for both heteroskedasticity and error autocorrelation.

White's Estimator

Consider again the model $\boldsymbol{y} = \boldsymbol{X\beta} + \boldsymbol{u}$, where

$$E(\boldsymbol{uu}') = \boldsymbol{\Omega} = \text{diag}(\omega_1^2, \ldots, \omega_T^2). \tag{4.49}$$

Finding a convergent estimator of the T diagonal elements of $\boldsymbol{\Omega}$ is, at first sight, an impossible task, since we have the same number of observations. So, we have to change our method.

Since

$$\hat{\boldsymbol{\beta}} = (\boldsymbol{X}'\boldsymbol{X})^{-1}\boldsymbol{X}'(\boldsymbol{X\beta} + \boldsymbol{u}) = \boldsymbol{\beta} + (\boldsymbol{X}'\boldsymbol{X})^{-1}\boldsymbol{X}'\boldsymbol{u},$$

the variance–covariance matrix of the ordinary least squares estimator $\hat{\boldsymbol{\beta}}$ is equal to

$$\begin{aligned}\operatorname{Var}(\hat{\boldsymbol{\beta}}) &= E\left[(\hat{\boldsymbol{\beta}}-\boldsymbol{\beta})(\hat{\boldsymbol{\beta}}-\boldsymbol{\beta})'\right] = E\left[(\boldsymbol{X}'\boldsymbol{X})^{-1}\boldsymbol{X}'\boldsymbol{u}\boldsymbol{u}'\boldsymbol{X}(\boldsymbol{X}'\boldsymbol{X})^{-1}\right] \\ &= (\boldsymbol{X}'\boldsymbol{X})^{-1}\boldsymbol{X}'\boldsymbol{\Omega}\boldsymbol{X}(\boldsymbol{X}'\boldsymbol{X})^{-1}. \end{aligned} \tag{4.50}$$

In an asymptotic framework, the limit equivalent of (4.50), i.e. the variance–covariance matrix of $\sqrt{T}(\hat{\boldsymbol{\beta}}-\boldsymbol{\beta})$, is given by

$$\operatorname{plim}\left(\frac{1}{T}\boldsymbol{X}'\boldsymbol{X}\right)^{-1} \cdot \operatorname{plim}\frac{1}{T}\boldsymbol{X}'\boldsymbol{\Omega}\boldsymbol{X}\cdot \operatorname{plim}\left(\frac{1}{T}\boldsymbol{X}'\boldsymbol{X}\right)^{-1}. \tag{4.51}$$

Under the usual assumptions, we can be sure that the extreme terms of the product in (4.51) tend towards a finite deterministic matrix. But what about the central term?

In fact, White (1980), preceded by Eicker, showed that under certain conditions it was possible to find a **convergent** estimator of

$$(1/T)\boldsymbol{X}'\hat{\boldsymbol{\Omega}}\boldsymbol{X},$$

even if $\hat{\boldsymbol{\Omega}}$ is **not** a convergent estimator of $\boldsymbol{\Omega}$. In the present case, where the matrix $\boldsymbol{\Omega}$ is diagonal—and therefore heteroskedastic of an unknown form—the estimator of the t-th diagonal element is provided by $\hat{u}_t^2$, i.e. by the square of the t-th residual obtained by OLS.

Alternatively, the (i, j) element of plim $(1/T)$ $\boldsymbol{X}'\boldsymbol{\Omega}\boldsymbol{X}$, i.e.

$$\operatorname{plim}\left((1/T)\sum_{t=1}^{T}\omega_t^2 x_{ti}x_{tj}\right)$$

can be estimated convergently by

$$(1/T)\sum_{t=1}^{T}\hat{u}_t^2 x_{ti}x_{tj}.$$

This result and the assumptions on which it is based are the subject of the following proposal:

Proposition 4.3 *Let be* $\mathbf{y} = \boldsymbol{X}\boldsymbol{\beta}+\boldsymbol{u}$, *where* $E(\boldsymbol{u}\boldsymbol{u}') = \boldsymbol{\Omega} = \operatorname{diag}(\omega_1^2, \ldots, \omega_T^2)$, and $\mathbf{W}$ *the White matrix defined by*

$$\mathbf{W} = \left(\frac{\boldsymbol{X}'\boldsymbol{X}}{T}\right)^{-1}\left(\frac{1}{T}\sum_{t=1}^{T}\hat{u}_t^2\boldsymbol{x}_t\boldsymbol{x}'_t\right)\left(\frac{\boldsymbol{X}'\boldsymbol{X}}{T}\right)^{-1},$$

with $\hat{u}_t^2$ *the squared residuals from the OLS regression of* $\mathbf{y}$ *on* $\mathbf{X}$.

If

(i) $E(\boldsymbol{u}|\,\boldsymbol{X}) = \mathbf{0}$;

(ii) $(1/T)(\boldsymbol{X}'\boldsymbol{X}) \xrightarrow{P} \boldsymbol{Q}$, $\mathbf{Q}$ *being a finite non-singular matrix;*
(iii) *when* $T \rightarrow +\infty$, *the average of* ω_t^2, $t = 1, \ldots, T$, *exists and is equal* to σ^2,

then **W** ***converges in probability*** *to the asymptotic variance–covariance matrix* of $\sqrt{T}(\hat{\boldsymbol{\beta}} - \boldsymbol{\beta})$.

This result has important practical implications. In particular, when the number of observations T is sufficiently large, the White matrix, i.e. the heteroskedastic convergent standard errors (HCSE), can be used to implement the tests usually applied to the linear model estimated by OLS, even in the presence of heteroskedastic errors.

The Newey–West Estimator

Certain economic series sometimes exhibit a marked form of heteroskedasticity: this is particularly true of financial series. Time series are almost always characterised by the presence of autocorrelation of a definite order.

The problem is then to generalise White's analysis to variance–covariance matrices with non-zero elements outside the main diagonal. However, it is clear that it is not possible to estimate all the non-diagonal elements of a matrix of dimension $T \times T$.. Moreover, we wish to obtain a convergent estimator of the matrix in question.

In other words, we need to restrict the range of possible autocorrelations between the values of the disturbances: in this way, if we designate the autocorrelation of order $t - s$ by $E(u_t u_{t-s})$, $s \neq 0$, we need to determine the maximum acceptable deviation $p = t - s$.

Let's now return to Eq. (4.50), but this time estimated as follows

$$\widehat{\text{Var}(\hat{\boldsymbol{\beta}})} = (\boldsymbol{X}'\boldsymbol{X})^{-1}\boldsymbol{X}'\hat{\boldsymbol{\Omega}}\boldsymbol{X}(\boldsymbol{X}'\boldsymbol{X})^{-1}.$$

The problem is to use the OLS residuals $\hat{\boldsymbol{u}}$, to obtain a convergent estimator of

$$\hat{\boldsymbol{\Sigma}} = \boldsymbol{X}'\hat{\boldsymbol{\Omega}}\boldsymbol{X}.$$

It can be shown that under certain conditions $(1/T)\,\hat{\boldsymbol{\Sigma}}$ is a convergent estimator of the variance–covariance matrix of the K-vector $(x_{t1}u_t, \ldots, x_{tK}u_t)$.

The Newey–West (1987) matrix—named HAC, *Heteroskedasticity and Autocorrelation Convergent*—is calculated using the weighted sums

$$\hat{\boldsymbol{\Sigma}} = \hat{\boldsymbol{\Gamma}}(0) + \sum_{j=1}^{p} w\,(j,p)\left[\hat{\boldsymbol{\Gamma}}(j) + \hat{\boldsymbol{\Gamma}}'(j)\right],$$

where $w\,(j,p)$ is the weight assigned to offset j and $\hat{\boldsymbol{\Gamma}}(j)$ is the dimensional matrix $K \times K$ given, for $j \geq 0$, by

$$\hat{\boldsymbol{\Gamma}}(j) = \sum_{t=j}^{p} \hat{u}_t \hat{u}_{t-j} \boldsymbol{x}_t \hat{\boldsymbol{x}}_{t-j}.$$

However, there are still two problems: how to determine the weights w and the maximum length p? The answer to the first question makes use of what is known as the time window. From this point of view, the most widely used window is the Bartlett window defined by

$$w(j, p) = 1 - \frac{j}{p+1}, \qquad j = 1, \ldots, p.$$

so that the weights decrease linearly as j increases. (There are several other windows, including the Parzen window and the spectral quadratic window).

The answer to the second question is trickier. Generally speaking, the maximum length p should grow in proportion to a fractional power of T. Stock and Watson (2007) suggest taking $p = 0,75T^{1/3}$. Other choices are, of course, possible, but none of them is obvious, which does not suit the applied econometrician.

From a practical point of view, it is essential to remember that the Newey–West estimator, like the White estimator, is convergent when the size of the sample T tends towards infinity. As a result, the estimator in question is only valid for samples with a large number of observations.

It should also be remembered that the presence of autocorrelation in the residuals is above all a signal that the estimated model is probably misspecified and that, consequently, it is this misspecification that needs to be remedied before anything else!

4.8 Application

We will re-examine the AR(2) model of French real GDP, which is widely used throughout this chapter. Essentially, we will add several binary variables in order to take account of the *outliers* present in the series. We will then compare this new model with the simple AR(2). Finally, we will subject it to some of the tests studied in this chapter.

The Model with Dummy Variables

In Example 4.2, we showed that the equation relating the variable ΔPIB_t to its values lagged was associated with significant coefficients, both individually and globally. Figure 4.3 provides a plot of the estimation residuals.

Clearly, the residuals make a big difference during the recession linked to the pandemic: briefly, there is a significant fall in GDP in the first and second quarters

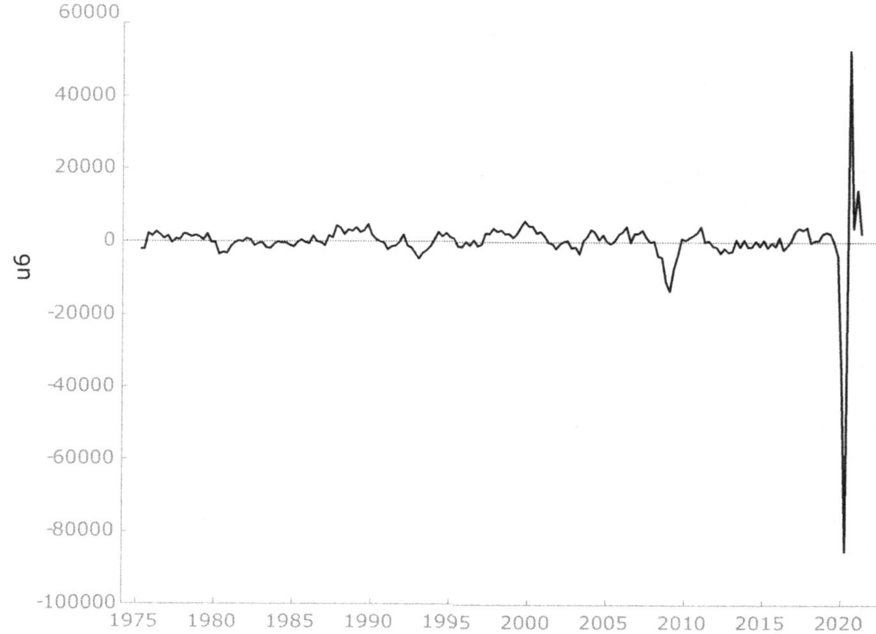

Fig. 4.3 Residuals of the estimated AR(2)

of 2020, followed by a significant recovery in the third quarter of the same year. Consequently, we will now introduce two *dummy* variables into the AR(2) model:

- one (denoted Bin1) will take the value 0 throughout the sampling period (1975 T2–2021 T2), except in 2020 T1 and T2, when it will take the value 1;
- the other (denoted Bin2) will wear the unit exclusively in quarter 3 of 2020 and zero everywhere else.

The result is an ARX model, with "X" denoting the two **exogenous** binary variables.

Here is an initial table showing the estimation results—by exact maximum likelihood—for the two models selected (Table 4.1):

Table 4.1 Estimation and significance of coefficients

Model	Constant	ΔPIB_{t-1}	ΔPIB_{t-2}	Bin 1	Bin2
AR (2)	1659***	– 0.33***	– 0.22***	–	–
ARX (2)	1816***	– 0.15*	0.33***	– 52,842***	76,035***

***Significant at 1%; *significant at 5.16%

Table 4.2 Selection criteria

Model	*AIC*	*BIC*	*HQC*	R^2
AR (2)	3869.9	3882.8	3375.1	0.11
ARX (2)	3458.5	3477.8	3466.3	0.91

Selecting a Model

As can be seen, the introduction of the two binary variables considerably modifies the estimators obtained, even though they are all significantly different from zero.

Note in particular that the sign of the GDP coefficient (in first differences) lagged by two quarters becomes positive in the second model and that the Bin1 and Bin2 coefficients take on a very large value, of opposite sign. This fact alone testifies to the depth of the decline in GDP recorded during the initial phases of the pandemic caused by SARS-COV2.

To select one or other model, we will refer, as we have explained on several occasions, to the three information criteria—Akaike, Schwarz Bayesian and Hannan-Quinn—as well as to the coefficient R^2 (Table 4.2).

The values of the three criteria are all significantly lower for the ARX model, leading to the same result: the AR(2) model with binary variables should be selected. For its part, the coefficient R^2 increases sharply from 0.1 to 0.9, leading to the same conclusion.

Hypothesis Testing

Once the ARX model has been selected, we will apply a number of validation tests, bearing in mind that the choice of tests is reduced in the case of autoregressive models.

Let's start with an autocorrelation test up to order 4. The Lagrange multiplier statistic is equal to

$$LM = 9.632,$$

while the corresponding p-value is given by

$$P\left(\chi^2(2) > 9.632\right) = 0.008.$$

We therefore have to reject the null hypothesis that there is no autocorrelation.

The fourth-order ARCH test also leads us to reject the null hypothesis of residual homoskedasticity, since the $ARCH = 69.56$ statistic is associated with a p-value equal to $P\left(\chi^2(4) > 69.56\right) < 0.0001$. The variance of the residuals is therefore heteroskedastic.

Finally, the Bera–Jarque statistic, under the null hypothesis of normality of the residuals, follows a Chi-square distribution with two degrees of freedom. In this case, this statistic is equal to $BJ = 494.68$ and the corresponding p-value is given by

$$P\left(\chi^2(2) > 494.68\right) = 0.$$

Here too, we have to reject the null hypothesis that the residuals are normally distributed.

Overall, the validation tests for the ARX (2) model are hardly conclusive, given that the estimated residuals are autocorrelated, heteroskedastic, and not normally distributed. (The absence of normality has important consequences in this case, since the model was estimated by maximum likelihood).

These observations demonstrate the inadequacy of simple time series models. It is clear that GDP is economically dependent on variables other than its past. In other words, we need to move towards full linear models and take into account the non-stationarity of economic series.

Estimation of the HAC Matrix

We will now estimate the variance–covariance matrix of the errors, both conventionally and using the Newey–West estimator. We will then compare the results obtained in each case.

To do this, we will estimate an AR(2)-type model, in which the variable ΔPIB_t in period t is simply related to its previous values in periods $t - 1$ and $t - 2$.

Firstly, it should be noted that the HAC standard errors were estimated using the Bartlett window, while the maximum length was equal to 4. This latter value corresponds to the criterion put forward by Stock and Watson.

Secondly, Table 4.3 shows, first and foremost, that the use of HAC estimators leads to higher standard errors than in the "classic" case. As a result, the t-test statistic takes on lower values while remaining largely significant.

Table 4.3 Comparison of standard error estimates

AR(2) model	Coefficient	Standard errors	Statistic t	HAC stan. errors	Statistic t
Const	732. 05	184.35	3.971	265.28	2.760
Δ $PIB(-1)$	0.428	0.076	5.664	0.103	4.162
Δ $PIB(-2)$	0.171	0.075	2.273	0.078	2.188

4.9 Exercises

4.9.1 Let $(X_1, \ldots, X_T)$ be a random sample, where for all t, $X_t \sim N(\mu, 1)$, μ being unknow. We want to test the null hypothesis $H_0 : \mu = \mu_0$ against the alternative hypothesis $H_1 : \mu \neq \mu_0$, using a likelihood ratio test. To do this, please proceed as follows:

1. Estimate $\hat{\mu}_{ML}$.
2. Use this estimator to determine the value of the likelihood function.
3. Then establish what the value of this likelihood function is under null hypothesis $H_0 : \mu = \mu_0$.
4. Take the ratio of the maximum likelihoods to obtain the LR statistic.

4.9.2 In the proof of Theorem 4.1, we introduced the matrix **Q** such that

$$\boldsymbol{Q} = \boldsymbol{X}(\boldsymbol{X}'\boldsymbol{X})^{-1}\boldsymbol{R}'\left[\boldsymbol{R}(\boldsymbol{X}'\boldsymbol{X})^{-1}\boldsymbol{R}'\right]^{-1}\boldsymbol{R}(\boldsymbol{X}'\boldsymbol{X})^{-1}\boldsymbol{X}'.$$

You are asked to show that this matrix is idempotent, i.e. that

$$\boldsymbol{Q}\boldsymbol{Q} = \boldsymbol{Q}.$$

4.9.3 Consider the following linear regression with two explanatory variables and $T = 232$:

$$\begin{array}{llll} y_t = 5.063 & -\ 0.335\ x_{1t} & +\ 0.553\ x_{2t}, & R^2 = 0.697 \\ \quad\ (0.586) & \quad (0.025) & \quad (0.045) & \end{array} \tag{4.52}$$

Use Eq. (4.20) to establish the value of $F(2, 229)$ and thus test the null hypothesis

$$H_0 : \beta_2 = \beta_3 = 0.$$

4.9.4 Let $(X_1, \ldots, X_T)$ be a random sample, where for all t, $X_t \sim N(\mu, 4)$. We ask you to derive the three statistics for the likelihood ratio, the Wald test and *the Lagrange multiplier test, in order to test the null hypothesis* $H_0 : \mu = 2$ against the alternative hypothesis $H_1 : \mu \neq 2$.

4.9.5 Let us refer to Eq. (4.52) in exercise 4.9.3 and apply three tests to the residuals of this estimated equation, the null hypothesis of which is respectively the absence of autocorrelation, the homoskedasticity and the normality of these residuals.

What conclusions would you draw from analysing this Table 4.4.

Table 4.4 Statistics and p-values

	Stat.	p-values
Autocorrelation (Lagrange multiplier)	$LMF = 60.238$	< 0.0001
Heteroskedasticity (Breusch–Pagan)	$BP = 10.705$	0.004
Normality (Bera–Jarque)	$BJ = 28.532$	< 0.0001

Reference

Bismans, F. (2016). *Probabilités et statistique inférentielle. Prélude à l'économétrie.* Ellipses.

Newey, W., & West, K. (1987). A simple positive semi-definite, heteroskedastic and autocorrelation consistent covariance matrix. *Econometrica, 55*, 703–708.

Stock, J. M., & Watson, M. H. (2007). *Introduction to econometrics* (2nd ed.): Pearson Education.

White, H. (1980). A heteroskedasticity-consistent covariance matrix estimator and a direct test for heteroscedasticity. *Econometrica, 48*, 827–838.

Chapter 5
Non-stationarity and Cointegration

5.1 Spurious Regressions

A spurious regression is simply, as its name suggests, a relationship between variables that is apparently statistically significant, but whose significance is only due to the presence of non-stationarity in these variables.

A Theoretical Example

To illustrate the phenomenon, let's consider two independent random walks

$$\begin{cases} y_t = y_{t-1} + u_t, & u_t \sim \text{IND}(0, \sigma_u^2) \\ x_t = x_{t-1} + v_t, & v_t \sim \text{IND}(0, \sigma_v^2). \end{cases}$$

Given this assumption of independence, the covariances between the random terms are obviously zero: for all t, s, $E(u_t u_s) = 0$. $E(u_t u_s) = 0$.

Under these conditions, regressing y_t on x_t leads to fallacious regression:

$$y_t = m + \beta x_t + \varepsilon_t. \tag{5.1}$$

Logically, the coefficient β in (5.1) should converge in probability to zero, which would simply reflect the absence of a relationship between the two variables. However, this is not the case.

Supplementary Information The online version contains supplementary material available at https://doi.org/10.1007/978-3-031-72910-2_5.

F. J. Bismans and O. Damette, *Dynamic Econometrics*,
https://doi.org/10.1007/978-3-031-72910-2_5

To show this, we will need the following result, which we will present in the form of an unproved lemma.

Lemma 5.1 *Let be the driftless* ***random walk*** $z_t = z_{t-1} + \varepsilon_t$, *where* $z_0 = 0$ *and* $\varepsilon_t \sim \text{IID}(0, \sigma^2)$. *Then*

(i) *if* $\bar{z} = (1/T)\sum z_t, T^{-1/2}\bar{z} \xrightarrow{D} \sigma \cdot \int_0^1 B(r)\,dr,$

(ii) $(1/T^2)\sum_{t=1}^{T} y_{t-1}^2 \xrightarrow{D} \sigma^2 \cdot \int_0^1 [B(r)]^2 dr,$

where $B(r)$ *represents* ***a standard Brownian motion*** (see Definition 2.9, "Dickey–Fuller Test" section).

This lemma is immediately used to prove Theorem 5.1, which can be stated as follows:

Theorem 5.1 *The least squares estimator of* β_2 *in* (5.1) *converges in distribution to the non-degenerate distribution:*

$$\frac{\sigma_u\sigma_v \int_0^1 B_u(r)B_v(r)\,dr - \sigma_u\sigma_v \int_0^1 B_u(r)\,dr \int_0^1 B_v(r)\,dr}{\sigma_v^2 \int_0^1 B_v(r)\,dr - \sigma_v^2\left[\int_0^1 B_v(r)\,dr\right]^2} = \sigma_u\sigma_v^{-1}\xi. \tag{5.2}$$

Demonstration

The least squares estimator for β_2—which you can easily check—is given by

$$\hat{\beta} = \frac{\sum\limits_{t=1}^{T} x_t y_t - T\,\overline{x}\overline{y}}{\sum\limits_{t=1}^{T} x_t^2 - T\overline{x}^2}, \tag{5.3}$$

where $\overline{x}$ and $\overline{y}$ represent the empirical means of the series (x_t) and (y_t). respectively. (See "The Ordinary Least Squares (OLS)" section.)

Dividing (5.3), top and bottom, by T^2, gives

$$\begin{aligned}\hat{\beta} &= \left[T^{-2}\sum_{t=1}^{T} x_t y_t - T^{-1}\,\overline{x}\overline{y}\right]\left[T^{-2}\sum_{t=1}^{T} x_t^2 - T^{-1}\overline{x}^2\right]^{-1} \\ &= \left[T^{-2}\sum_{t=1}^{T} x_t y_t - (T^{-1/2}\,\overline{x})(T^{-1/2}\overline{y})\right]\left[T^{-2}\sum_{t=1}^{T} x_t^2 - T^{-1}\overline{x}^2\right]^{-1}.\end{aligned} \tag{5.4}$$

Given Lemma 5.1-(i), we obtain

$$\begin{cases} T^{-1/2}\overline{y} \xrightarrow{D} \sigma_u \cdot \int_0^1 B_u(r)\, dr \\ T^{-1/2}\overline{x} \xrightarrow{D} \sigma_v \cdot \int_0^1 B_v(r)\, dr. \end{cases} \tag{5.5}$$

Furthermore, Lemma 5.1-(ii) leads to

$$T^{-2}\sum_{t=1}^{T} x_t^2 \xrightarrow{D} \sigma_u^2 \cdot \int_0^1 [B(r)]^2 dr. \tag{5.6}$$

In light of this convergence, we have

$$T^{-2}\sum_{t=1}^{T} y_t x_t \xrightarrow{D} \sigma_u \sigma_v \int_0^1 B_u(r) B_v(r)\, dr. \tag{5.7}$$

Transferring (5.5)–(5.7) in (5.4), we obtain

$$\hat{\beta} \xrightarrow{D} \frac{\sigma_u \sigma_v \int_0^1 B_u(r) B_v(r)\, dr - \sigma_u \sigma_v \int_0^1 B_u(r)\, dr \int_0^1 B_v(r)\, dr}{\sigma_v^2 \int_0^1 B_v(r)\, dr - \sigma_v^2 \left[\int_0^1 B_v(r)\, dr\right]^2}.$$

This theorem shows that $\hat{\beta}$ does not converge in probability to zero, but that on the contrary, this estimator asymptotically admits a non-degenerate distribution. It follows that the regression in question is spurious.

Similarly, it could be shown that the test statistic $T^{-1/2}\ t$, used to assess the nullity of $\hat{\beta}$, also has a non-degenerate asymptotic distribution. It follows that when $T \to +\infty$, the probability of this statistic being non-zero tends towards unity. This result also demonstrates the fallacy of regression (5.1).

What this theoretical example shows is that the non-stationarity of the variables under consideration induces a relationship, the significance of which is only apparent.

How Can Spurious Regressions be Remedied?

The question now is how to avoid such "*spurious*" regressions. One answer lies in the introduction of time lags for both the dependent variable and the explanatory variable.

Applying this methodological rule to the model (5.1) would lead, for example, to the estimation by ordinary least squares of the dynamic relationship

$$y_t = m + \alpha y_{t-1} + \beta x_t + \delta x_{t-1} + \varepsilon_t. \tag{5.8}$$

Looking at this last regression, it is clear that there are values for the estimated coefficients—in particular, $\alpha = 1$ and $\beta = \delta = 0$—such that the residuals $\hat{\varepsilon}_t$ are stationary, i.e. integrated of order zero, $I(0)$. $I(0)$.

We can then show (i) that the estimators of the parameters of (5.8) obtained by OLS are convergent; (ii) that the coefficients $\hat{\beta}_T$ and $\hat{\delta}_T$ converge towards a normal distribution, at a rate equal to $T^{1/2}$; (iii) that the t-tests of the null hypotheses $\hat{\beta} = 0$ and $\hat{\delta} = 0$ are asymptotically distributed according to a normal distribution $N(0, 1)$.

An inferential difficulty remains, however, insofar as testing the joint hypothesis $\hat{\beta} = \hat{\delta} = 0$, results in an F statistic that follows a non-standard distribution.

Notwithstanding this reservation, the fact remains that the procedure for boosting the estimated relationship resolves most of the problems raised by the existence of spurious relationships and should therefore be particularly recommended.

A second possible approach is based on a theorem by Granger, which essentially states that, under certain conditions, there can be a linear combination of unit root processes, i.e. all $I(1)$,, which is itself stationary.

This type of combination is called a **cointegration relationship** and is interpreted as a long-term equilibrium relationship between the variables $I(1)$ concerned.

Such an approach is quite fruitful, but it needs to be generalised in several directions, in particular by considering a set of k non-stationary variables.

5.2 The Concept of Cointegration

Chronologically, the first method for estimating a cointegrating relationship was proposed by Engle and Granger (1987) (See also Granger, 1983, 1991).

It does, however, have a few shortcomings, as we shall see. Nevertheless, it is still worth presenting it, without going into too much technical details.

The Engle–Granger Approach

Consider the variables x_t and y_t, both non-stationary and integrated of order one, i.e. $I(1)$. In general, the linear combination

$$z_t = x_t - \alpha y_t$$

will also be $I(1)$. In some cases, however, this combination may be stationary, i.e.

$$z_t \sim I(0).$$

By generalising this possibility to a m-vector $\boldsymbol{x}_t$, we get the definition:

Definition 5.1 If all the components of the vector $\boldsymbol{x}_t$ are $I(1)$ and there is a m-vector $\boldsymbol{\alpha}' \neq \boldsymbol{0}$ such that $\boldsymbol{\alpha}'\boldsymbol{x}_t \sim I(\boldsymbol{0})$, then these components are said to be ***cointegrated*** and the row vector $\boldsymbol{\alpha}'$ is called the ***cointegrating vector.***

Engle and Granger (1987) then suggested a two-stage procedure for testing cointegration between the components of the vector $\boldsymbol{x}_t = (x_{1t}, \ldots, x_{mt})$:

(i) first, the linear regression of x_{1t} on $x_{2t}, \ldots, x_{mt}$, is estimated by least squares, which provides the residuals of this regression;
(ii) in the second step, an Augmented Dickey–Fuller (ADF) test is applied to these residuals, assuming that there is no trend and no constant. The null hypothesis is therefore the existence of a unit root in the residuals.

Consequently, if we reject the null hypothesis of a unit root in these residuals, this rejection will be interpreted as evidence in favour of the existence of a cointegrating relationship between the variables under consideration. Of course, the other tests detailed in "Other Tests of Stationarity" section can be used in the same spirit.

However, it should not be forgotten that under the null hypothesis that the x_{it} components are integrated of order one, the asymptotic distribution of the ADF test based on the residuals will differ from that obtained when a unit root is present in a time series. It follows that the critical values associated with each of these two cases will differ.

It will suffice here to point out that the critical values originally obtained by simulation at Engle–Granger have been the subject of several improvements, in particular with MacKinnon (1991) who uses the response surface technique.

However, such tests using estimation residuals have several shortcomings. Firstly, at finite distance, the results will be strongly determined by the choice of dependent variable in the first stage of the Engle–Granger (EG) procedure. Another shortcoming is that the method does not allow the estimation and testing of several cointegrating relationships.

Lastly, the associated tests generally have very low power, which means that the probability of second-grade error, i.e. wrongly accepting the null hypothesis, is increased.

In an attempt to overcome these limitations, other procedures have been implemented. They are also based on least squares estimation of the cointegration relationship and are called dynamic or fully modified least squares.

Alternative Methods

In principle, **the Dynamic Ordinary Least Squares** (DOLS) procedure, developed by Stock and Watson (1993), involves estimating the following dynamic equation:

$$y_t = m + \beta x_t + \sum_{j=-p}^{p} \delta_j \, \Delta x_{t-j} + u_t. \tag{5.9}$$

In practice, the regressors in this equation are, on the one hand, x_t and, on the other hand, $\Delta x_{t+p}, \ldots, \Delta x_{t-p}$. The DOLS estimator of β is precisely the least squares estimator of β in (5.9).

If y_t and x_t are cointegrated, the DOLS estimator is efficient in large samples. Furthermore, it can also be shown that the t-statistic calculated using the DOLS estimator of β and the corresponding standard errors, whether or not they are heteroskedastic and autocorrelated (HAC), converges asymptotically to a standard normal distribution.

Another procedure is associated with the names Phillips and Hansen (1990) and Phillips (1995). It is called **Fully Modified Ordinary Least Squares** (FM-OLS).

We begin by considering the linear regression model

$$y_t = m + \boldsymbol{\beta}'_K \boldsymbol{x}_t + u_t, \quad t = 1, 2, \ldots, T, \tag{5.10}$$

where the components of the K-vector $\boldsymbol{x}_t$ are all integrated of order one, so $I(1)$, but not cointegrated. Since this is the case, $\boldsymbol{x}_t$ expressed in first differences leads to the stationary vector equation

$$\Delta \boldsymbol{x}_t = \boldsymbol{\gamma} + \boldsymbol{v}_t, \quad t = 2, 3, \ldots, T, \tag{5.11}$$

where $\boldsymbol{\gamma}$ is a K-vector of drift parameters and $\boldsymbol{v}_t$ is a K-vector of variables $I(0)$.

It will also be assumed that the vector $\boldsymbol{\theta}_t = (u_t, \boldsymbol{v}'_t)'$ is strongly stationary with a zero mathematical expectation and a positive definite variance–covariance matrix $\boldsymbol{\Sigma}$.

The FM-OLS estimators are calculated in two main stages. In the first step, y_t is corrected for the long-run correlation between u_t and $\boldsymbol{v}_t$. To do this, we denote $\hat{u}_t, \ t = 1, 2, \ldots, T$, the vector of residuals obtained from the OLS estimation of (5.10).

He then

$$\boldsymbol{\theta}_t = (\hat{u}_t, \hat{\boldsymbol{v}}'_t)', \quad t = 2, 3, \ldots, T, \tag{5.12}$$

where $\hat{\boldsymbol{v}}_t = \Delta \boldsymbol{x}_t - \hat{\boldsymbol{\gamma}}, \ t = 2, 3, \ldots, T$, with $\hat{\boldsymbol{\gamma}} = (1/(T-1)) \sum_{t=2}^{T} \Delta \boldsymbol{x}_t$.

A convergent estimator of the long-term variance of $\boldsymbol{\theta}_t$ leads us to the partitioned matrix

$$\hat{\boldsymbol{\Omega}} = \hat{\boldsymbol{\Sigma}} + \hat{\boldsymbol{\Lambda}} + \hat{\boldsymbol{\Lambda}}' = \begin{pmatrix} \hat{\Omega}_{11} & \hat{\boldsymbol{\Omega}}'_{12} \\ \hat{\boldsymbol{\Omega}}_{21} & \hat{\boldsymbol{\Omega}}_{22} \end{pmatrix}, \tag{5.13}$$

where $\hat{\boldsymbol{\Sigma}}$ is the estimator of the variance–covariance matrix $\boldsymbol{\Sigma}$ and $\hat{\boldsymbol{\Lambda}}$ is the Newey-West HAC estimator (see Sect. 4.7), which corrects the variance–covariance matrix using an estimator that is convergent and robust to heteroskedasticity and autocorrelation of the residuals. Moreover, $\hat{\Omega}_{11}$ is a scalar, $\hat{\boldsymbol{\Omega}}'_{12}$ a K-row vector, $\hat{\boldsymbol{\Omega}}_{21}$ the transpose of the previous one and $\hat{\boldsymbol{\Omega}}_{22}$ a submatrix of dimension $K \times K$.

Before we move on to the second stage, let's start by writing

$$\hat{\boldsymbol{\Delta}} = \hat{\boldsymbol{\Sigma}} + \hat{\boldsymbol{\Lambda}} = \begin{pmatrix} \hat{\Delta}_{11} & \hat{\boldsymbol{\Delta}}'_{12} \\ \hat{\boldsymbol{\Delta}}_{21} & \hat{\boldsymbol{\Delta}}_{22} \end{pmatrix},$$

$$\hat{\boldsymbol{Z}} = \hat{\boldsymbol{\Delta}}_{21} - \hat{\boldsymbol{\Delta}}_{22}\hat{\boldsymbol{\Omega}}_{22}^{-1}\hat{\boldsymbol{\Omega}}_{21},$$

$$\hat{y}_t^* = y_t - \hat{\boldsymbol{\Omega}}'_{12}\hat{\boldsymbol{\Omega}}_{22}^{-1}\hat{\boldsymbol{v}}_t,$$

as well as

$$\boldsymbol{G}_{(K+1)\times K} = \begin{pmatrix} \boldsymbol{0}'_{1\times K} \\ \boldsymbol{I}_{K\times K} \end{pmatrix}.$$

In the second step, the FM-OLS estimator of $\boldsymbol{\beta}$ is then given by

$$\hat{\boldsymbol{\beta}}_{FM} = (\boldsymbol{W}'\boldsymbol{W})^{-1}(\boldsymbol{W}'\hat{\boldsymbol{y}}^* - T\boldsymbol{G}\hat{\boldsymbol{Z}}), \tag{5.14}$$

where $\hat{\boldsymbol{y}}^* = (\hat{y}_1^*, \ldots, \hat{y}_T^*)'$ and $\boldsymbol{W} = \big((1, 1, \ldots, 1)', \boldsymbol{X}\big)$.

A convergent estimator of the variance–covariance matrix of $\hat{\boldsymbol{\beta}}_{FM}$ is given by

$$\widehat{\text{Var}}(\hat{\boldsymbol{\beta}}_{FM}) = \hat{\omega}_{11.2}(\boldsymbol{W}'\boldsymbol{W})^{-1}, \tag{5.15}$$

with $\hat{\omega}_{11.2} = \hat{\Omega}_{11} - \hat{\boldsymbol{\Omega}}'_{12}\hat{\boldsymbol{\Omega}}_{22}^{-1}\hat{\boldsymbol{\Omega}}_{21}$.

The methods we have just described have one thing in common with the Engle–Granger procedure: they are all based on the estimation of the static cointegration relationship alone.

5.3 ARDL Models and Error Correction

We introduced autoregressive distributed lags (ARDL) models in "Estimation of an ARDL Model" section. We will now go one step further and show that these models can be used to associate an error correction mechanism with a cointegrating relationship.

A Simple Example

Consider the ARDL(1,1) model

$$y_t = m + \alpha_1 y_{t-1} + \beta_0 x_t + \beta_1 x_{t-1} + u_t, \quad t = 1, \ldots, T, \quad u_t \sim BB(0, \sigma^2), \quad |\alpha_1| < 1. \tag{5.16}$$

and suppose that x follows the random walk

$$x_t = x_{t-1} + \varepsilon_t \sim I(1).$$

Since the perturbations u_t are $I(0)$, it follows that y_t is a combination of variables $I(1)$ and $I(0)$. Therefore, in general, $y_t \sim I(1)$ is non-stationary, given that linear combinations of variables $I(1)$ are also $I(1)$.

However—as we have seen—this conclusion can be invalidated, and it is therefore possible for such a combination to be stationary.

Consider Eq. (5.16) again and reparametrise it using the first differences as follows:

$$\begin{cases} \Delta y_t = y_t - y_{t-1} \\ \Delta x_t = x_t - x_{t-1}. \end{cases}$$

We can then write that

$$\begin{aligned} \Delta y_t &= m - (1 - \alpha_1) y_{t-1} + \beta_0 (\Delta x_t + x_{t-1}) + \beta_1 x_{t-1} + u_t \\ &= \beta_0 \Delta x_t - (1 - \alpha_1) y_{t-1} + m + (\beta_0 + \beta_1) x_{t-1} + u_t \\ &= \beta_0 \Delta x_t - (1 - \alpha_1) \left[y_{t-1} - \frac{m}{1 - \alpha_1} - \frac{\beta_0 + \beta_1}{1 - \alpha_1} x_{t-1} \right] + u_t. \end{aligned}$$

Next comes

$$\Delta y_t = \beta_0 \Delta x_t - (1 - \alpha_1) \left[y_{t-1} - a - \gamma x_{t-1} \right] + u_t, \tag{5.17}$$

with $a = m/(1 - \alpha_1)$ and $\gamma = (\beta_0 + \beta_1)/(1 - \alpha_1)$.

This last formulation is an example of an ARDL model, which combines both the cointegration relation in square brackets in (5.17) and an error correction mechanism in the context of a link between Δy_t and Δx_t.

More precisely, Eq. (5.17) is said to be an **error correction representation of** the ARDL(1,1) model as expressed in Eq. (5.16). (This is also referred to as the Error Correction Mechanism, ECM).

In this case, the current variation in y is actually the sum of two components:

- the first is proportional to the current variation in x;
- the second is a partial correction which takes account of the extent to which y_{t-1} deviates from its equilibrium value corresponding to x_{t-1}.

We can of course generalise this analysis to the example of a model with more lags, both for the dependent variable and for the explanatory variable(s).

Generalisation

A first generalisation consists of adding lags to (5.16). We then have the ARDL(p, q) model

$$\begin{aligned} y_t = m + \alpha_1 y_{t-1} + \alpha_2 y_{t-2} \cdots + \alpha_p y_{t-p} + \beta_0 x_t \\ + \beta_1 x_{t-1} + \cdots + \beta_q x_{t-q} + u_t, \end{aligned} \tag{5.18}$$

which, using the delay operator, can still be expressed in the alternative form

$$\alpha(L) y_t = m + \beta(L) x_t + u_t. \tag{5.19}$$

Of course, it is possible to add a deterministic trend or binary variables to this model.

Such a model is **stable** if all the roots of the equation of degree p

$$P(z) = \alpha_p z^p + \alpha_{p-1} z^{p-1} + \cdots + \alpha_1 z + 1 = 0$$

are outside the unit circle, in other words, if $|z| > 1$. (The explosive case that occurs when one or more roots of this equation are inside the unit circle, is irrelevant for economic variables).

Based on Granger's representation theorem, which essentially states that

any set of cointegrated variables has a representation
with error correction,

the model (5.18), by analogy with (5.17), becomes

$$\begin{aligned} \Delta y_t = \sum_{i=1}^{p} \alpha_i \Delta y_t + \beta_0 \Delta x_t + \beta_1 \Delta x_{t-1} + \cdots + \beta_q \Delta x_{t-q} \\ - (1 - \alpha_1 - \cdots - \alpha_p)\left[y_{t-1} - a - \gamma x_{t-1}\right] + u_t, \end{aligned} \tag{5.20}$$

with $a = m/(1 - \alpha_1 - \cdots - \alpha_p)$ and $\gamma = (\beta_0 + \beta_1 + \cdots + \beta_q)/(1 - \alpha_1 - \cdots - \alpha_p)$.

A second possible generalisation arises when we introduce K explanatory variables into (5.16) rather than just one. In this way, we obtain a model ARDL($p, k_1, \ldots, k_K$).

The equivalent of (5.18) in level is then

$$y_t = m + \sum_{i=1}^{p} \alpha_i y_{t-i} + \sum_{j=0}^{k_1} \beta_{1j} x_{1,t-j} + \cdots + \sum_{j=0}^{k_K} \beta_{Kj} x_{K,t-j} + u_t \tag{5.21}$$

and under the error correction representation, the equivalent of (5.20) is given by

$$\Delta y_t = \sum_{i=1}^{p} \alpha_i \Delta y_{t-i} + \sum_{j=0}^{k_1} \beta_{1j} \Delta x_{1,t-j} + \cdots + \sum_{j=0}^{k_K} \beta_{Kj} \Delta x_{K,t-j} - (1 - \alpha_1 - \cdots - \alpha_p)\left[y_{t-1} - a - \sum_{r=1}^{K} \gamma_r x_{r,\,t-1} \right] + u_t, \tag{5.22}$$

where $\forall r,\ \ \gamma_r = \beta_r / (1 - \alpha_1 - \cdots - \alpha_p)$, a preserving, of course, the same expression as that associated with (5.20).

Obviously, as we have already pointed out, the stability of these general ARDL models depends entirely on the fact that the associated polynomial equation has all its roots outside the unit circle.

The question now is how to estimate such models and reveal the associated cointegration relationship.

Estimation and Testing of ARDL Models

Basically, there are two main approaches to estimating ARDL models: one involves checking that all the variables under consideration are integrated of the same order (in practice of order one); the other does not prejudge the stationary or non-stationary nature of these variables and is linked to the work of Pesaran, Shin and Smith (2001).

The first method therefore begins by checking that the dependent variable and the K explanatory variables are all $I(1)$. The parameters of (5.22) are then estimated using ordinary least squares. The long-run coefficients of the cointegration relationship are then obtained simply by running the regression of Δy_t on the different variables in (5.22).

Alternatively, it is also possible to estimate Eq. (5.21) by least squares; the coefficients γ_r of the cointegration relationship are then derived from the estimators of the regression (5.13) according to the formula:

$$\forall r = 1, \ldots, K, \qquad \gamma_r^* = \frac{\hat{\beta}_r}{1 - \sum_{i=1}^{p} \hat{\alpha}_i}.$$

The two procedures must give identical results.

The second method—abbreviated as PESS, for PEsaran, Shin and Smith—does not, as we have said, require us to check that the variables in the ARDL model are all $I(1)$.. This is its comparative advantage, based on the fact that unit root tests sometimes have fairly low power and can therefore cast doubt on the character $I(1)$ or $I(0)$ of the different series used.

To see what the PESS method consists of, we will start with a simple ARDL model, in which only three variables are taken into account, namely: y_t and $x_{2t},\ x_{3t}$. For the rest, we will consider that the dependent variable is lagged by p periods, while the two explanatory variables will incorporate a specific number of lags.

The model will therefore be written as

$$\Delta y_t = m + \sum_{i=1}^{p} \alpha_i \Delta y_{t-i} + \sum_{j=0}^{k_2} \beta_{2j} \Delta x_{2,t-j} + \sum_{j=0}^{k_3} \beta_{3j} \Delta x_{3,t-j} + \gamma_1 y_{t-1} + \gamma_2 x_{2,t-1} + \gamma_3 x_{3,t-1} + v_t. \tag{5.23}$$

The PESS procedure has three steps.

Step 1: Estimate the model (5.23) by least squares.

Step 2: Calculate the Wald statistic or the F statistic to test the null hypothesis

$$H_0 : \gamma_1 = \gamma_2 = \gamma_3 = 0$$

in (5.23) estimated.

The distribution of each of these statistics is non-standard. However, they have been tabulated by PESS. Here is a table showing the critical values of the two distributions at 95% and 99%, for the case where there is a constant in the regression. Note also that k denotes the number of variables taken into account in the cointegration relationship (Table 5.1).

On reading this table, it is clear that we are given two limits for the critical values of the two distributions: an upper limit $I(1)$, noted W_U or F_U depending on the statistic used; a lower limit $I(0)$, noted W_L or F_L.

Step 3: Compare the W and F statistics calculated in the previous step with the upper and lower bounds given in the table, for a given level of significance. There are then three possibilities, limited here to the Wald statistic:

- If $W < W_L$, we accept the null hypothesis $\gamma_1 = \gamma_2 = \gamma_3 = 0$ and conclude that, potentially, there is no cointegrating relationship between the three variables;

Table 5.1 Critical value limits

k	Stat. wald (W)				Stat. F			
	95%		99%		95%		99%	
	$I(0)$	$I(1)$	$I(0)$	$I(1)$	$I(0)$	$I(1)$	$I(0)$	$I(1)$
2	9.687	11.528	14.114	15.630	4.934	5.764	7.057	7.815
3	11.380	14.506	15.864	18.926	3.793	4.855	5.288	6.309
4	12.875	17.512	17.540	22.460	3.219	4.378	4.385	5.615
5	14.252	20.247	19.085	25.612	2.850	4.049	3.817	5.122

Source Pesaran, Shin and Smith (2001)

- If $W > W_U$, we reject the null hypothesis and deduce that there is a cointegrating relationship between y_t, x_{2t} and x_{3t}.
- Finally, if $W_L < W < W_U$, there is indecision as to whether to reject or accept the null hypothesis.

In the latter case, the order of integration of the different variables needs to be checked using the usual unit root tests.

5.4 Cointegration and Systems of Equations

It goes without saying that if a cointegrating relationship comprises more than two variables, say $n > 2$,, then there can be $(n - 1)$ cointegrating relationships between these variables. This leads us to study true systems of dynamic equations.

Vector Autoregressions

Let's consider a simple example of a **Vector Autoregressive** model (VAR), that of a model with two variables and a single lag:

$$\begin{pmatrix} y_{1t} \\ y_{2t} \end{pmatrix} = \begin{pmatrix} m_1 \\ m_2 \end{pmatrix} + \begin{pmatrix} a_{11} & a_{12} \\ a_{21} & a_{22} \end{pmatrix} \begin{pmatrix} y_{1,t-1} \\ y_{2,t-1} \end{pmatrix} + \begin{pmatrix} u_1 \\ u_2 \end{pmatrix},$$

or in matrix form,

$$\boldsymbol{y}_t = \boldsymbol{m} + \boldsymbol{A}\boldsymbol{y}_{t-1} + \boldsymbol{u}_t. \tag{5.24}$$

The time trajectory of y_t will depend entirely on the properties of the matrix **A**, in particular its eigenvalues and eigenvectors. Remember that an **eigenvalue** of this matrix is a scalar $\lambda \in \mathbb{R}$, associated with $\boldsymbol{x} \in \mathbb{R}^2$, such that

$$\boldsymbol{A}\boldsymbol{x} = \lambda \boldsymbol{x},$$

where **x** is an **eigenvector** of **A**. (See Appendix B, Eq. 52.)

Now suppose that the eigenvalues and eigenvectors of $\boldsymbol{A}_{2\times 2}$ are given by

$$\boldsymbol{\Lambda} = \begin{pmatrix} \lambda_1 & 0 \\ 0 & \lambda_2 \end{pmatrix}, \qquad \boldsymbol{B}_{2\times 2} = \begin{pmatrix} \boldsymbol{b}_1 & \boldsymbol{b}_2 \end{pmatrix}.$$

To progress with the analysis of this VAR(1), we will need several mathematical results, which we will present in the general form of a reminder.

Reminder

Let **A** be a ***symmetric*** matrix of dimension n-squared

Proposition P1. The eigenvalues of **A** are all ***real***

Proposition P2. If all the eigenvalues of **A** are ***distinct***, then there exists a ***non-singular*** matrix **Q** composed of n ***linearly independent*** eigenvectors such that

$Q^{-1}AQ = \Lambda = \text{diag}(\lambda_1, \ldots, \lambda_n)$,

where diag(·) denotes the ***diagonal matrix***, the elements of which are the n eigenvalues of **A**

Applying this last proposition to our matrix $\boldsymbol{A}_{2\times 2}$, it follows that if the eigenvalues of **A** are all distinct, there exists an invertible matrix $\boldsymbol{B}$ such that

$$\boldsymbol{B}^{-1}\boldsymbol{A}\boldsymbol{B} = \Lambda = \text{diag}(\lambda_1, \lambda_2). \tag{5.25}$$

Next, let's define the vector z_t by

$$z_t = \boldsymbol{B}^{-1}\boldsymbol{y}_t. \tag{5.26}$$

Given (5.25), pre-multiplying (5.24) by $\boldsymbol{B}^{-1}$ gives

$$\begin{aligned} \boldsymbol{B}^{-1}\boldsymbol{y}_t &= \boldsymbol{B}^{-1}\boldsymbol{m} + \boldsymbol{B}^{-1}\boldsymbol{A}\boldsymbol{y}_{t-1} + \boldsymbol{B}^{-1}\boldsymbol{u}_t \\ &= \boldsymbol{B}^{-1}\boldsymbol{m} + \boldsymbol{B}^{-1}\boldsymbol{B}\boldsymbol{\Lambda}\boldsymbol{B}^{-1}\boldsymbol{y}_{t-1} + \boldsymbol{B}^{-1}\boldsymbol{u}_t \\ &= \boldsymbol{m}^* + \boldsymbol{\Lambda}\boldsymbol{B}^{-1}\boldsymbol{y}_{t-1} + \boldsymbol{v}_t. \end{aligned} \tag{5.27}$$

Using (5.26), we can write (5.27) as

$$z_t = \boldsymbol{m}^* + \boldsymbol{\Lambda} z_{t-1} + \boldsymbol{v}_t. \tag{5.28}$$

From the above equation, we deduce that

$$\begin{cases} z_{1t} = m_1^* + \lambda_1 z_{1,t-1} + v_{1t} \\ z_{2t} = m_2^* + \lambda_2 z_{2,t-1} + v_{2t}. \end{cases}$$

In other words, each variable z follows an autonomous process which is totally dependent on the values of λ_1 or λ_2. There are four main possibilities:

(i) the eigenvalues λ_1 and λ_2 are both less than unity in modulus, in which case z_{1t} and z_{2t} follows a specific AR(1) process and are stationary.

(ii) If one of the two eigenvalues is equal to unity and the other is less than unity, then the corresponding equation in (5.28) represents a non-stationary random walk with drift, i.e. $I(1)$.

(iii) The two eigenvalues of **A** are equal to unity, so that z_{1t} and z_{2t} follow a random walk with drift, which is then non-stationary and therefore $I(1)$.

(iv) If λ_1 and λ_2 are greater than unity in modulus, then (5.28) follows an explosive trajectory, which obviously does not fit with our economic reality. Consequently, we can dismiss this possibility as highly unlikely.

In view of what we are interested in this chapter, i.e. the modelling of non-stationary processes, in the weak sense, i.e. integrated of order one, it is obvious that we will only consider cases (ii) and (iii) which have just been described.

Non-stationary VAR Models

To analyse the two cases in more detail, we need a number of results from matrix algebra. First, here is an example of a particular, quasi-diagonal matrix:

$$\boldsymbol{J} = \begin{pmatrix} \lambda_1 & 1 & 0 \\ 0 & \lambda_2 & 0 \\ 0 & 0 & \lambda_3 \end{pmatrix}.$$

Such a matrix is diagonal in blocks: in this case, the first block consists of the submatrix $\boldsymbol{J}^{(1)}_{2\times 2}$ and the second of $\boldsymbol{J}^{(2)} = (\lambda_3)$. It is called a **Jordan matrix.**

We will then also use the following three proposals, again presented as a reminder.

Reminder

Proposition P3. The ***rank*** of **A**, rk(A),, is equal to the ***number of non-zero*** eigenvalues of **A**

Proposition P4. Let **A** be a symmetric n-squared matrix. Then the ***eigenvectors*** of **A** and $\boldsymbol{\Pi} = \boldsymbol{I}_n - \boldsymbol{A}$ are ***identical***

Proposition P5. If **J** is a ***Jordan matrix***, then there exists a ***non-singular*** matrix **P** such that

$P^{-1}AP = J$.

Case (ii): $\lambda_1 = 1,\ \lambda_2 < 1$.

Let's write the model (5.24) in first differences, which gives

$$\Delta \boldsymbol{y}_t = \boldsymbol{y}_t - \boldsymbol{y}_{t-1} = \boldsymbol{m} + \boldsymbol{A}\boldsymbol{y}_{t-1} - \boldsymbol{y}_{t-1} + \boldsymbol{u}_t$$

$$= \boldsymbol{m} - (\boldsymbol{I}_2 - \boldsymbol{A})\boldsymbol{y}_{t-1} + \boldsymbol{u}_t$$
$$= \boldsymbol{m} - \boldsymbol{\Pi}\, \boldsymbol{y}_{t-1} + \boldsymbol{u}_t, \tag{5.29}$$

where $\boldsymbol{\Pi} = (\boldsymbol{I}_2 - \boldsymbol{A})$.

The eigenvalues of $\boldsymbol{\Pi}$ are equal to $1 - \lambda_1 = 0$ and $(1 - \lambda_2)$. Furthermore, by virtue of Proposition P4, the eigenvectors of **A** and $\boldsymbol{\Pi}$ are identical. Therefore, in view of Proposition P3, the rank of this matrix $\boldsymbol{\Pi}$ is equal to one, which implies that it is singular. Furthermore, by proposition P2, since the eigenvalues are distinct, we have

$$\boldsymbol{\Pi} = \boldsymbol{B}\begin{pmatrix} 0 & 0 \\ 0 & 1 - \lambda_2 \end{pmatrix}\boldsymbol{B}^{-1}$$
$$= \begin{pmatrix} b_{11} & b_{12} \\ b_{21} & b_{22} \end{pmatrix}\begin{pmatrix} 0 & 0 \\ 0 & 1 - \lambda_2 \end{pmatrix}\begin{pmatrix} b_{11}^* & b_{12}^* \\ b_{21}^* & b_{22}^* \end{pmatrix}$$
$$= \begin{pmatrix} b_{12}(1 - \lambda_2)b_{21}^* & b_{12}(1 - \lambda_2)b_{22}^* \\ b_{22}(1 - \lambda_2)b_{21}^* & b_{22}(1 - \lambda_2)b_{22}^* \end{pmatrix}. \tag{5.30}$$

Given (5.30), the matrix $\boldsymbol{\Pi}$ can be factored as a product of the column vector $\boldsymbol{\alpha} = \big(b_{11}(1 - \lambda_2)\; b_{21}(1 - \lambda_2)\big)'$ and the row vector $\boldsymbol{\beta}' = \big(b_{21}^*\; b_{22}^*\big)$, so that

$$\boldsymbol{\Pi} = \boldsymbol{\alpha}\boldsymbol{\beta}'. \tag{5.31}$$

The vector $\boldsymbol{\beta}'$ is precisely the **cointegrating vector**, while $\boldsymbol{\alpha}$ provides the weights with which the cointegrating relationship intervenes in each of the two VAR(1) equations.

Case (iii): $\lambda_1 = \lambda_2 = 1$.

As the eigenvalues are equal to unity, there are no two linearly independent eigenvectors associated with this eigenvalue. Consequently, there is no matrix **B** that allows the diagonalisation of **A**.

However, using proposition P5, it is possible to find a regular matrix **P** such that

$$\boldsymbol{P}^{-1}\boldsymbol{A}\boldsymbol{P} = \boldsymbol{J}, \tag{5.32}$$

where **J** is the Jordan matrix given by

$$\boldsymbol{J} = \begin{pmatrix} \lambda & 1 \\ 0 & \lambda \end{pmatrix}.$$

Now consider the vector **z** defined by

$$z_t = \boldsymbol{P}^{-1}\boldsymbol{y}_t. \tag{5.33}$$

Pre-multiplying (5.24) by $\boldsymbol{P}^{-1}$ renders

$$\boldsymbol{P}^{-1}\boldsymbol{y}_t = \boldsymbol{P}^{-1}\boldsymbol{m} + \boldsymbol{P}^{-1}\boldsymbol{P}\boldsymbol{J}\boldsymbol{P}^{-1}\boldsymbol{y}_{t-1} + \boldsymbol{P}^{-1}\boldsymbol{u}_t,$$

which leads, taking into account (5.33), to the system of two equations

$$\begin{cases} z_{1t} = m_1^* + \lambda z_{1,t-1} + z_{2,t-1} + w_{1t} \\ z_{2t} = m_2^* + \lambda z_{2,t-1} + w_{2t}. \end{cases}$$

Since the eigenvalue λ is equal to unity, this system is written:

$$\begin{cases} (1-L)z_{1t} = m_1^* + z_{2,t-1} + w_{1t} \\ (1-L)z_{2t} = m_2^* + w_{2t}. \end{cases} \tag{5.34}$$

Let's multiply the first equation of (5.34) by $(1-L)$, in order to eliminate $z_{2,t-1}$:

$$\begin{aligned}(1-L)^2 z_{1t} &= (1-L)z_{2,t-1} + (1-L)w_{1t} \\ &= m_2^* + w_{2,t-1} + (w_{1t} - w_{1,t-1}).\end{aligned}$$

The equation transformed in this way tells us that z_{1t} is integrated of order two, given that it has to be differentiated twice in order to be $I(0)$. On the other hand, the second equation in (5.34) is integrated of order one. It follows that the y's are $I(2)$.

It would therefore appear that there is no cointegrating relationship. Such a statement is inaccurate, simply because it ignores the general definition of cointegration, which is as follows:

Definition 5.2 A vector $\mathbf{x}$ composed of variables all $I(d)$ is said to be ***cointegrated*** of order (d, b), if there exists a linear combination of the components of $\mathbf{x}$ which is $I(d-b)$, with $b \in \mathbb{N}^*$.

We then note

$$\boldsymbol{x} \sim CI(d, b).$$

In our case, we have

$$\boldsymbol{y} \sim CI(2, 1).$$

The cointegrating relationship is easily isolated from (5.32) and the matrix $-\boldsymbol{\Pi} = \boldsymbol{A} - \boldsymbol{I}_2$:

$$\boldsymbol{\Pi} = \boldsymbol{P}\begin{pmatrix} 0 & 1 \\ 0 & 0 \end{pmatrix}\boldsymbol{P}^{-1} = \begin{pmatrix} p_{11} & p_{12} \\ p_{21} & p_{22} \end{pmatrix}\begin{pmatrix} 0 & 1 \\ 0 & 0 \end{pmatrix}\begin{pmatrix} p_{11}^* & p_{12}^* \\ p_{21}^* & p_{22}^* \end{pmatrix}$$

$$= \begin{pmatrix} p_{11}p_{21}^* & p_{11}p_{22}^* \\ p_{21}p_{21}^* & p_{21}p_{22}^* \end{pmatrix}.$$

The cointegration relationship we are looking for is given by $\boldsymbol{\beta}' = \begin{pmatrix} p_{21}^* & p_{22}^* \end{pmatrix}$. It therefore allows us to linearly combine variables y which are, individually, $I(2)$, the combination itself being $I(1)$.

The VAR(p) *Model*

We begin by considering k variables and two time lags. Under these conditions, the VAR(2) model, in which we will not include any constants, is as follows:

$$\boldsymbol{y}_t = \boldsymbol{A}_1 \boldsymbol{y}_{t-1} + \boldsymbol{A}_2 \boldsymbol{y}_{t-2} + \boldsymbol{u}_t, \quad t = 1, \ldots, T, \tag{5.35}$$

where $\boldsymbol{u}_t$ is the column vector with k components satisfying the hypotheses

H1: $E(\boldsymbol{u}_t) = \boldsymbol{0}, \quad \forall t$.
H2: $E(\boldsymbol{u}_t \boldsymbol{u}'_t) = \boldsymbol{\Omega}$, with $\boldsymbol{\Omega}$ a positive definite matrix of dimension $k \times k$.
H3: $E(\boldsymbol{u}_s \boldsymbol{u}'_t) = \boldsymbol{0}, \quad \forall s \neq t$.

By subtracting $\boldsymbol{y}_{t-1}$ from the two members of (5.35), the VAR is written in first differences

$$\Delta \boldsymbol{y}_t = (\boldsymbol{A}_1 - \boldsymbol{I}_k)\boldsymbol{y}_{t-1} + \boldsymbol{A}_2 \boldsymbol{y}_{t-2} + \boldsymbol{u}_t. \tag{5.36}$$

If we add and then subtract $\boldsymbol{A}_2 \boldsymbol{y}_{t-1}$ in the right-hand side of (5.36), we get

$$\Delta \boldsymbol{y}_t = -\boldsymbol{A}_2 \Delta \boldsymbol{y}_{t-1} - (\boldsymbol{I}_k - \boldsymbol{A}_1 - \boldsymbol{A}_2)\boldsymbol{y}_{t-1} + \boldsymbol{u}_t, \tag{5.37}$$

which leads to

$$\Delta \boldsymbol{y}_t = -\boldsymbol{A}_2 \Delta \boldsymbol{y}_{t-1} - \boldsymbol{\Pi} \boldsymbol{y}_{t-1} - \boldsymbol{u}_t, \tag{5.38}$$

with $\boldsymbol{\Pi} = (\boldsymbol{I}_k - \boldsymbol{A}_1 - \boldsymbol{A}_2)$.

We can go one step further by formulating a VAR(p) as follows:

$$\boldsymbol{y}_t = \boldsymbol{A}_1 \boldsymbol{y}_{t-1} + \boldsymbol{A}_2 \boldsymbol{y}_{t-2} + \cdots + \boldsymbol{A}_p \boldsymbol{y}_{t-p} + \boldsymbol{u}_t, \quad \boldsymbol{u}_t \sim IID(\boldsymbol{0}, \boldsymbol{\Omega}). \tag{5.39}$$

This model, inspired by (5.38), can also be written as

$$\Delta \boldsymbol{y}_t = \sum_{j=1}^{p-1} \boldsymbol{\Phi}_j \Delta \boldsymbol{y}_{t-j} - \boldsymbol{\Pi} \boldsymbol{y}_{t-1} + \boldsymbol{u}_t, \tag{5.40}$$

where $\boldsymbol{\Pi} = (\boldsymbol{I}_k - \boldsymbol{A}_1 - \boldsymbol{A}_2 - \cdots - \boldsymbol{A}_p)$ and $\boldsymbol{\Phi}_j = -\sum_{i=j+1}^{p} \boldsymbol{A}_i,\ \forall j = 1, 2, \ldots, p-1$.

If $\boldsymbol{y}_t \sim I(1)$, the left-hand side of (5.40) is $I(0)$, while in the right-hand side, $\Delta \boldsymbol{y}_{t-j}$ and $\boldsymbol{u}_t$ are also stationary. It follows that (5.40) can only hold if $-\boldsymbol{\Pi}\, \boldsymbol{y}_{t-1} \sim I(0)$, in other words, if the linear combinations of the k variables are stationary.

The trajectory of the k components of the vector $\mathbf{y}$ will depend crucially on the values of λ which are solutions of

$$\det(\lambda^p \boldsymbol{I}_k - \lambda^{p-1}\boldsymbol{A}_1 - \quad \cdots \quad - \lambda \boldsymbol{A}_{p-1} - \boldsymbol{A}_p) = 0. \tag{5.41}$$

Immediately ruling out the case of explosive roots—in this case, greater than unity—there are still two main possibilities for the stationarity of (5.40) to be ensured:

(i) $\mathrm{rk}(\boldsymbol{\Pi}) = k$. Each root of (5.41) is less than 1, so the matrix $\boldsymbol{\Pi}$ has full rank. All the variables y in the VAR are then $I(0)$.
(ii) $\mathrm{rk}(\boldsymbol{\Pi}) = r < k$. This case arises when there is a unit root of multiplicity $(k - r)$ and the r other roots are all less than unity, so $I(0)$.

In the latter case, the matrix $\boldsymbol{\Pi}$ from (4.40) can be factored, so that we obtain

$$\boldsymbol{\Pi} = \boldsymbol{\alpha}\boldsymbol{\beta}' \sim I(0),$$

where $\boldsymbol{\alpha}$ and $\boldsymbol{\beta}$ are matrices of dimension $(k \times r)$. (A clarification is necessary here. Throughout this book, matrices are represented by upper case letters and vectors by lower case letters, both in bold. It follows that, according to this logic, $\boldsymbol{\alpha}$ and $\boldsymbol{\beta}$,, which are matrices, should have been capitalised. But it is now well-established practice to use lower case. We'll go along with it).

At this point, considering a VAR(p), we can then state a more general definition of cointegration.

Definition 4.3 If $\boldsymbol{y}_{t-1} \sim I(1)$ and if the linear combinations $\boldsymbol{\Pi}\, \boldsymbol{y}_{t-1}$ are $I(0)$, then the VAR(p) model is said to be ***cointegrated***. If $\mathrm{rk}(\boldsymbol{\Pi}) = r < k$, r is the ***cointegrating rank*** and represents the ***dimension of*** the cointegrating subspace.

5.5 Johansen's Approach

Johansen's (1988; 1995) methodology is based on the representation of a VAR model in error correction form. It also makes extensive use of maximum likelihood estimation. It is closely related to canonical correlations.

On Canonical Correlations

The correlation coefficient between two random variables is well known and is equal to the covariance between these variables divided by the product of their standard deviations.

The canonical correlation coefficient generalises the concept to the correlation between two random vectors. From this point of view, let us take the vector **x** of dimension $(p \times 1)$, with $E(\boldsymbol{x}) = \mathbf{0}$ and $\text{Var}(\boldsymbol{x}) = \boldsymbol{\Sigma}$, the latter being positive. Let's divide **x** into two subsets of variables, so that

$$\boldsymbol{x} = \begin{pmatrix} y \\ z \end{pmatrix} \quad \text{and} \quad \begin{pmatrix} \boldsymbol{\Sigma}_{11} & \boldsymbol{\Sigma}_{12} \\ \boldsymbol{\Sigma}_{21} & \boldsymbol{\Sigma}_{22} \end{pmatrix},$$

where **y** is a p_1-column vector and **z** is a p_2-vector, with $p_1 + p_2 = p$.

Let $\boldsymbol{\delta}_{p_1 \times 1}$ and $\boldsymbol{\gamma}_{p_2 \times 1}$ be two unknown vectors to be determined under the condition that the correlation between $\boldsymbol{\delta}'\boldsymbol{y}$ and $\boldsymbol{\gamma}'\boldsymbol{z}$ is as high as possible. Let $U_1 = \boldsymbol{\delta}'\boldsymbol{y}$ and $V_1 = \boldsymbol{\gamma}'\boldsymbol{z}$, and note $\rho(U_1, V_1)$ the correlation coefficient between U_1 and V_1. Furthermore, given that the correlations are invariant up to a linear application, we can obtain a unique solution by imposing the condition that $\text{Var}(U_1) = \text{Var}(V_1) = 1$:

$$\begin{aligned} \text{Var}(U_1) &= \boldsymbol{\delta}'\left[\text{Var}(\boldsymbol{y})\right]\boldsymbol{\delta} = \boldsymbol{\delta}'\boldsymbol{\Sigma}_{11}\boldsymbol{\delta} = 1 \\ \text{Var}(V_1) &= \boldsymbol{\gamma}'[\text{Var}(\boldsymbol{z})]\boldsymbol{\gamma} = \boldsymbol{\gamma}'\boldsymbol{\Sigma}_{22}\boldsymbol{\gamma} = 1 \end{aligned} \tag{5.42}$$

The problem is now to maximise, under the constraints (5.42),

$$\rho(U_1, V_1) = \frac{E(U_1 V_1)}{\sqrt{\text{Var}(U_1)\text{Var}(V_1)}} = \frac{\boldsymbol{\delta}'\boldsymbol{\Sigma}_{12}\boldsymbol{\gamma}}{\left[(\boldsymbol{\delta}'\boldsymbol{\Sigma}_{11}\boldsymbol{\delta})(\boldsymbol{\gamma}'\boldsymbol{\Sigma}_{22}\boldsymbol{\gamma})\right]^{1/2}}.$$

As a result, we form the Lagrangian

$$L(\boldsymbol{\delta}, \boldsymbol{\gamma}) = \boldsymbol{\delta}'\boldsymbol{\Sigma}_{12}\boldsymbol{\gamma} - \mu_1\left(\boldsymbol{\delta}'\boldsymbol{\Sigma}_{11}\boldsymbol{\delta}' - 1\right) - \mu_2\left(\boldsymbol{\gamma}'\boldsymbol{\Sigma}_{22}\boldsymbol{\gamma} - 1\right),$$

where μ_1 and μ_2 are Lagrange multipliers.

The first-order conditions then give

$$\begin{cases} \partial L/\partial \boldsymbol{\delta} = \boldsymbol{\Sigma}_{12}\boldsymbol{\gamma} - 2\mu_1 \boldsymbol{\Sigma}_{11}\boldsymbol{\delta} = 0 \\ \partial L/\partial \boldsymbol{\gamma} = \boldsymbol{\Sigma}_{21}\boldsymbol{\delta} - 2\mu_2 \boldsymbol{\Sigma}_{22}\boldsymbol{\gamma} = 0. \end{cases}$$

This system, given what

$$\lambda = 2\mu_1 = 2\mu_2 = \rho(U_1, V_1), \tag{5.43}$$

can also be written as

$$\begin{pmatrix} -\lambda\boldsymbol{\Sigma}_{11} & \boldsymbol{\Sigma}_{12} \\ \boldsymbol{\Sigma}_{21} & -\lambda\boldsymbol{\Sigma}_{22} \end{pmatrix} \begin{pmatrix} \boldsymbol{\delta} \\ \boldsymbol{\gamma} \end{pmatrix} = \mathbf{0}. \tag{5.44}$$

For the system of homogeneous Eqs. (5.44) to admit non-trivial solutions, it is necessary that

$$\begin{vmatrix} -\lambda\boldsymbol{\Sigma}_{11} & \boldsymbol{\Sigma}_{12} \\ \boldsymbol{\Sigma}_{21} & -\lambda\boldsymbol{\Sigma}_{22} \end{vmatrix} = 0. \tag{5.45}$$

Using a theorem on the determinant of partitioned matrices (see Appendix B, Theorem 41), if $\boldsymbol{\Sigma}_{22}$ is non-singular, the determinant in (5.45) is equivalent to

$$\det(\boldsymbol{\Sigma}_{11}).\det(\boldsymbol{\Sigma}_{12}\boldsymbol{\Sigma}_{22}^{-1}\boldsymbol{\Sigma}_{21} - \lambda^2\boldsymbol{\Sigma}_{11}) = \det(\boldsymbol{\Sigma}_{12}\boldsymbol{\Sigma}_{22}^{-1}\boldsymbol{\Sigma}_{21} - \lambda^2\boldsymbol{\Sigma}_{11}) = 0 \tag{5.46}$$

and the system

$$(\boldsymbol{\Sigma}_{12}\boldsymbol{\Sigma}_{22}^{-1}\boldsymbol{\Sigma}_{21} - \lambda^2\boldsymbol{\Sigma}_{11})\boldsymbol{\delta} = \mathbf{0} \tag{5.47}$$

has p_1 solutions for both λ^2 and $\boldsymbol{\delta}$.

Similarly, if $\boldsymbol{\Sigma}_{11}$ is non-singular, then

$$\det(\boldsymbol{\Sigma}_{21}\boldsymbol{\Sigma}_{11}^{-1}\boldsymbol{\Sigma}_{12} - \lambda^2\boldsymbol{\Sigma}_{22}) = 0 \tag{5.48}$$

and the homogeneous system

$$(\boldsymbol{\Sigma}_{21}\boldsymbol{\Sigma}_{11}^{-1}\boldsymbol{\Sigma}_{12} - \lambda^2\boldsymbol{\Sigma}_{22})\boldsymbol{\gamma} = \mathbf{0} \tag{5.49}$$

admits p_2 solutions for λ^2 and as many for $\boldsymbol{\gamma}$. In (5.46) to (5.49), λ is an eigenvalue, while $\boldsymbol{\delta}$ and $\boldsymbol{\gamma}$ are the associated eigenvectors.

We will call $U_1 = \boldsymbol{\delta}'_1\boldsymbol{y}$ and $V_1 = \boldsymbol{\gamma}'_1\boldsymbol{z}$ **the first canonical variables**, and the correlation $\lambda_1 = \rho(U_1, V_1)$ will be called the **first canonical correlation**.

Next, let's set $U_2 = \boldsymbol{\delta}'_2\boldsymbol{y}$ and $V_2 = \boldsymbol{\gamma}'_2\boldsymbol{z}$, and normalise their respective variances. Let us also add that U_2 and V_2 are not correlated with U_2 and V_2. The problem is now to maximise $\rho(U_2, V_2)$ in the same way as above.

The result of maximisation under the normalisation constraints on the variances is $\rho(U_2, V_2) = \lambda_2$, the latter being the second largest root of (5.44). If $\boldsymbol{\delta}_2$ and $\boldsymbol{\gamma}_2$

are solutions of (5.47) and (5.49) or, equivalently, of (5.44), associated with λ_2, then U_2 and V_2 are the **second canonical variables**, while λ_2 is the **second canonical correlation.**

We can continue this procedure in an iterative way, while specifying that the number of pairs of canonical variables is limited to p_1. We would then obtain:

$$\lambda_1 \geq \lambda_2 \geq \cdots \geq \lambda_{p_1} \geq 0.$$

It would be possible to group all these results together in matrix form, but we have said enough to approach Johansen's approach in full knowledge of the facts.

Reduced Rank Regression

Consider the VAR(p) model as formulated in (5.39), with no constant term or deterministic trend, where all variables are $I(1)$. Such a model can be represented in the error-corrected form (5.40), i.e.

$$\Delta \boldsymbol{y}_t = \sum_{j=1}^{p-1} \boldsymbol{\Phi}_j \Delta \boldsymbol{y}_{t-j} - \boldsymbol{\Pi} \boldsymbol{y}_{t-1} + \boldsymbol{u}_t.$$

Next, let's define the $H(r)$ model as follows:

Definition 5.4 The first-order integrated model $H(r)$ is determined as the sub-model of the VAR(p) obtained under the **reduced rank** condition

$$\boldsymbol{\Pi} = \boldsymbol{\alpha}\boldsymbol{\beta}',$$

where $\boldsymbol{\alpha}$ and $\boldsymbol{\beta}$ are matrices $p \times r$, with $\text{rk}(\boldsymbol{\Pi}) \leq r$.

The **error correction** model is then written as

$$\Delta \boldsymbol{y}_t = \sum_{j=1}^{p-1} \boldsymbol{\Phi}_j \, \Delta \boldsymbol{y}_{t-j} + \boldsymbol{\alpha}\boldsymbol{\beta}' \boldsymbol{y}_{t-1} + \boldsymbol{u}_t, \quad t = 1, \ldots, T, \tag{5.50}$$

where the parameters $\left(\boldsymbol{\alpha}, \boldsymbol{\beta}, \boldsymbol{\Phi}_1, \ldots, \boldsymbol{\Phi}_{p-1}, \boldsymbol{\Omega}\right)$ vary freely.

This model $H(r)$ is formulated under the condition that the rank of $\boldsymbol{\Pi}$ is less than or equal to r. We can then introduce a series of models, nested according to the diagram:

$$H(0) \subset \cdots \subset H(r) \subset \cdots \subset H(p),$$

where $H(p)$ represents the unconstrained model, i.e. $I(0)$, and $H(0)$ is the model for which the rank of $\boldsymbol{\Pi}$ is equal to zero, in which case the VAR process comprises only first differences.

Consequently, in what follows we will focus on the only models $H(1), \ldots, H(p-1)$, that can address the question of the existence of cointegrating relationships in a system of equations.

Let us now consider the model (5.50) without constant term or deterministic trend and, following Juselius (2006), introduce the following notations:

$$\mathbf{Z}_{0t} = \Delta \boldsymbol{y}_t, \ \mathbf{Z}_{1t} = \boldsymbol{y}_{t-1}, \ \mathbf{Z}_{2t} = (\Delta \boldsymbol{y}_{t-1}, \ldots, \Delta \boldsymbol{y}_{t-k+1}).$$

Taking into account these writing conventions and the fact that $\boldsymbol{\Psi} = (\boldsymbol{\Phi}_1, \ldots, \boldsymbol{\Phi}_{k-1})$, (5.50) is written as

$$\mathbf{Z}_{0t} = \boldsymbol{\alpha}\boldsymbol{\beta}'\mathbf{Z}_{1t} + \boldsymbol{\Psi}\mathbf{Z}_{2t} + u_t, \quad t = 1, \ldots, T. \tag{5.51}$$

We then eliminate the second term in (5.51) and carry out a **reduced rank regression** (RRR), which is applied when the rank r of $\boldsymbol{\Pi}$ is less than k. Ultimately, as we shall see, this method leads to a canonical correlation analysis between the constituent variables of the two vectors $\Delta \boldsymbol{y}_t$ and $\boldsymbol{y}_{t-1}$, in this case between $\mathbf{Z}_{0t}$ and $\mathbf{Z}_{1t}$.

The first step in determining the estimators of $\boldsymbol{\alpha}$, $\boldsymbol{\beta}$ and $\boldsymbol{\Omega}$ is to carry out the two regressions using least squares:

$$\begin{aligned} \mathbf{Z}_{0t} &= \hat{\boldsymbol{B}}'_1\mathbf{Z}_{2t} + \boldsymbol{R}_{0t} \\ \mathbf{Z}_{1t} &= \hat{\boldsymbol{B}}'_2\mathbf{Z}_{2t} + \boldsymbol{R}_{1t}, \end{aligned}$$

after which we will use the residuals $\boldsymbol{R}_{0t}$ and $\boldsymbol{R}_{1t}$ from the two previous regressions to estimate

$$\boldsymbol{R}_{0t} = \boldsymbol{\alpha}\boldsymbol{\beta}'\boldsymbol{R}_{1t} + \boldsymbol{u}_t, \quad t = 1, \ldots, T, \quad \boldsymbol{u}_t \sim IND(\mathbf{0}, \boldsymbol{\Omega}). \tag{5.52}$$

Such a procedure is known as reduced rank regression and proceeds in two stages.

Stage 1: Since the matrix $\boldsymbol{\beta}$ is considered as data, it is easy to estimate $\boldsymbol{\alpha}$: in fact, by post-multiplying (5.52) by $\boldsymbol{R}'_{1t}\boldsymbol{\beta}$ while neglecting the error term, we get

$$\boldsymbol{R}_{0t}\boldsymbol{R}'_{1t}\boldsymbol{\beta} = \boldsymbol{\alpha}\boldsymbol{\beta}'\boldsymbol{R}_{1t}\boldsymbol{R}'_{1t}\boldsymbol{\beta}.$$

Summing the previous expression and dividing by T gives

$$S_{01}\boldsymbol{\beta} = \boldsymbol{\alpha}\boldsymbol{\beta}' \boldsymbol{S}_{11}\boldsymbol{\beta},$$

where $\boldsymbol{S}_{ij} = T^{-1}\sum_t \boldsymbol{R}_{it}\boldsymbol{R}_{jt},\quad i=0,\,1$ and $j=1$.

The least squares estimator of $\boldsymbol{\alpha}$ as a function of $\boldsymbol{\beta}$ is therefore given by

$$\hat{\boldsymbol{\alpha}}(\boldsymbol{\beta}) = S_{01}\boldsymbol{\beta}(\boldsymbol{\beta}'\boldsymbol{S}_{11}\boldsymbol{\beta})^{-} \quad . \tag{5.53}$$

Stage 2: We have assumed that $\boldsymbol{u}_t$, in (5.52), follows a multivariate normal distribution with zero expectation and a variance–covariance matrix $\boldsymbol{\Omega}$. The log-likelihood function will then take the form

$$\begin{aligned} \ln L_c(\boldsymbol{\beta}\,;\,r) = & -\frac{Tk}{2}\ln(2\pi) - \frac{T}{2}\ln\left[\det\left(\hat{\boldsymbol{\Omega}}(\boldsymbol{\beta})\right)\right] \\ & -\frac{1}{2}\sum_{t=1}^{T}\boldsymbol{u}'(\hat{\alpha})\left[\hat{\boldsymbol{\Omega}}(\boldsymbol{\beta})\right]^{-1}\boldsymbol{u}(\hat{\alpha}). \end{aligned} \tag{5.54}$$

Straight away, we obtain the estimator of the matrix $\boldsymbol{\Omega}$ as a function of $\boldsymbol{\alpha}$ and $\boldsymbol{\beta}$:

$$\begin{aligned} \hat{\boldsymbol{\Omega}}(\boldsymbol{\alpha},\boldsymbol{\beta}) &= T^{-1}\sum_{t=1}^{T}(\boldsymbol{R}_{0t}-\boldsymbol{\alpha}\boldsymbol{\beta}'\boldsymbol{R}_{1t})(\boldsymbol{R}_{0t}-\boldsymbol{\alpha}\boldsymbol{\beta}'\boldsymbol{R}_{1t})' \\ &= T^{-1}\left(\sum \boldsymbol{R}_{0t}\boldsymbol{R}'_{0t} - \sum \boldsymbol{R}_{0t}\boldsymbol{R}'_{1t}\boldsymbol{\beta}\boldsymbol{\alpha}' - \boldsymbol{\alpha}\boldsymbol{\beta}'\sum \boldsymbol{R}_{1t}\boldsymbol{R}'_{0t} + \boldsymbol{\alpha}\boldsymbol{\beta}'\sum \boldsymbol{R}_{1t}\boldsymbol{R}'_{1t}\boldsymbol{\beta}\boldsymbol{\alpha}'\right) \\ &= S_{00} - S_{01}\boldsymbol{\beta}\boldsymbol{\alpha}' - \boldsymbol{\alpha}\boldsymbol{\beta}'\boldsymbol{S}_{10} + \boldsymbol{\alpha}\boldsymbol{\beta}'\boldsymbol{S}_{11}\boldsymbol{\beta}\boldsymbol{\alpha}'. \end{aligned}$$

Substituting (5.53) into the previous expression gives the estimator of the variance–covariance matrix, which depends exclusively on $\boldsymbol{\beta}$:

$$\begin{aligned} \hat{\boldsymbol{\Omega}}(\boldsymbol{\beta}) &= S_{00} - 2\,S_{01}\boldsymbol{\beta}(\boldsymbol{\beta}'\boldsymbol{S}_{11}\boldsymbol{\beta})^{-1}\boldsymbol{\beta}'\boldsymbol{S}_{10} + S_{01}\boldsymbol{\beta}(\boldsymbol{\beta}'\boldsymbol{S}_{11}\boldsymbol{\beta})^{-1}\boldsymbol{\beta}'\boldsymbol{S}_{11}\boldsymbol{\beta}(\boldsymbol{\beta}'\boldsymbol{S}_{11}\boldsymbol{\beta})^{-1}\boldsymbol{\beta}'\boldsymbol{S}_{10} \\ &= S_{00} - S_{01}\boldsymbol{\beta}(\boldsymbol{\beta}'\boldsymbol{S}_{11}\boldsymbol{\beta})^{-1}\boldsymbol{\beta}'\boldsymbol{S}_{10}. \end{aligned}$$

It then follows that

$$\hat{\boldsymbol{\Omega}}(\boldsymbol{\beta}) = T^{-1}\sum \boldsymbol{u}_t(\hat{\alpha})\boldsymbol{u}'_t(\hat{\alpha}) = S_{00} - S_{01}\boldsymbol{\beta}(\boldsymbol{\beta}'\boldsymbol{S}_{11}\boldsymbol{\beta})^{-1}\boldsymbol{\beta}'\boldsymbol{S}_{10}. \tag{5.55}$$

Taking into account (5.53) and (5.55), the log-likelihood function given in (5.54) becomes

$$\ln L_c(\boldsymbol{\beta}\,;\,r) = -\frac{Tk}{2}(1+\ln 2\pi) - \frac{T}{2}\left[\det(S_{00} - S_{01}\boldsymbol{\beta}(\boldsymbol{\beta}'\boldsymbol{S}_{11}\boldsymbol{\beta})^{-1}\boldsymbol{\beta}'\boldsymbol{S}_{10})\right]. \tag{5.56}$$

The Maximum Likelihood Estimator

In view of (5.56), maximising the log-likelihood with respect to $\boldsymbol{\beta}$ is in fact equivalent to minimising the determinant in it. Given that the value of T is high and therefore the matrices $\boldsymbol{S}_{00}$ and $\boldsymbol{S}_{11}$ are non-singular, we can again use Theorem 41 (Appendix B, Theorem 41) by applying it to the partitioned matrix

$$\boldsymbol{A}(\boldsymbol{\beta}) = \begin{pmatrix} \boldsymbol{S}_{00} & \boldsymbol{S}_{01}\boldsymbol{\beta} \\ \boldsymbol{\beta}'\boldsymbol{S}_{10} & \boldsymbol{\beta}'\boldsymbol{S}_{11}\boldsymbol{\beta} \end{pmatrix},$$

which leads to

$$\begin{aligned} \det(\boldsymbol{A}(\boldsymbol{\beta})) &= \det(\boldsymbol{S}_{00}) \cdot \det\big(\boldsymbol{\beta}'\boldsymbol{S}_{11}\boldsymbol{\beta} - \boldsymbol{\beta}'\boldsymbol{S}_{10}\boldsymbol{S}_{00}^{-1}\boldsymbol{S}_{01}\boldsymbol{\beta}\big) \\ &= \det\big(\boldsymbol{\beta}'\boldsymbol{S}_{11}\boldsymbol{\beta}\big) \cdot \det\big(\boldsymbol{S}_{00} - \boldsymbol{S}_{01}\boldsymbol{\beta}(\boldsymbol{\beta}'\boldsymbol{S}_{11}\boldsymbol{\beta})^{-1}\boldsymbol{\beta}'\boldsymbol{S}_{10}\big). \end{aligned} \tag{5.57}$$

Taking into account (5.55), (5.57) can be written as

$$\begin{aligned} \det\big(\hat{\boldsymbol{\Omega}}(\boldsymbol{\beta})\big) &= \frac{\det(\boldsymbol{S}_{00}) \cdot \det\big(\boldsymbol{\beta}'\boldsymbol{S}_{11}\boldsymbol{\beta} - \boldsymbol{\beta}'\boldsymbol{S}_{10}\boldsymbol{S}_{00}^{-1}\boldsymbol{S}_{01}\boldsymbol{\beta}\big)}{\det\big(\boldsymbol{\beta}'\boldsymbol{S}_{11}\boldsymbol{\beta}\big)} \\ &= \frac{\det(\boldsymbol{S}_{00}) \cdot \det\big(\boldsymbol{\beta}'(\boldsymbol{S}_{11} - \boldsymbol{S}_{10}\boldsymbol{S}_{00}^{-1}\boldsymbol{S}_{01})\boldsymbol{\beta}\big)}{\det\big(\boldsymbol{\beta}'\boldsymbol{S}_{11}\boldsymbol{\beta}\big)}. \end{aligned}$$

However, as $\boldsymbol{S}_{00}$ does not depend on $\boldsymbol{\beta}$, it will suffice to minimise the determinant

$$\frac{\det\big(\boldsymbol{\beta}'(\boldsymbol{S}_{11} - \boldsymbol{S}_{10}\boldsymbol{S}_{00}^{-1}\boldsymbol{S}_{01})\boldsymbol{\beta}\big)}{\det\big(\boldsymbol{\beta}'\boldsymbol{S}_{11}\boldsymbol{\beta}\big)}. \tag{5.58}$$

To solve this optimisation problem, we need a lemma (see Johansen, 1995, Lemma A.8):

Lemma 5.2 *Let* **M** *and* **N** *be symmetric positive semidefinite and symmetric positive definite matrices respectively. The function*

$$f(\boldsymbol{x}) = \frac{\det(\boldsymbol{x}'\boldsymbol{M}\boldsymbol{x})}{\det(\boldsymbol{x}'\boldsymbol{N}\boldsymbol{x})}$$

is minimised among all matrices of dimension $(p \times K)$ *by* $\boldsymbol{x} = (\boldsymbol{v}_1, \ldots, \boldsymbol{v}_K)$ *and* $\Pi_{i=1}^{K} \rho_i$, *where* ρ_i *and* $\boldsymbol{v}_i$ *are solutions of the equation*

$$\det(\rho\boldsymbol{N} - \boldsymbol{M}) = \boldsymbol{0},$$

with the eigenvalues ordered according to $\rho_1 > \cdots > \rho_p \geq 0$.

By applying this lemma to (5.58), we get

$$\det\left(\rho S_{11} - S_{11} + S_{10}S_{00}^{-1}S_{01}\right) = \det\left(\lambda S_{11} - S_{10}S_{00}^{-1}S_{01}\right) = 0, \tag{5.59}$$

where $\lambda = 1 - \rho$.

Solving Eq. (4.59) will provide p eigenvalues $\hat{\lambda}_1, \ldots, \hat{\lambda}_p$, such that $1 > \hat{\lambda}_1 > \cdots > \hat{\lambda}_p > 0$ and the associated eigenvectors $\boldsymbol{V} = (\hat{\boldsymbol{v}}_1, \ldots, \hat{\boldsymbol{v}}_p)$, normalised by imposing the restrictions

$$\boldsymbol{V}'\boldsymbol{S}_{11}\boldsymbol{V} = \boldsymbol{I}_p. \tag{5.60}$$

Given that $\text{rk}(\boldsymbol{\Pi}) = r$, the r largest eigenvalues correspond to the r eigenvectors making up the matrix $\hat{\boldsymbol{\beta}} = (\hat{\boldsymbol{v}}_1, \ldots, \hat{\boldsymbol{v}}_r)$. The determinant of the variance–covariance matrix in (5.58) is therefore equal to

$$\det\left(\hat{\boldsymbol{\Omega}}(\hat{\boldsymbol{\beta}})\right) = \det(\boldsymbol{S}_{00})\, \Pi_{i=1}^{r}(1 - \lambda_i), \tag{5.61}$$

since $\hat{\boldsymbol{\beta}}'\boldsymbol{S}_{11}\hat{\boldsymbol{\beta}} = \boldsymbol{I}_r$, on the one hand, and $\hat{\boldsymbol{\beta}}'\boldsymbol{S}_{10}\boldsymbol{S}_{00}^{-1}\boldsymbol{S}_{01}\hat{\boldsymbol{\beta}} = \text{diag}(\hat{\lambda}_1, \ldots, \hat{\lambda}_r)$, on the other. The other parameters, i.e. the weight matrix, are easily obtained by inserting $\hat{\boldsymbol{\beta}}$ into (5.53).

A final remark: the estimators of $\boldsymbol{\alpha}$ and $\boldsymbol{\beta}$ are related to the **canonical variables $\boldsymbol{R}_{0t}$** and $\boldsymbol{R}_{0t}$, while the r highest eigenvalues are the squares of the **canonical correlations**. To this extent, estimating $\hat{\boldsymbol{\beta}}$ amounts to selecting r linear combinations of $\boldsymbol{y}_{t-1}$ which have the strongest correlations (squared) with the stationary process $\Delta\boldsymbol{y}_t$.

Determining the Cointegration Rank

So far, we have assumed that the cointegration rank, i.e. alternatively the number of cointegrating relationships, is known, whereas in reality it is not. Tests are therefore needed to find it. The tests we are going to develop are based on the consideration of the vector error correction model comprising k variables and p lags:

$$\Delta\boldsymbol{y}_t = \boldsymbol{\alpha}\boldsymbol{\beta}'\boldsymbol{y}_{t-1} + \sum_{j=1}^{p-1}\boldsymbol{\Phi}_j\Delta\boldsymbol{y}_{t-j} + \boldsymbol{u}_t. \tag{5.62}$$

Two tests are used to determine the number of cointegrating relationships, itself given by the number of significant canonical correlations: the trace test and the maximum eigenvalue test. Both are similar to likelihood ratio tests (see "Testing the Model" chapter, "Test on Constrained Estimators" section).

Let's start with the first of these and suppose that we are testing

$$H_r : \ \operatorname{rk}(\boldsymbol{\Pi}) = r < k$$

against the alternative

$$H_k : \ \operatorname{rk}(\boldsymbol{\Pi}) = k,$$

with $r = 0, 1, \ldots, k-1$, r being the number of cointegrating relationships.

The log-likelihood ratio statistic is then defined by

$$\lambda_{\text{trace}}(H_r/H_k) = 2\left[\ln L_c(\hat{\boldsymbol{\beta}}, k) - \ln L_c(\hat{\boldsymbol{\beta}}, r)\right].$$

Since the concentrated log-likelihood function is given for all r by

$$\ln L_c(r) = -\frac{Tk}{2}(1 + \ln 2\pi) - \frac{T}{2}\ln(\det(\boldsymbol{S}_{00})) - \frac{T}{2}\sum_{i=1}^{r}\ln(1 - \lambda_i), \tag{5.63}$$

where the $\lambda'_i s$ are the r highest eigenvalues of the matrix $\boldsymbol{S}_{10}\boldsymbol{S}_{00}^{-1}\boldsymbol{S}_{01}\boldsymbol{S}_{11}^{-1}$, the **trace statistic** is written as

$$\lambda_{\text{trace}}(H_r/H_k) = -T\sum_{i=r+1}^{k}\ln(1 - \lambda_i), \tag{5.64}$$

λ_i being the $(k - r)$ lowest eigenvalues of the matrix $\boldsymbol{S}_{10}\boldsymbol{S}_{00}^{-1}\boldsymbol{S}_{01}\boldsymbol{S}_{11}^{-1}$.

We now turn to the **maximum eigenvalue** test, where the null hypothesis is

$$H_r : \ \operatorname{rg}(\boldsymbol{\Pi}) = r < k$$

against the alternative

$$H_{r+1} : \ \operatorname{rg}(\boldsymbol{\Pi}) = r + 1, \qquad r = 0, 1, \ldots, k-1,$$

which leads, given (5.63), to the statistic

$$\lambda_{\max}(H_r/H_{r+1}) = -T\ln(1 - \hat{\lambda}_{r+1}), \tag{5.65}$$

where $\hat{\lambda}_{r+1}$ is an eigenvalue solution of $\boldsymbol{S}_{10}\boldsymbol{S}_{00}^{-1}\boldsymbol{S}_{01}\boldsymbol{S}_{11}^{-1}$.

Here is a table summarising the sequential approach to implementing each of these tests, assuming that there are four variables in the VAR model (Table 5.2).

This goes without saying, but it can happen that the conclusions drawn from the two tests do not completely coincide.

Table 5.2 Comparison of assumptions

Rank	λ trace		λ max	
	H_0	H_1	H_0	H_1
0	$r = 0$	$r = 4$	$r = 0$	$r = 1$
1	$r = 1$	$r = 4$	$r = 1$	$r = 2$
2	$r = 2$	$r = 4$	$r = 2$	$r = 3$
3	$r = 3$	$r = 4$	$r = 3$	$r = 4$

Before looking at how the critical values for these statistics are established, we should point out that we have introduced neither a constant nor a trend in (5.62). From this point of view, we can establish five main possible cases by distinguishing the series in level y_t from cointegration relations (CR):

(i) No constant or trend in y_t and CR;
(ii) No trend in y_t; a constant in CR;
(iii) Trend linear in y_t; a constant in CR;
(iv) Trend linear in both y_t and CR;
(v) Quadratic trend in y_t; linear trend in CR.

Next, we know that a likelihood ratio statistic asymptotically follows a Chi-square distribution with a number of degrees of freedom corresponding to the number of restrictions tested.

However, this is not the case here, as the two cointegration rank statistics asymptotically follow a non-standard distribution. More precisely, the distributions in question are vector Brownian motion functionals and can therefore only be obtained by simulation.

In total, taking into account the five cases distinguished for trend and constant, the critical values for the two statistics have been tabulated, for example, by Johansen (1995, Chapter 15).

5.6 Applications

We will develop three applications: firstly, we will give an example of spurious regression between two simulated random walks; secondly, we will compare different methods for estimating a cointegrating relationship between first-order integrated series.

Finally, we will fully develop Pesaran's approach by estimating and testing a complete ARDL model. (In the latter case, we will use the *Microfit* 5.0 software developed by Bahram and Hashem Pesaran, 2009).

A Spurious Regression

We have generated a sample of size 100 of the two random walks

$$\begin{cases} y_t = y_{t-1} + u_t, & u_t \sim \text{IND}(0, 1) \\ x_t = x_{t-1} + v_t, & v_t \sim \text{IND}(0, 1). \end{cases} \tag{5.66}$$

The independently distributed perturbations were produced using a pseudo-random number generator. The two series were then temporally structured (1er quarter 1995–4ème quarter 2019).

Here is a graph of these two series (Fig. 5.1):

It is clear from their respective graphs that the behaviour of these two series is totally independent and that there is no link between them.

But this is not what the linear regression shows

$$y_t = \alpha x_t + \varepsilon_t, \quad \varepsilon_t \sim N(0, \sigma^2),$$

where $y_t = rwalk1$ and $x_t = rwalk2$ (Table 5.3).

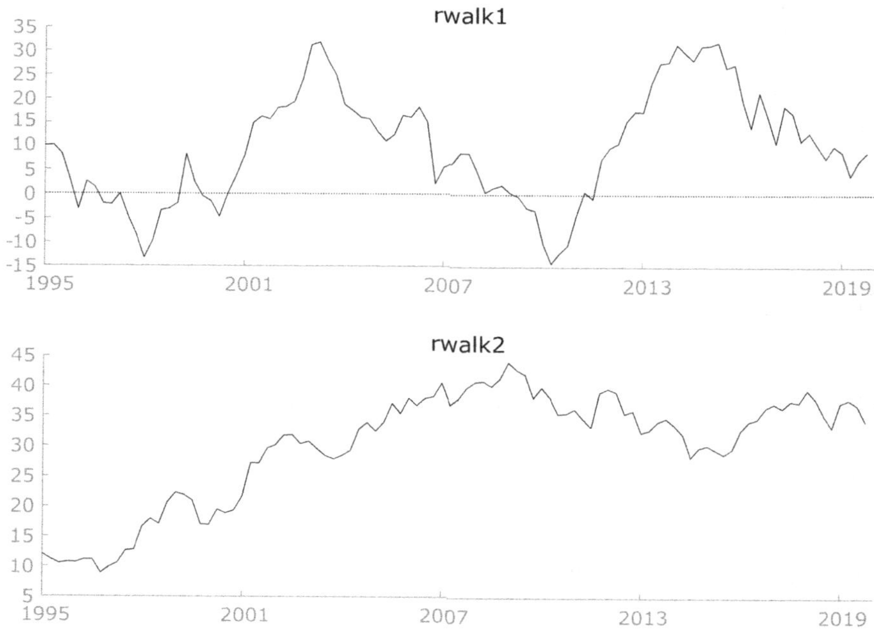

Fig. 5.1 Two simulated random walks

Table 5.3 Linear regression

	Coefficient	Stand. error	Stat. t	p-value
x_t	0.325	0.071	4.55	0.0000
$R^2 = 0.449$	$F(1, 99) = 20.698$		p-value (F)= 0.00001	

As can be seen, there is a positive relationship between the two random walks; moreover, this relationship is highly significant with a very low *p-value*. All in all, we have a spurious regression due to the non-stationary nature of the two variables used.

Estimation of a Cointegrating Relationship

We now examine the long-term determinants of aggregate electricity demand in France (period 1990-Q1–2015-Q3). We have selected three of these, namely an electricity price index (EPI), the price index of a possible substitute, gas (GPI) and real GDP (GDPr). All these variables are expressed in logarithms.

The purpose of this section is therefore to establish the existence of a cointegrating relationship between the four variables, with electricity consumption (EC) being the dependent variable. We begin by examining how many cointegrating relationships there could be (Table 5.4).

Both tests lead to the same conclusion: at 5%, we must reject the null hypothesis of no cointegrating relationship and conclude that there is only one cointegrating relationship. Examination of the p-values for $k = 1$ confirms this choice.

Table 5.4 Cointegration rank test

Rank	Stat Trace	p-values	Stat max EV	p-values
0	61.106	0.0022	31.729	0.011
1	29.377	0.0633	20.903	0.0523
2	0.085	0.434	8.3882	0.3485
3	0.001	0.7732	0.086	0.7694

(1) The p-values associated with the trace statistic are corrected according to the sample size. (2) The calculation of the statistics corresponds to the presence of a constant in the cointegration relationship

Let us now estimate the cointegrating relationship between consumption and the explanatory variables by applying the various estimation methods detailed in the previous sections: DOLS, FM-OLS and Johansen's method (Table 5.5).

Table 5.5 Comparison of results

Coefficients	DOLS	FM-OLS	JOHANSEN
Constant	4.185**	4.48**	7.109**
EPI	−0.090*	−0.12**	−0.24**
GPI	−0.243**	−0.20**	0.0115
GDPr	0.701**	0.67**	0.429**

* and ** denote coefficients significant at 5% and 1% respectively

Several points can be made from this table:

(i) The first two methods lead to very similar estimates; this is not the case for the estimators obtained using the Johansen procedure, which differ quite markedly from the previous ones: in particular, the elasticity of electricity consumption with respect to the price of gas is not significantly different from zero and even has a sign opposite to that of the other two estimates;
(ii) On the other hand, the three variables selected have a differentiated effect on electricity demand; in particular, real GDP has a strong influence on electricity consumption, given that a one percent increase in GDP translates into a 0.7% increase in consumption (remember that all these variables are expressed in logarithms); on the other hand, the price of electricity has little influence on demand, with an elasticity of less than ten percentage points;
(iii) It should also be noted that the various elasticities have the sign expected by theory, with the exception, of course, of that relating to the price of gas; we must conclude from this that electricity and gas are not substitutable energies.

Pesaran Estimates

As we saw in "Estimation and Testing of ARDL Models" section of this chapter, Pesaran's approach—we have named it PESS to take account of his co-authors—amounts to estimating an ARDL model while not, in principle, requiring us to check whether the variables involved are $I(0)$ or $I(1)$. We continue to use the four variables presented in the previous section, although we will refer to EC as C, GDPr as Y, EPI as E and GPI as G.

We will work with the

$$\Delta C_t = m + \sum_{i=1}^{4} \alpha_i \Delta C_{t-i} + \sum_{i=0}^{4} \beta_{Yi} \Delta Y_{t-i} + \sum_{i=0}^{4} \beta_{Ei} \Delta E_{t-i}$$

$$+\sum_{i=0}^{4}\beta_{Gi}\Delta G_{t-i}+\gamma_1 C_{t-1}+\gamma_2 Y_{t-1}+\gamma_3 E_{t-1}+\gamma_4 G_{t-1}+u_t \tag{5.66}$$

and go through three steps.

1. The first consists of estimating the model (5.66) using ordinary least squares. This linear regression is not interesting in itself, but it does allow us to calculate the Wald statistic or the F statistic, under the null hypothesis

$$H_0:\ \gamma_1=\gamma_2=\gamma_3=\gamma_4=0,$$

   assumption, which amounts to saying that there is no cointegrating relationship between the four variables selected. (The distribution of these statistics is not that of the usual Fisher-Snedecor distribution and must be tabulated by simulation, as we saw again in "Estimation and Testing of ARDL Models" section of this chapter.)
2. The second step consists of interpreting the F statistic, for example, the one corresponding first to the model where electricity consumption is the dependent variable. This statistic is, in this case, equal to

$$F(4,77)=4.369.$$

   As in the model with constant and $k=4$, at a level of 95%, the interval determined by the lower and upper bounds of the critical values is given by $[2.850\ ,\ 4.049]$ and $F(4,77)=4.369>4.049$, it follows that we reject the null hypothesis and accept the alternative of the existence of a cointegrating relationship.

   On the other hand, if ΔY, ΔE or ΔG is the dependent variable in (5.66), then the F statistic is equal to 0.753, 2.189 and 1.317 respectively, all values below the lower bound. We are therefore led to accept the null hypothesis that there is no cointegrating relationship.

   In total, the various tests indicate that there is one cointegrating relationship and only one, when C is the dependent variable, which is consistent with the conclusion drawn from the application of the two Johansen tests (see "Estimation of a Cointegrating Relationship" section).
3. The third step involves estimating the long-term relationship isolated in the previous step. The final ARDL model is determined by using the information criteria, either Akaike (AIC), Bayesian Schwarz (BIC) or Hannan-Quinn (HQC). In the latter two cases, the same ARDL model is selected and the estimators of the long-term parameters are the same (Table 5.6).

As can be seen from the table, all the long-term coefficients are highly significant and the estimators obtained are very close to each other, regardless of which of the information criteria is used.

Table 5.6 Estimators of the cointegrating relationship

	AIC-ARDL(3,0,0,1)		BIC/HQC-ARDL(3,0,0,0)	
	Coefficients	*p*-values*	Coefficients	*p*-values*
Cst	5.0489	0.000	5.2057	0.000
Y	0.63474	0.000	0.62626	0.000
E	−0.14315	0.002	−0.16485	0.000
G	−0.19374	0.000	−0.18248	0.000

The asterisk (*) indicates asymptotic *p*-values. The asterisk (*) indicates asymptotic *p*-values

Finally, if we refer to the coefficients resulting from the application of the Bayesian Schwarz or Hannan-Quinn criteria, we obtain the following values for the corresponding F or Wald statistics:

$$\begin{aligned} \text{Stat } F &= 11.1882 \quad [3.3367;\ 4.4408] \\ \text{Stat } W &= 44.7526 \quad [13.3469;\ 17.7632]. \end{aligned}$$

In both cases, the F or W statistic is well above the upper limit of the interval and, consequently, the null hypothesis of no cointegrating relationship between the variables concerned must be rejected.

The same conclusion would be reached if we were to consider Akaike's information criterion.

5.7 Exercises

5.7.1 Consider the following linear regression

$$y_t = m + \beta x_t + \varepsilon_t, \qquad t = 1, \ldots, T,$$

where ε_t is a Gaussian white noise, and proof that the OLS estimators of the parameters m and β are given by

$$\begin{cases} \hat{m} = \overline{y} - \hat{\beta}\overline{x} \\ \hat{\beta} = \dfrac{\sum\limits_{t=1}^{T} x_t y_t - T\,\overline{x}\,\overline{y}}{\sum\limits_{t=1}^{T} x_t^2 - T\,\overline{x}^2}, \end{cases}$$

with $\overline{y} = (1/T)\sum_{t=1}^{T} y_t$ and $\overline{x} = (1/T)\sum_{t=1}^{T} x_t$.

5.7.2 The file (pandemic.gdt) contains three series, namely the number of hospitalisations due to COVID (denoted *HOSP*), the total number of people fully vaccinated (i.e. vaccinated with two doses) against the pandemic, this variable being denoted

VAC, and finally the number of contaminations, denoted *CONT*. These data, which concern Belgium exclusively, are daily (7 days a week) and cover the period from 1 February 2021 to 20 September of the same year.

You are asked

1. to verify that these series are integrated of order 1;
2. to estimate the cointegration relationship

$$HOSP = \alpha + \beta_1 \, VAC + \beta_2 \, CONT,$$

   by using DOLS and FM-OLS.
3. to compare and interpret the estimators thus obtained.

5.7.3 Let the ARDL $(p, k_1, \ldots, k_K)$ model be

$$\Delta y_t = \sum_{i=1}^{p} \alpha_i \Delta y_{t-i} + \sum_{j=0}^{k_1} \beta_{1j} \Delta x_{1,t-j} + \ldots + \sum_{j=0}^{k_K} \beta_{Kj} \Delta x_{K,t-j}$$
$$- (1 - \alpha_1 - \ldots - \alpha_p) \left[y_{t-1} - a - \sum_{r=1}^{K} \gamma_r x_{r,t-1} \right] + \varepsilon_t .$$

We ask you to explain the statistical and economic significance of this ARDL model.

5.7.4 Let's go back to the pandemic.gdt file, expressing the three variables in logarithmic form, all $I(1)$. We then obtain the following final estimates for an ARDL model of hospitalisations (D is, of course, the first difference of the variables under consideration.) (Table 5.7):

Table 5.7 Parameters estimation

Variables	Coefficients	t Statistic	p-value
LHOSP (−1)	−0.252	−11.11	< 0.01
LVAC (−1)	−0.061	−10.40	< 0.01
LCONT (−1)	0.265	11.69	< 0.01
DLHOSP (−1)	−0.497	−9.72	< 0.01
DLHOSP (−2)	−0.317	−5.66	< 0.01
DLHOSP (−3)	−0.139	−2.07	0.03
DLCONT (−2)	−0.135	−5.61	< 0.01
DLCONT (−3)	−0.127	−5.74	< 0.01
DLCONT (−4)	−0.178	−5.63	< 0.01
DLCONT (−5)	−0.108	−4.71	< 0.01
DLCONT (−6)	−0.118	−5.97	< 0.01
BIN1	−0.976	−28.92	< 0.01

(continued)

Table 5.7 (continued)

Variables	Coefficients	t Statistic	p-value
BIN2	0.563	9.97	< 0.01
BIN3	−0.173	−26.06	< 0.01
$R^2 = 0.75$		$F(14, 211) = 2.84e^{+16}$, p-valeur = 0.000000	

We ask:

(1) interpret all the statistical results obtained;

(2) establish the cointegration relationship between *LHOSP* and the other two variables;

(3) interpret the elasticities of *LHOSP* in relation to the number of people vaccinated and the number of people infected by the virus.

5.7.5 As in the previous exercise, we'll use the pandemic.gdt file, but we'll use the PESS (Pesaran, Shin and Smith) procedure. Let's now consider the ARDL model

$$\Delta LHOSP_t = m + \sum_{i=1}^{7} \alpha_i \Delta LHOSP_{t-i} + \sum_{i=0}^{7} \beta_{Vi} \Delta VAC_{t-i} + \sum_{i=0}^{7} \beta_{Ci} \Delta CONT_{t-i} + \gamma_1 LHOSP_{t-1} + \gamma_2 LVAC_{t-1} + \gamma_3 LCONT_{t-1} + u_t$$

You are required to

1. estimate this model by ordinary least squares.
2. calculate the F statistic under the null hypothesis

$$H_0 : \gamma_1 = \gamma_2 = \gamma_3 = 0.$$

3. use the critical values given in Table 5.1 to reject or accept H_0.
4. interpret the result obtained in the previous point.

5.7.6 Using the data in the following table

Rank	Trace Stat.	p-value	Eigenvalue max Stat	p-value
0	120.52	0.0000	98.88	0.0000
1	21.64	0.0009	19.23	0.0012
2	2.41	0.1416	2.41	0.1429

determine the rank of cointegration and therefore the number of cointegrating relationships between the variables *LHOSP*, *LVAC* and *LCONT*.

5.7.7 Use Johansen's multivariate procedure to

1. estimate the cointegration relationship

$$LHOSP = \hat{m} + \hat{\gamma}_1 LVAC + \hat{\gamma}_2 LCONT.$$

2. interpret this relationship.

References

Engle, R. F., & Granger, C. W. J. (1987). Co-integration and error-correction: Representation, estimation and testing. *Econometrica, 55*, 251–276.

Granger, C. W. J. (1983). *Cointegrated variables and error-correcting models*. USCD Discussion Paper, 83–13.

Granger, C. W. J. (1991). Developments in the study of cointegrated economic variables. In R.F. Engle & C. W. J.Granger (Eds.), *Long-run economic relationships. Readings in cointegration* (Chapter 4). Oxford University Press.

Johansen, S. (1988). Statistical analysis of cointegration vectors. *Journal of Economic Dynamics and Control, 12*, 231–254.

Johansen, S. (1995). *Likelihood-based inference in cointegrated vector autoregressive models*. Oxford University Press.

Juselius, K. (2006). *The cointegrated VAR model: methodology and applications*. Oxford University Press.

Mackinnon, J. G. (1991). Critical values for cointegration tests. In R. F. Engle & C. W. J. Granger (Eds.), *Long-run economic relationship. Readings in cointegration*. Oxford University Press.

Pesaran, B., & Pesaran, M. H. (2009). *Time series econometrics using Microfit 5.0*. Oxford University Press.

Pesaran, M. H., Shin, Y., & Smith, R. J. (2001). Bounds testing approaches to the analysis of level relationships. *Journal of Applied Econometrics, 16*, 289–326.

Phillips, P. C. B. (1995). Fully modified least squares and vector autoregressions. *Econometrica, 63*, 1023–1078.

Phillips, P. C. B., & Hansen, B. E. (1990). Statistical Inference in Instrumental variable regression with $I(1)$ processes. *Review of Economic Studies, 57*, 99–125.

Stock, J. M., & Watson, M. H. (1993). A simple estimator of cointegrating vectors in higher order integrated systems. *Econometrica, 61*, 783–820.

Chapter 6
Specification of the ARDL Model

6.1 Why Start from a General Specification?

In the period before the proliferation of personal computers, the econometric approach consisted of starting with a simple equation containing few variables, and then possibly adding other variables or time lags. In reality, the opposite approach is recommended, starting with a very general model and then simplifying it.

Theoretical Justification

Let's start with an extremely simple model

$$y_t = \alpha x_t + \varepsilon_t, \qquad t = 1, \ldots, T, \tag{6.1}$$

assuming that the correct model is given by the equation

$$y_t = \alpha x_t + \beta z_t + \nu_t, \qquad \nu_t \sim IND(0, \sigma_\nu^2). \tag{6.2}$$

The least squares estimator of α in (6.1) is

$$\tilde{\alpha} = \sum_{t=1}^{T} y_t x_t / \sum_{t=1}^{T} x_t^2,$$

Supplementary Information The online version contains supplementary material available at https://doi.org/10.1007/978-3-031-72910-2_6.

F. J. Bismans and O. Damette, *Dynamic Econometrics*, https://doi.org/10.1007/978-3-031-72910-2_6

which, after substituting (6.2), leads to

$$\tilde{\alpha} = \alpha + \beta \frac{\sum_t z_t x_t}{\sum_t x_t^2} + \frac{\sum_t v_t x_t}{\sum_t x_t^2}.$$

Taking the expectation of $\tilde{\alpha}$, he comes

$$E(\tilde{\alpha}) = \alpha + \beta \frac{\sum_t z_t x_t}{\sum_t x_t^2} = \alpha + \beta\, b_{zx}, \tag{6.3}$$

where b_{zx} is the coefficient resulting from the regression of z on x.

In view of (6.3), if this coefficient b_{zx} is not zero, i.e. if there is an empirical relationship between the observations relating to the two variables z and x, it follows that the estimator $\tilde{\alpha}$ will be biased. Consequently, the Gauss-Markov theorem will be invalidated.

Furthermore, given the "true" model (6.2), the variance $\tilde{\alpha}$ is given by

$$\begin{aligned} \mathrm{Var}(\tilde{\alpha}) &= \mathrm{Var}\left[\sum_t \frac{v_t x_t}{x_t^2}\right] = \sum_t \mathrm{Var}\left[\frac{v_t x_t}{x_t^2}\right] \\ &= \left(1/\sum_t x_t^2\right)\mathrm{Var}(v_t) \\ &= \left(1/\sum_t x_t^2\right)\sigma_v^2. \end{aligned} \tag{6.4}$$

Since, once again with regard to the "true" model, the disturbance term in (6.1) is in fact equal to $\varepsilon_t = \beta z_t + v_t$, it follows that the variance estimated from this Eq. (6.1) will probably overestimate σ_v^2.

It follows from these developments that omitting a variable in the specification of a model can lead to statistically inefficient estimators. However, this proposition is no longer true if a superfluous variable is included in the model.

Adding an Irrelevant Variable

Suppose that Eq. (6.1) correctly specifies the model, but that for the rest it is Eq. (6.2) that we wish to estimate by least squares.

To determine the estimators α and β, we need to invert the observation matrix (**X'X**) (see Appendix B, Definition 21), i.e.

$$\boldsymbol{X}'\boldsymbol{X} = \begin{pmatrix} x_1 & \cdots & x_T \\ z_1 & \cdots & z_T \end{pmatrix} \begin{pmatrix} x_1 & z_1 \\ \vdots & \vdots \\ x_T & z_T \end{pmatrix} = \begin{pmatrix} \sum x_t^2 & \sum x_t z_t \\ \sum x_t z_t & \sum z_t^2 \end{pmatrix}.$$

Inverting the latter matrix gives

$$(\boldsymbol{X}'\boldsymbol{X})^{-1} = \frac{1}{\det(\boldsymbol{X}'\boldsymbol{X})} [\text{Adj } (\boldsymbol{X}'\boldsymbol{X})]' = \frac{1}{\underbrace{\sum x_t^2 z_t^2 - \left(\sum x_t z_t\right)^2}_{D}} \begin{pmatrix} \sum z_t^2 & -\sum x_t z_t \\ -\sum x_t z_t & \sum x_t^2 \end{pmatrix}.$$

Then we have the OLS estimators

$$\begin{aligned} \begin{pmatrix} \hat{\alpha} \\ \hat{\beta} \end{pmatrix} &= \frac{1}{D} \begin{pmatrix} \sum z_t^2 & -\sum x_t z_t \\ -\sum x_t z_t & \sum x_t^2 \end{pmatrix} \begin{pmatrix} x_1 & \cdots & x_T \\ z_1 & \cdots & z_T \end{pmatrix} \begin{pmatrix} y_1 \\ \vdots \\ y_T \end{pmatrix} \\ &= \frac{1}{D} \begin{pmatrix} \sum z_t^2 & -\sum x_t z_t \\ -\sum x_t z_t & \sum x_t^2 \end{pmatrix} \begin{pmatrix} \sum x_t y_t \\ \sum y_t z_t \end{pmatrix} \\ &= \frac{1}{D} \begin{pmatrix} \sum z_t^2 \sum x_t y_t - \sum x_t z_t \sum y_t z_t \\ \sum x_t^2 \sum y_t z_t - \sum x_t z_t \sum x_t y_t \end{pmatrix}. \end{aligned} \tag{6.5}$$

If (6.1) is the correct model, it follows that $\sum y_t x_t = \alpha \sum x_t^2 + \sum \varepsilon_t x_t$, from which we derive that

$$E\left(\sum y_t x_t\right) = \alpha \sum x_t^2$$

and as $\sum y_t z_t = \alpha \sum x_t z_t + \sum \varepsilon_t z_t$, it also comes

$$E(y_t z_t) = \alpha \sum x_t z_t.$$

Taking the expectations from (6.5) and using the previous expectations, we obtain

$$\begin{aligned} E(\hat{\alpha}) &= (1/D)\left[E\left(\sum z_t^2\right)E\left(\sum x_t y_t\right) - E\left(\sum x_t z_t\right)E\left(\sum y_t z_t\right)\right] \\ &= (1/D)\left[\sum z_t^2 \alpha \sum x_t^2 - \sum x_t z_t \alpha \sum x_t z_t\right] \\ &= \alpha. \end{aligned}$$

Similarly, the expectation of $\hat{\beta}$ is given by

$$E(\hat{\beta}) = (1/D)\left[E\left(\sum x_t^2\right)E\left(\sum y_t z_t\right) - E\left(\sum x_t z_t\right)E\left(\sum x_t y_t\right)\right]$$

$$
\begin{aligned}
&= (1/D)\left[\sum x_t^2\alpha \sum x_t z_t - \sum x_t z_t \alpha \sum x_t^2\right] \\
&= 0.
\end{aligned}
$$

In this case, the application of least squares leads to estimators that are unbiased. It could also be shown that the residuals from the regression would yield unbiased estimators of the variance of the disturbances.

From General to Specific

The main conclusion to be drawn from the above developments is that we need to start with a very general model containing a large number of explanatory variables and time lags. This *General Unconstrained Model* (GUM) is then subjected to a series of reduction operations leading to the final model, which is simple, parsimonious and correctly adjusted to the observed data.

The reduction operations in question are carried out through a series of hypothesis tests applied to each stage of the reduction as well as to the final model.

This methodological approach can be summed up by the following formula, associated with David Hendry's name: from the general to the specific or particular. (GETS stands for *GEneral To Specific methodology*).

It is convenient to represent this methodology in a diagram (see Hendry and Doornik, 2014). To do this, a number of terminological clarifications are essential. The *process* underlying the societal set under study will be referred to as *the Global Data Generating Process* (GDGP). The *Local Data Generating Process* (LDGP) is a subset of the previous one, which generates the data relating to the variables selected in the initial model—known as the GUM model, as we have already explained.

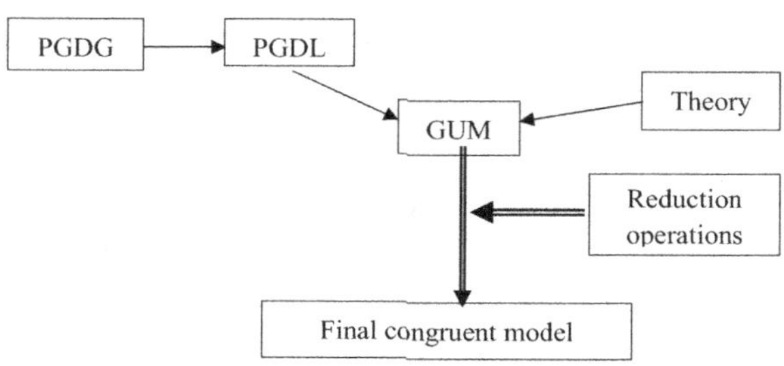

The final model is said to be *congruent with* the data when it is not contradicted by these empirical observations.

6.2 Specification Tests

In our case, the first specification test to be carried out obviously concerns the linear or non-linear nature of the ARDL model obtained after the various reduction operations.

RESET Test

For simplicity, let's assume that the regression model is given by

$$y = X\beta + u,$$

where **X is** a deterministic matrix of format $T \times K$ and rank K. Let us also assume that $u \sim MN(\mathbf{0}, \sigma^2 I_T)$. Let us then assume that the choice of the linear functional form is wrong and that, as a result, we have:

$$E(u|\, X) = \xi \neq \mathbf{0}.$$

The basis of the RESET linearity test—an acronym for "*REgression Specification Error Test*"—is to approximate ξ by the vector $Z\gamma$ by estimating the model

$$y = X\beta + Z\gamma + u, \tag{6.6}$$

where **Z** is a matrix of dimension $T \times P$. To construct this last matrix, we use the T row vectors

$$z'_t = \left((x'_t \hat{\beta})^2, \ldots, (x'_t \hat{\beta})^P\right),$$

where $\hat{\beta}$ is the usual ordinary least squares estimator. The scalar $x'_t \hat{\beta}$ is consequently the t-th element of the vector of fitted values $\hat{y} = X\hat{\beta}$ (In practice, we take $P = 3$, which implies that **Z** is a matrix of dimension $T \times 2$.)

The null hypothesis of linearity then consists of testing $\gamma = \mathbf{0}$ in the model (6.6). Rejecting H_0 leads to the conclusion that the functional form chosen, linear in this case, is inadequate. Acceptance implies that the linearity hypothesis is not invalidated.

Clearly, testing the nullity of γ in (6.6) amounts to testing the nullity of a subset of the coefficients of this model. It follows that a F-test is entirely appropriate.

By simply applying Theorem 4.1 and formula (4.12), the test statistic takes the form

$$\text{RESET} = \frac{(\hat{u}'_C \hat{u}'_C - \hat{u}'_F \hat{u}'_F)/P}{\hat{u}'_F \hat{u}'_F/(T - K - P)} \sim F(P, T - K - P), \tag{6.7}$$

where the index F denotes the estimation residuals of the free, unconstrained model and the index C those of the model constrained by the null hypothesis $\boldsymbol{\gamma} = \mathbf{0}$.

In other words, the RESET statistic is therefore distributed according to a Fisher-Snedecor distribution with P and $T–K–P$ degrees of freedom, P being equal to the number of powers of $(\boldsymbol{x}'_t \hat{\boldsymbol{\beta}})$ introduced into the augmented regression.

As usual, the null hypothesis (of linearity in this case) will be rejected if the p-value associated with the statistic (6.7) is, for example, less than a threshold of five per cent.

The Chow Test

Even when the estimated relationship has passed a linearity test, there may still be a break in the relationship at some point. Testing for the existence of such a break is possible using the Chow test.

Let's assume that a break in time occurs at a known date. Let's split the T observations into two parts, the first containing T_1 observations and the second the remaining $T_2 = T - T_1$ observations.

The null hypothesis of the absence of a break, as opposed to the alternative constituted by the presence of a structural break, is then

$$H_0 : \boldsymbol{\beta}_1 = \boldsymbol{\beta}_2 \text{ versus } H_0 : \boldsymbol{\beta}_1 \neq \boldsymbol{\beta}_2,$$

in the

$$\begin{pmatrix} \boldsymbol{y}_1 \\ \boldsymbol{y}_2 \end{pmatrix} = \begin{pmatrix} \boldsymbol{X}_1 & \mathbf{0} \\ \mathbf{0} & \boldsymbol{X}_2 \end{pmatrix} \begin{pmatrix} \boldsymbol{\beta}_1 \\ \boldsymbol{\beta}_2 \end{pmatrix} + \begin{pmatrix} \boldsymbol{u}_1 \\ \boldsymbol{u}_2 \end{pmatrix}, \tag{6.8}$$

where $\boldsymbol{y}_1$ and $\boldsymbol{y}_2$ are respectively the $(T_1 \times 1)$ and $(T_2 \times 1)$ vectors of observations on the dependent variable and $\boldsymbol{X}_1, \boldsymbol{X}_2$ respectively the $(T_1 \times K)$ and $(T_2 \times K)$ matrices of the explanatory variables.

Of course, the model (6.8) is assumed to satisfy all the usual assumptions (on the perturbations **u** and the matrices **X**), which are necessary to apply least squares.

In this situation, as with the RESET test developed in the previous section, an F-test is appropriate, so the Chow test statistic is written as follows:

$$\text{CH}=\frac{(\hat{\boldsymbol{u}}'_C\hat{\boldsymbol{u}}_C - \hat{\boldsymbol{u}}'_{F_1}\hat{\boldsymbol{u}}_{F_1} - \hat{\boldsymbol{u}}'_{F_2}\hat{\boldsymbol{u}}_{F_2})/K}{(\hat{\boldsymbol{u}}'_{F_1}\hat{\boldsymbol{u}}_{F_1} + \hat{\boldsymbol{u}}'_{F_2}\hat{\boldsymbol{u}}_{F_2})/(T_1 + T_2 - 2K)} \sim F(K, T_1 + T_2 - 2K),$$

where $\hat{\boldsymbol{u}}'_C\hat{\boldsymbol{u}}_C$ is the sum of the squares of the residuals from the model constrained by the null hypothesis. For their part, $\hat{\boldsymbol{u}}'_{F_1}\hat{\boldsymbol{u}}_{F_1}$ and $\hat{\boldsymbol{u}}'_{F_2}\hat{\boldsymbol{u}}_{F_2}$ represent respectively the sums of the squares of the free residuals obtained by running the two regressions shown in (6.8).

A word of warning, however: under the alternative hypothesis, both regressions require the restrictions

$$T_1 \geq K \text{ and } T_2 \geq K,$$

i.e. the number of observations in each of the two sub-samples must be at least as large as the number of parameters to be estimated.

The Quandt Test

The Chow test assumes that the date on which a structural break occurs is known to the econometrician. Sometimes, however, such a date is not identifiable or is identifiable only within a time interval and not at a precise moment in time.

Let's assume that a break occurs between the dates τ_1 and τ_2. The Chow test must then be modified to test for potential breaks occurring on all possible dates τ within the interval $[\tau_1, \tau_2]$.

Modified in this way, the test is obtained by maximum likelihood and is due to Quandt. Hence the name: *Quandt Likelihood Ratio* Test, abbreviated to QLR test.

The QLR statistic is then the largest of the F statistics, i.e.

$$\text{QLR} = \max[F(\tau_1), F(\tau_1 + 1), \ldots, F(\tau_2)]. \tag{6.9}$$

However, the distribution of (6.9) is different from that of the individual F statistics that make it up. In fact, this distribution depends essentially on two factors:

- the number of restrictions to be introduced into the test, in other words, the number of coefficients (including the constant) likely to break under the alternative hypothesis;
- τ_1/T and τ_2/T, ratios, i.e. τ_1 and τ_2 limits expressed as fractions of the total size of the sample under consideration.

On a practical level, to approximate the distribution of the QLR statistic in large samples, the points τ_1 and τ_2 cannot be too close to the beginning or end of the sample.

From this point of view, the most common choice is to use a ***trimming*** of 15%, in other words, to put $\tau_1 = 0.15T$ and $\tau_2 = 0.85T$, rounded to the nearest integer.

Here is a table of the critical values of the distribution of the QLR statistic calculated with a "cut-off" of 15% (Table 6.1).

The interpretation of these critical values is straightforward: if the p-value associated with the QLR statistic is less than 5% or 1%, taking into account the number of restrictions introduced, then we reject the null hypothesis of no structural break and conclude that such a break exists.

Table 6.1 Critical values (15% cut-off)

Number of restrictions	5%	1%
1	8.68	12.16
2	5.86	7.78
3	4.71	6.02
4	4.09	5.12
5	3.66	4.53
6	3.37	4.12
7	3.15	3.82
8	2.98	3.57

6.3 Stability Tests

These tests are the work of the trio Brown, Durbin, and Evans (1975). They are based on the use of a technique known as recursive least squares. There are two main varieties: *Cumulative Sums* of residuals of recursive least squares (CUSUM) and *Cumulative Sums* of *Squares* of residuals of recursive least squares (CUSUMSQ).

Recursive Least Squares

The recursive residuals of a regression are obtained using a procedure that can be summarised as follows:

1. Select the first $t > K$ observations in the sample and then calculate the estimated vector

$$\hat{\boldsymbol{\beta}}_t = (\boldsymbol{X}'_t\boldsymbol{X}_t)^{-1}\boldsymbol{X}'_t\boldsymbol{y}_t,$$

   where $\boldsymbol{X}_t$ denotes the matrix of dimension $t \times K$ of the t observations on the K variables determined and where the column vector $\boldsymbol{y}_t$ is given by $\boldsymbol{y}_t = (y_1, \ldots, y_t)$. The recursive residuals are then the standardised one-step forecast residuals such that

$$w_{t+1} = \frac{(y_{t+1} - \boldsymbol{x}'_{t+1}\boldsymbol{\beta}_t)}{\sqrt{1 + \boldsymbol{x}'_{t+1}(\boldsymbol{X}'_t\boldsymbol{X}_t)^{-1}\boldsymbol{x}_{t+1}}}. \tag{6.10}$$

2. Add the $(t + 1)$-th observation to the data set to obtain

$$\hat{\boldsymbol{\beta}}_{t+1} = (\boldsymbol{X}'_{t+1}\boldsymbol{X}_{t+1})^{-1}\boldsymbol{X}'_{t+1}\boldsymbol{y}_{t+1}.$$

   Calculate w_{t+2}.
3. Repeat step 2, adding one observation for each date.

In total, we have the sequence of recursive residuals $(w_{K+1}, \ldots, w_T)$. Brown et al. (1975) have also shown that, calibrated in this way, these residuals are such that

$$\boldsymbol{w} \sim MN(\boldsymbol{0}, \sigma^2 \boldsymbol{I}_{T-K}).$$

The CUSUM Test

As already noted, the recursive residuals constructed in this way will be used to test the hypothesis of the existence of structural change over time.

The null hypothesis is then

$$H_0 : \begin{cases} \boldsymbol{\beta}_t = \boldsymbol{\beta}_{t+1} = \cdots = \boldsymbol{\beta}_T \\ \sigma_t^2 = \sigma_{t+1}^2 = \cdots = \sigma_T^2, \end{cases}$$

where $\boldsymbol{\beta}_t$ is the vector of coefficients to be estimated at date t, and σ_t^2 is the variance of the perturbations at that same date.

The CUSUM test is based on the statistic

$$W_r = \frac{1}{s_w} \sum_{t=K+1}^{r} w_t, \qquad r = K+1, \ldots, T, \tag{6.11}$$

where

$$s_w^2 = \frac{1}{T-K-1} \sum_{t=K+1}^{T} (w_t - \overline{w})^2,$$

with $\overline{w}$ the arithmetic mean of these residuals.

From (6.11), each W_r represents a **cumulative sum** (CUSUM), which can be plotted as a function of r. If the estimated parameters of the model are temporally stable, i.e. if the null hypothesis holds, then $E(W_r) = 0$. However, if there is a break in stability, the sums W_r will tend to deviate from zero, and hence from the abscissa of the graph (Fig. 6.1).

The two lines are symmetrical with respect to $W_r = 0$ and have the equation

$$W = \pm\left[a\sqrt{T-K} + 2a(r-K)/(T-K)^{1/2}\right],$$

where, for example, $a = 0.948$ for a test level $\alpha = 0.05$.

They are constructed in such a way that the probability of crossing them on either side is equal to the chosen significance level. Consequently, the null hypothesis

Fig. 6.1 The CUSUM test

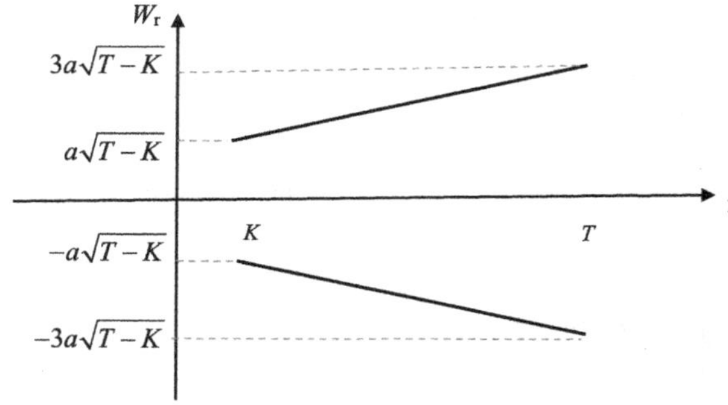

of parameter constancy will be rejected when the sequence of cumulative sums intersects one of these lines.

The CUSUMSQ Test

Brown et al. (1975) also proposed another test based on the cumulative sums of **squares** of recursive residuals (CUSUMSQ). The corresponding statistic is given by

$$W_r^* = \frac{\sum_{t=K+1}^{r} w_t^2}{\sum_{t=K+1}^{T} w_t^2}, \qquad r = K+1, \ldots, T. \tag{6.12}$$

Under the null hypothesis of constant parameters, W_r^* follows a Beta distribution (see Bismans, 2016, p. 271 et seq.), with an expectation equal to

$$E\left(W_r^*\right) = (r-K)/(T-K),$$

which varies from 0 for $r = K$ to 1 for $r = T$.

Here, too, we can plot the evolution of the sums of squares W_r^* as a function of r (Fig. 6.2):

Fig. 6.2 The CUSUMSQ test

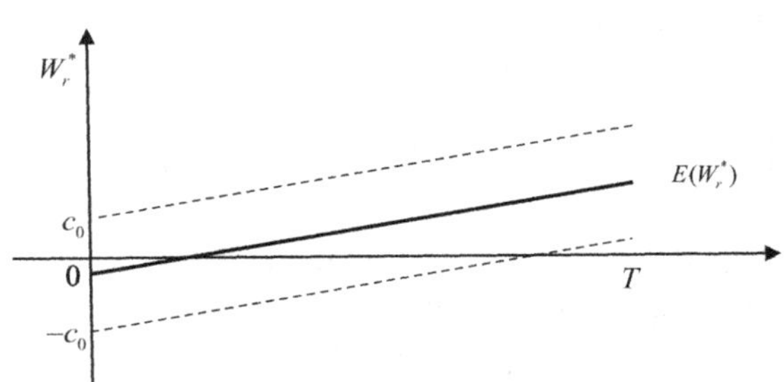

On this graph, the lines parallel to the mean value line are defined by the equation

$$W_r^* = \pm\ c_0 + \frac{r-K}{T-K},$$

where the value of c_0 is determined in such a way that the probability of crossing one of these lines, under the null hypothesis that the parameters are constant, is equal to the chosen significance threshold α. (For example, for $K = 5$ and $T = 55$, if $\alpha = 0.05$, then $c_0 = 0.18677$.)

Finally, the CUSUMSQ test will lead to the rejection of the null hypothesis of stability of estimates, when the sequence of cumulative sums of squares intersects one of these two lines, thus determining a date at which a break occurs in the constancy of the parameters.

A final clarification on these tests is useful from an empirical point of view. While the CUSUM test mainly detects *systematic* deviations from the stability of the estimated parameters, the CUSUMSQ test tends to detect *accidental* deviations. They therefore complement each other well in this respect.

6.4 Co-linearity and Encompassing

We pointed out (see "Model Assumptions and Their Interpretation" section) that the application of least squares was based on the assumption that the matrix $\mathbf{X}$ of the observations was of full rank, in other words, the K columns of X were linearly independent. What happens when this assumption is not verified?

Detecting Multicollinearity

A priori, the presence of some form of linear relationship within a set of explanatory variables is not surprising, insofar as there is necessarily a general interdependence within any economic system. This is all the more true when we consider that macroeconomic variables such as GDP, aggregate consumption, etc., incorporate a common trend.

The problem is more serious when independent variables show a high degree of intercorrelation. From this point of view, let the model

$$y_t = \beta_1 + \beta_2 x_{t2} + \beta_3 x_{t3} + u_t, \qquad t = 1, \ldots, T. \tag{6.13}$$

Let us then assume that the two explanatory variables follow the same two-dimensional distribution with a correlation coefficient of

$$\rho = \frac{\mathrm{Cov}(x_{t2}, x_{t3})}{(\mathrm{Var}(x_{t2})\mathrm{Var}(x_{t3}))^{1/2}}.$$

If ρ is equal to unity, it is impossible to estimate the two parameters β_2 and β_3 separately. In this case, the two variables are linked by the function $x_{t3} = \delta x_{t2}$ and, as a result, Eq. (6.13) becomes

$$y_t = \beta_1 + (\beta_2 + \beta_3\delta)x_{t2} + u_t.$$

Only the estimation of $\beta_2 + \beta_3\delta$ is now possible and, all in all, we are dealing with a case of perfect multicollinearity.

From the point of view of hypothesis testing, multicollinearity leads to a real difficulty when the separate hypotheses $\beta_2 = 0$ and $\beta_3 = 0$ cannot be rejected, while the joint hypothesis $\beta_2 = \beta_3 = 0$, on the other hand, is rejected.

There are several procedures for attempting to detect the presence of (strong) multicollinearity between the explanatory variables of a linear model. The first of these is based on the analysis of the eigenvalues of the matrix $(\boldsymbol{X}'\boldsymbol{X})$. More precisely, we begin by recalibrating the data so that the columns of $\mathbf{X}$ are all of length equal to unity and then we calculate an ***index*** (known as the *condition index*) defined as the square root of the ratio between the largest and smallest eigenvalue of the matrix $(\boldsymbol{X}'\boldsymbol{X})$.

The higher this index, the more problematic it becomes to calculate the inverse matrix $(\boldsymbol{X}'\boldsymbol{X})^{-1}$. A value of 30 is generally considered to indicate the presence of strong multicollinearity between the explanatory data.

Another approach, which is easier to implement, involves calculating the *Variance-Inflation Factor* (*VIF*), which is defined as follows:

$$VIF_k = (1 - R_k^2)^{-1},$$

where $R_k^2, \quad k = 1, \ldots, K$, is the squared coefficient of correlation of the regression of x_{tk} on all the other independent variables included in this regression.

A high value of the *VIF* indicates that R_k^2 is close to unity and that, as a result, the variable x_{tk} is highly correlated with the rest of the explanatory variables. In general, a $VIF_k > 10$ indicates the presence of severe multicollinearity within the observation data.

The question then arises as to how to remedy such a problem. Several remedies have been proposed, such as the use of factor analysis of multivariate statistics or *ridge regression*. However, we are sceptical about these procedures. With Pesaran (2015, p. 71), we believe that the real solution to the problem is "to increase the information on the basis of which the regression is estimated", in other words, to have additional data.

Encompassing Tests

First, a word about terminology. The term "encompassing" conveys the idea of envelopment, *i.e.* the fact that a well-specified model accounts for the main results obtained by rival models.

Consider the two regression models

$$\mathrm{M}_1 : \boldsymbol{y} = \boldsymbol{X\beta}_1 + \boldsymbol{u}_1, \qquad \boldsymbol{u}_1 \sim MN(\boldsymbol{0}, \sigma^2 \boldsymbol{I}_T) \tag{6.14}$$

$$\mathrm{M}_2 : \boldsymbol{y} = \boldsymbol{Z\beta}_2 + \boldsymbol{u}_2, \qquad \boldsymbol{u}_2 \sim MN(\boldsymbol{0}, \omega^2 \boldsymbol{I}_T), \tag{6.15}$$

where $\mathbf{y}$ is the vector $(T \times 1)$ of observations on the dependent variable, $\mathbf{X}$ and $\mathbf{Z}$ are matrices $(T \times K_1)$ and $(T \times K_2)$, $\boldsymbol{\beta}_1$ and $\boldsymbol{\beta}_2$ the column vectors of the regression coefficients of dimensions K_1 and K_2 respectively, and finally $\boldsymbol{u}_1, \boldsymbol{u}_2$ are the vectors of the T-component perturbations.

Models M_1 and M_2 are said to be ***non-nested*** if the regressors of M_1 cannot be expressed as a linear combination of those of M_2 and vice versa.

We will now present two tests that allow us to compare the two models.

1. *Cox's N-test*

   The Cox statistic under the null hypothesis that M_1 encompasses M_2, is given by

$$N_1 = \left[(T/2)\ln(\hat{\omega}^2/\hat{\omega}_*^2)\right]/\hat{V}_1 \sim N(0, 1), \tag{6.16}$$

   with

$$\begin{aligned}
\hat{\omega}^2 &= \hat{\boldsymbol{u}}'_2\hat{\boldsymbol{u}}_2/T,\\
\hat{\omega}_*^2 &= (\hat{\boldsymbol{u}}'_1\hat{\boldsymbol{u}}_1 + \hat{\boldsymbol{\beta}}'_1\boldsymbol{X}'\boldsymbol{M}_2\boldsymbol{X}\hat{\boldsymbol{\beta}}_1)/T,\\
\hat{V}_1^2 &= (\hat{\sigma}^2/\hat{\omega}_*^4)\hat{\boldsymbol{\beta}}'_1\boldsymbol{X}'\boldsymbol{M}_2\boldsymbol{M}_1\boldsymbol{M}_2\boldsymbol{X}\hat{\boldsymbol{\beta}}_1,\\
\hat{\sigma}^2 &= \hat{\boldsymbol{u}}'_2\hat{\boldsymbol{u}}_2/T, \qquad \hat{\boldsymbol{\beta}}_1 = (\boldsymbol{X}'\boldsymbol{X})^{-1}\boldsymbol{X}'\boldsymbol{y}\\
\boldsymbol{M}_1 &= \boldsymbol{I}_T - \boldsymbol{X}(\boldsymbol{X}'\boldsymbol{X})^{-1}\boldsymbol{X}', \qquad \boldsymbol{M}_2 = \boldsymbol{I}_T - \boldsymbol{Z}(\boldsymbol{Z}'\boldsymbol{Z})^{-1}\boldsymbol{Z}'.
\end{aligned}$$

   In a similar way, we can establish the Cox statistic N_2 for the null hypothesis test that M_2 encompasses M_1.

2. *The Mizon-Richard F-test*

   When it comes to testing the M_1 model against the M_2 model, the encompassing test is equivalent to a conventional F-test. This involves calculating the F_1 statistics to test the $H_0 : \boldsymbol{\delta} = \boldsymbol{0}$ hypothesis in the augmented regression.

$$\boldsymbol{y} = \boldsymbol{X}a_0 + \boldsymbol{Z} * \boldsymbol{\delta} + \boldsymbol{u},$$

where $Z*$ denotes the variables of M_2 which cannot be expressed as exact linear combinations of the regressors of M_1. Of course, it is also possible to establish the F_2 statistic to test M_2 against M_1.

6.5 Application: Electricity Consumption

This involves fitting an ARDL-type model to electricity consumption in France using the General to Specific (GETS) methodology as summarised in "From General to Specific" section of this chapter.

First, we need to go into the details of the reduction operations that enable us to prune the initial very general dynamic model—the GUM—so as to obtain a final structure that is both parsimonious and congruent with the observation data.

On the Reduction Process

The starting point of the whole approach is, as we said, to start from a general unconstrained or free model with a large number of variables and time lags, and then to reduce the number of variables according to a number of criteria which we will now develop in full.

In this case, the sequential reduction procedure is based on an examination of the following four main elements:

1. Is the F statistic after the elimination of each set of variables significant?
2. Are the three information criteria (AIC, BIC, and HQC) improved at each stage of the process of reducing the number of variables?
3. At each iteration, do the coefficients of the parameters of the reduced model pass the t-test at a level of at least 10 per cent?
4. Finally, but this criterion is secondary to the previous ones, does the corrected R^2 improve with each successive reduction?

On the basis of the satisfaction of these criteria—essentially, in this case, the first three—we end up with a final model at the end of the various operations to eliminate irrelevant variables. This model must then satisfy all the validation and specification tests detailed above.

The Initial Unrestricted Model

Our aim at this stage is therefore to establish a very general model of overall electricity demand in France—overall because it includes demand from households, public administrations and businesses.

We have already presented—at least for the long term—the determinants of demand used (see "Estimation of a Cointegrating Relationship" section), i.e. an index of the price of electricity, an index of the price of a possible substitute, gas, and finally, real Gross Domestic Product, i.e. excluding price increases. All these series are quarterly and cover the period 1990 Q1–2015 Q3.

Using both Johansen's multidimensional analysis and Pesaran's procedure (see "Pesaran Estimates" section), we found that these four variables were all integrated of order 1 and that there was only one cointegrating relationship.

To complete the stationarity analysis, here is a table showing the results of the various unit root tests applied to these variables. (The letter L means that we have taken the logarithm of the variables in question; moreover, these variables are all expressed in levels; finally, the three central columns give the values of the corresponding test statistics) (Table 6.2).

From this table, it is clear that all the tests lead to one and the same conclusion: our four variables are integrated into order one, which confirms the results obtained in the previous chapter.

The autoregressive distributed lag (ARDL) model that we are going to fit to these data must include not only the four variables expressed in levels but also their equivalent in first differences.

In addition, *in the short term,* two other variables are likely to influence aggregate electricity consumption:

- Temperature is certainly one of the factors to be taken into account; logically, the lower the temperature, the more electricity will be consumed, at least for households.
- From a macroeconomic point of view, cyclical fluctuations in the economy are bound to affect electricity consumption, so we need to take into account the occurrence of recessions to complete the modelling.

We will therefore introduce these additional determinants in the form of two binary variables. One, denoted REC, has a value of 1 during periods of recession and 0 during phases of expansion in the French economy.

Table 6.2 Unit root test results

Variables	ADF test	PP test	ERS test	Conclusion
LCONSE	3.2 (1)	2.9 (1)	0.1 (3)	I(1)
LIPCE	2.2 (3)	1.1 (3)	−1.6 (3)	I(1)
LIPCG	2.1 (3)	3.4 (3)	10.8 (1)	I(1)
LPIBr	3.5 (1)	−1.6 (2)	0.9 (2)	I(1)

The number in brackets at the back of the test statistics varies from one to three: (1) indicates a test with no constant or trend; (2) a test with constant but no trend, and (3) a test with constant and trend.

The other, known as TEMP, is based on records of the average temperature in Paris during the months of February, May, August and November each year, and systematically takes the value 1 when the average temperature is below 8 °C. Obviously, the value 0 prevails in the other 2 months of the year. The aim of this specific binary variable is to see whether there is a "winter effect" on overall electricity demand.

Since the GETS methodology recommends that we start from a dynamic model with many lags, we will take into account three lags, which, in first differences, implies five quarters in the calculation. On the other hand, for the binary variable REC, a single lag will suffice to account for the recessions.

As we know, the initial ARDL model includes both a long-term relationship and an error correction mechanism and will therefore be presented as follows, with the "L" for logarithm omitted:

$$\begin{aligned} \Delta CONSE_t = & \sum_{i=1}^{3} \alpha_i \, \Delta CONSE_{t-i} + \sum_{j=0}^{3} \beta_{1j} \, \Delta PIB_{t-j} + \sum_{j=0}^{3} \beta_{2j} \, \Delta IPE_{t-j} \\ & + \sum_{j=0}^{3} \beta_{3j} \, \Delta IPG_{t-j} + \delta \, TEMP + \sum_{k=0}^{1} \tau_k \, REC_{t-k} \\ & + \left[\gamma_0 CONSE_{t-1} - \gamma_1 - \gamma_2 PIBr_{t-1} - \gamma_3 IPE_{t-1} - \gamma_4 IPG_{t-1} \right] + u_t . \end{aligned} \tag{6.17}$$

Note that the expression in square brackets represents the long-term equilibrium relationship and that the term u_t is assumed to satisfy the usual properties.

Estimation of Eq. (6.17) by OLS yields the results shown in Table 6.3, where the letter **L** denotes the logarithm and the letter **d** the first difference. In addition, the lags taken into account are indicated by the number next to the variable.

What is particularly striking in this table is the lack of statistical significance of the short-term variables, with the exception of the recession indicator lagged by one or two quarters.

In any case, the following table summarises the main features of this initial model (Table 6.4).

The F-test, whose corresponding p-value is less than one per cent, assures us of the overall significance of the estimated parameters. As for the corrected R^2, i.e. the correlation coefficient corrected by the number of variables, it is equal to 0.44.

Table 6.3 The initial model

	Coefficient	p-value
Constant	3.530	< 0.01
LCONSE(−1)	−0.573	< 0.01
LIPE(−1)	−0.127	< 0.01
LIPG(−1)	−0.070	0.058
GDPr(−1)	0.316	< 0.01
dLIPE	−0.092	0.472
dLIPE(−1)	−0.108	0.413
dLIPE(−2)	0.030	0.838
dLIPE(−3)	−0.114	0.418
dLIPG	−0.052	0.316
dLIPG(−1)	−0.035	0.560
dLIPG(−2)	0.045	0.362
dLIPG(−3)	−0.085	0.203
dLGDP	−0.016	0.960
dLGDP(−1)	−0.031	0.932
dLGDP(−2)	−0.203	0.457
dLGDP(−3)	−0.201	0.556
Temp	0.004	0.104
Temp(−1)	−0.002	0.405
REC	0.006	0.529
REC(−1)	−0.016	0.066
REC(−2)	−0.011	0.011
REC(−3)	0.001	0.830
dLCONSE(−1)	−0.320	0.021
dLCONSE(−2)	−0.266	0.016
dLCONSE(−3)	−0.046	0.651

Table 6.4 Some statistical characteristics

F(25, 73)	AIC	BIC	HQC	R^2
20.71 (< 0.01)	−581.22	−513.75	−553.92	0.444

The First Stage

We now need to reduce the model (6.17). We will do this in four successive stages. Other paths are possible. The first is to test all the restrictions

$$dLIPCE(-3) = dLIPCG(-3) = dLPIB(-3) = dLCONSE(-3) = REC(-3) = 0,$$

while estimating the corresponding model.

Table 6.5 provides information for assessing whether the reduction is statistically and economically justified.

Comparing this last table with the previous one, we can see that we end up with a new model that outperforms the old one in every respect. Thus, the F-test leads us to accept the non-nullity of the parameters of this new model resulting from reduction n°1. Next, the three information criteria are improved, while the coefficient R^2 is also improved.

Here are the parameter estimators obtained after the first reduction step, with the p-values in the third column (Table 6.6).

The table shows that the constant and the four cointegrated variables are fully significant, as are the non-lagged temperature and the recession indicator with one

Table 6.5 Reduction (stage 1)

F(20, 78)	AIC	BIC	HQC	R^2
25.69 (< 0.01)	−586.26	−531.76	−564.21	0.453

Table 6.6 The model at reduction stage 1

	Coefficient	p-value
Constant	3.507	< 0.01
LCONSE(−1)	−0.605	< 0.01
LIPE(−1)	−0.118	< 0.01
LIPG(−1)	−0.091	< 0.01
GDPr(−1)	0.351	< 0.01
dLIPE	−0.057	0.468
dLIPE(−1)	−0.126	0.323
dLIPE(−2)	0.018	0.90
dLIPG	−0.059	0.223
dLIPG(−1)	−0.043	0.445
dLIPG(−2)	0.027	0.503
dLGDP	0.053	0.860
dLGDP(−1)	−0.123	0.742
dLGDP(−2)	−0.387	0.126
Temp	0.005	0.043
Temp(−1)	−0.002	0.539
REC	0.007	0.447
REC(−1)	−0.017	0.047
REC(−2)	−0.009	0.015
dLCONSE(−1)	−0.283	0.026
dLCONSE(−2)	−0.241	0.020

or two lags. All the other regression coefficients resulting from the use of ordinary least squares are not significantly different from zero.

Reducing Again and Again

We are now going to carry out further reductions, of which there are three in this case: stages 2, 3 and 4, which will lead to the final model. These final stages will successively test the three sets of restrictions.

- $dLIPC(-1) = dLIPC(-2) = dLIPG(-1) = dLIPG(-2) = dLPIB(-1) = dLPIB(-2) = 0$
- $dLIPE = dLIPG = dLPIB = 0$
- $Temp(-1) = REC = 0$.

Here is a table showing how the main statistics behave at the end of each stage of the reduction procedure (Table 6.7).

On reading this table, the first thing to notice is that at each stage of the reduction, the p-value associated with the F statistic is less than one per cent, so the hypothesis of joint nullity of the estimated parameters is systematically rejected.

Secondly, there was an improvement in each of the three information criteria, which justifies the reductions made in stages 2 to 4. Finally, even if only slightly, the coefficient of determination improves at each stage.

All in all, the sequential reduction procedure for the initial model fully satisfies the selected criteria.

Let's now look at the final model, which has been estimated using ordinary least squares.

All the estimated parameters are significant at at least one per cent, with the exception of $REC(-2)$, which is significant at 4 per cent, and especially temperature, which is significant at 7 per cent. (Obviously, if all the coefficients were required to be significant at the five per cent level, the Temp variable would be eliminated from the model.)

We will defer a more detailed examination of the specification and interpretation of the final model to the next section, while giving a few details on the automation of sequential reduction.

Table 6.7 Reduction stage statistics

Steps	Stat F	AIC	BIC	HQC	R^2
2	32.4 (< 0.01)	−594.6	−555.7	−578.9	0.47
3	41.1 (< 0.01)	−599.0	−567.9	−586.4	0.48
4	43.2 (< 0.01)	−602.2	−576.2	−591.7	0.49

Note

The four-step procedure just described can be fully automated. In other words, the computer programme is responsible for producing the final result.

Thus, using GRETL, but without going into the programming details, the software delivers the estimates shown in Table 6.8 by proceeding as follows:

Table 6.8 The final model

	Coefficient	Stat *t*	*p*-value
Constant	3.24	7.3	< 0.01
LCONSE(−1)	−0.545	−6.63	< 0.01
LIPE(−1)	−0.111	−4.67	< 0.01
LIPG(−1)	−0.078	−3.54	< 0.01
LPIB(−1)	0.310	5.52	< 0.01
Temp	0.004	1.83	0.07
REC(−1)	−0.012	−3.96	< 0.01
REC(−2)	−0.007	−2.04	0.04
dLCONSE(−1)	−0.326	−3.38	< 0.01
dLCONSE(−2)	−0.284	−3.49	< 0.01

- At each stage, the variable with the highest *p*-value (and therefore not significantly different from zero) is eliminated, so that there are as many stages in the reduction as there are superfluous variables;
- the sequential procedure continues until all the variables have a *p*-value below a certain ceiling (by default, equal to 10% for GRETL);
- In our case, as the Temp variable has a *p*-value of 7 per cent, we will obviously choose an elimination ceiling with a value of 8 per cent.

6.6 Application: Testing and Interpreting the Final Model

We begin by presenting the various validation and specification tests for the final model of electricity consumption demand, after which we analyse and interpret in detail the econometric equation making up the ARDL model, i.e. its two components: the cointegration relationship and the error correction mechanism.

An Initial Series of Tests

The first step is to check that the assumptions underlying the linear model are respected in the case under consideration. As a reminder, the successive error terms

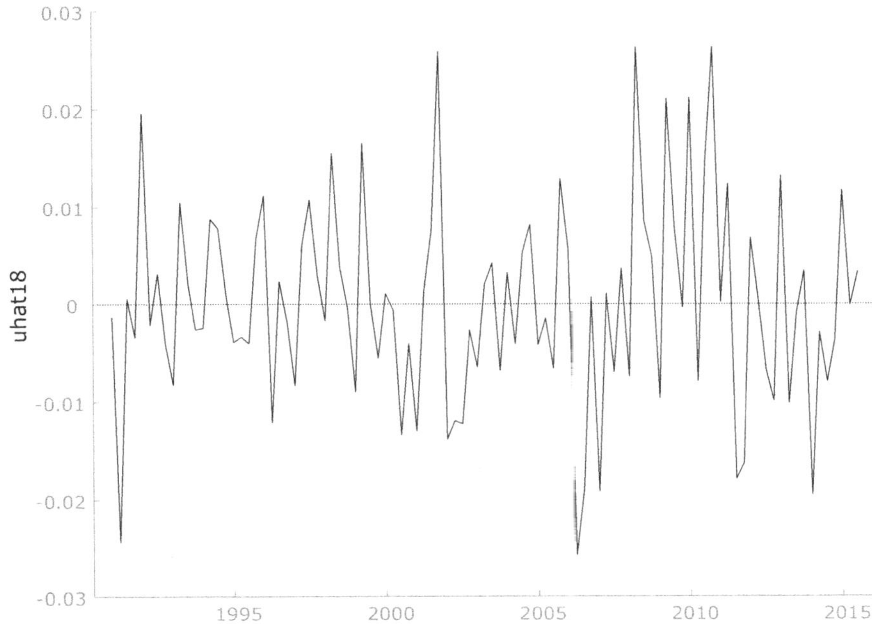

Fig. 6.3 Residuals of the ARDL model of electricity demand

in the ARDL equation to be estimated were assumed to be identically normally distributed with zero mean and constant variance.

We therefore need to examine whether the estimated residuals satisfy the following three conditions: absence of autocorrelation, homoscedasticity and normality.

The figure below provides a graphical representation of the residuals for the period from 1er quarter 1991 to 3ème quarter 2015 (Fig. 6.3).

Examination of this graph does not reveal the presence of autocorrelation in the residuals. At most, we would expect their variance not to be constant over time. However, it is important to be more precise and to use the following tests.

Turning first to the possible autocorrelation of these residuals, here are the statistics for the two tests described in "Breusch–Pagan Test" section, together with the associated p-values (Table 6.9).

Both tests lead to the same result, i.e. acceptance of the null hypothesis of no autocorrelation in the residuals. Examination of the corresponding autocorrelation function (not reproduced here) would lead to a similar conclusion.

Table 6.9 Residual autocorrelation tests (4 lags)

	Statistics	p-value
ML	0.050263	$P(\chi^2(4) > 0.050263) = 1$
MLF	0.010794	$P(F(4, 85) > 0.010794) = 1$

Table 6.10 Homoskedasticity and normality tests

	Statistics	*p*-value
ARCH	8.05399*	$P(\chi^2(4) > 8.05399) = 0.0837$
Bera-Jarque	$\chi^2(2) = 1,799$	0.40679

*4 time shifts

Now what about the behaviour of the variance of these residuals? The most appropriate test for this is ARCH, described in "ARCH Test" section. The Bera-Jarque test (see "Testing Normality" section) is used to assess their normality.

Table 6.10 uses these two tests and provides the corresponding statistics and *p*-values.

Given the *p*-values above the 5 per cent significance level, we accept the null hypothesis of no ARCH effects in the residuals, as well as their normality.

Specification Tests

Is the linear or more precisely log-linear nature of the final ARDL model justified? This question is answered by the RESET test developed in "RESET Test" section.

The test statistic is equal to

$$RESET = 0,7796,$$

while the associated *p*-value is

$$P(F(2,\ \ 87) > 0.7796) = 0.462.$$

This leads us to accept the null hypothesis that the ARDL equation is linear.

We now turn to the question of the stability of the estimated parameters. We have seen—"The CUSUM Test"–"Co-linearity and Encompassing" sections—that two tests based on sums of recursive least squares are available: the CUSUM and the CUSUM of squares (CUSUMSQ). Both are designed to test the existence or otherwise of structural change over time in the estimators.

Here is a graphical representation of the first of these, with a confidence band of 95 per cent (Fig. 6.4).

Since the null hypothesis is that of the temporal stability of the estimated parameters, the fact that the cumulative sums never cross the confidence band lines leads us to accept this temporal stability hypothesis.

Let's look at the second of these tests, the CUSUMSQ. Here too, a graphical representation of the cumulative sums is convenient (Fig. 6.5).

Examination of this graph confirms the existence of stability in the estimated parameters.

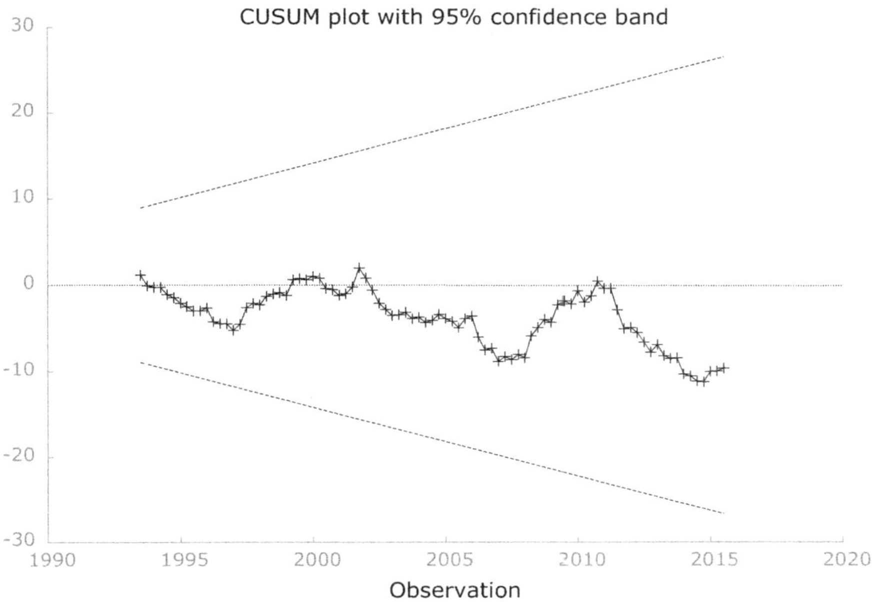

Fig. 6.4 Implementing the CUSUM test

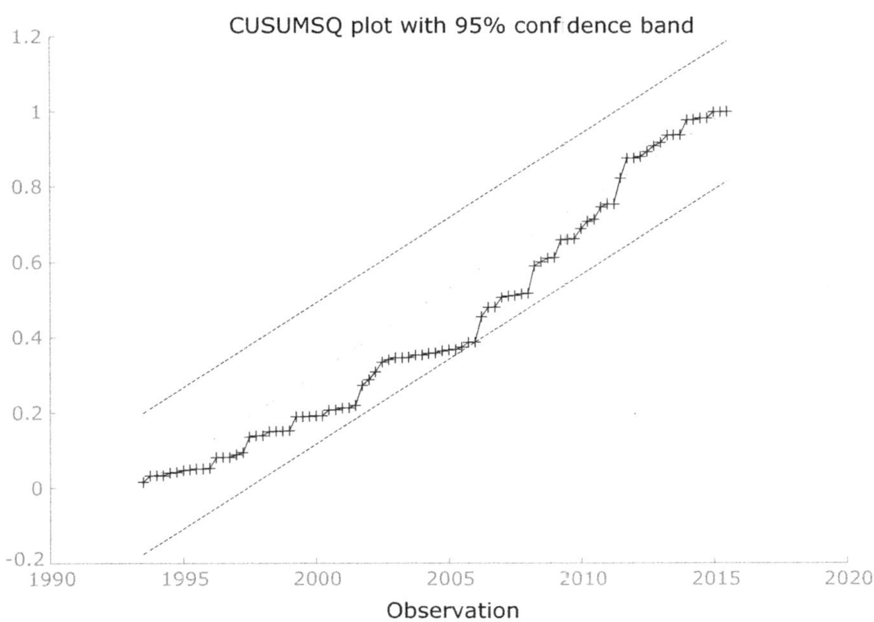

Fig. 6.5 Implementing the CUSUMSQ test

On the Cointegration Relationship

We first extract the cointegration relation from the ARDL equation, which leads, given that it is a static equilibrium relation, to the equation

$$3.24 - 0.545\ LCONSE - 0.111\ LIPE - 0.078\ LIPG \\ +0.31\ LPIB = 0,$$

equation which, after rewriting, gives the actual cointegrating relationship, i.e.

$$LCONSE = 5.945 - 0.204\ LIPE - 0.143\ LIPG + 0.569\ LPIB.$$

As all these variables are expressed in logarithms, the coefficients of the three explanatory variables represent elasticities, i.e. rates of relative change.

Three points need to be made about the interpretation of this static relationship:

1. The income elasticity of electricity demand is positive and equal to 0.569, which means that a 1% increase in real GDP *ultimately* translates into an increase in demand of almost 0.57%;
2. The price elasticity of overall electricity consumption is −0.111; in other words, if the electricity price index rises by 1%, its consumption will fall by around 0.11%;
3. The elasticity of demand for electricity in relation to the price of gas is also negative; in other words, a 1% increase in the price of gas leads to a 0.078% drop in electricity consumption; it follows that gas and electricity are complementary goods and that an increase in their price has a negative effect on electricity consumption. This result may seem surprising, since a priori we would expect these two goods to be substitutable; however, it should be noted that this result applies to consumption taken as a whole and that, moreover, it is certain that the price of electricity and that of gas evolve in parallel on a purely institutional level.

As we have seen ("Non-stationarity and Cointegration" chapter, "Estimation of a Cointegrating Relationship" section), other methods exist for estimating a cointegrating relationship, in particular those of Johansen and Pesaran. Table 6.11 compares our own results with those obtained by the two aforementioned authors.

Table 6.11 Comparison of three methods

	ARDL	PESARAN	JOHANSEN
Cste	5.945	5.049	7.109
LIPE	−0.204	−0.143	−0.24
LIPG	−0.143	−0.194	0.0115
LPIB	0.569	0.635	0.429

Apart from the constant, which is not of immediate interest if we want to compare the three methods, our own estimates are of the same sign as those obtained by Pesaran. Moreover, in both cases, all the estimated parameters are significant at less than one per cent.

This is not the case for the Johansen method: in the latter case, aggregate electricity demand is positively related to the gas price index, whereas in all other cases, it is negatively related. It should also be added that the coefficient in question is not even significantly different from zero.

These very different results are attributable to the fact that Johansen's methodology is based on the estimation of a VAR-type model—several equations—so that the parameters estimated in one equation are affected by the specifications adopted or obtained in the other equations of the system.

If we take a closer look at the comparative results, we can make at least two observations.

- The estimates made using our ARDL model are, in the end, fairly close to those of Pesaran, except that his method gives a lower elasticity of electricity consumption in relation to its own price than ours; on the other hand, the opposite is observed for the other two cointegrated variables, since the elasticities in relation to GDP or the price of gas are higher (in absolute terms, that is) than the values obtained from the estimation of our ARDL model;
- On the other hand, our own estimators differ substantially from those resulting from the application of Johansen's method; this is obvious for the price elasticity of electricity consumption in relation to gas, which is of the opposite sign to ours, but which is above all extremely low, i.e. of the order of one per cent compared with 14 per cent in our case; the other two elasticities—in relation to the price of electricity and real GDP—also diverge significantly depending on the approach used.

Short-Term Determinants

To facilitate the analysis, we will start by reproducing the estimates specific to the error correction mechanism (Table 6.12).

Table 6.12 Estimation of the ECM model

Variables	Coefficients
dLCONSE (−1)	−0.126
dLCONSE (−2)	−0.284
Temp	0.004
REC (−1)	−0.012
REC (−2)	−0.007

First, it should be noted that none of the variables in the cointegration relationship is a short-term determinant of electricity consumption. In particular, variations in the price of electricity do not influence its consumption in the short term. The same is true for changes in real GDP.

On the other hand, the occurrence of a recession, with a lag of one or two quarters, has a negative effect on electricity demand. The effects are not very significant quantitatively, but it is certain—the significance level of the REC (−1) variable is less than one per cent and that of REC (−2) less than five per cent. Moreover, these two effects are cumulative, which tends to show the negative nature of cyclical fluctuations on the dependent variable.

Such a result may come as a surprise. But if you think about it, there is nothing exceptional about it: variations in GDP do not describe the economic cycle as such—which everyone, following Burns and Mitchell, now describes as classic—but at most a possible growth cycle. On the other hand, the REC indicator does record the real movement of the economy. Hence its importance.

However, it is the variations in consumption itself, with a lag of one or two quarters, that are the main determinants. The corresponding coefficients have a negative sign, which means that a positive variation in electricity demand over two successive quarters results in a reduction in that demand. An endogenous self-correcting mechanism, as it were.

It remains to say a word about another short-term determinant of consumption, which is also significant at less than 8%: the dummy variable Temp. Its coefficient has a positive sign, which is quite logical given that harsh winter quarters result in higher electricity consumption.

6.7 Exercises

6.7.1 Let's suppose that the equation

$$y_t = \alpha x_t + \varepsilon_t, \qquad t = 1, \ldots, T$$

represents the "true" model under study, but that we estimate the equation

$$y_t = \alpha x_t + \beta z_t + v_t, \qquad v_t \sim IND(0, \sigma_v^2).$$

Demonstrate that the residuals resulting from the estimation by OLS of this last equation lead to an unbiased estimator of the variance σ_ε^2.

6.7.2 Using your own words, explain what the "general to specific" (GETS) methodology consists of and detail the conditions that dictate the reduction operations of the initial general model (GUM).

6.7.3 The file pandemic.gdt (already used in Chapter 5) contains three series, namely the number of hospitalisations due to COVID (denoted *HOP*), the total number of people fully vaccinated (i.e. vaccinated with two doses) against the pandemic, this variable being denoted *VAC*, and finally, the number of contaminations denoted *CONT*. These data, which concern Belgium exclusively, are daily (7 days a week) and cover the period from 1 February 2021 to 20 September of the same year.

You are required by using GETS methodology to estimate the following ARDL model after checking that all the variables are integrated of order 1

$$\Delta LHOP_t = \sum_{i=1}^{7} \alpha_i \Delta LHOP_{t-i} + \sum_{j=0}^{7} \beta_{1j} \Delta LVAC_{1,t-j} + \sum_{j=0}^{7} \beta_{2j} \Delta LCONT_{2,t-j} + \left[\gamma_0 LHOP_{t-1} - a - \gamma_1 LVAC_{1,t-1} - \gamma_2 LCONT_{2,t-1}\right] + \varepsilon_t,$$

where L denotes the logarithm base e and Δ the first difference of the considered variable.

1. reduce this GUM to obtain a final model, parsimonious and congruent with respect for the data.
2. test the autocorrelation, homoskedasticity, linearity, stability and specification in the final model.
3. determine the cointegration relation between hospitalisations and the two other variables.
4. interpret this last relation.

6.7.4 In "Stability Tests" section, we develop the Quanct Likelihood Ratio test (QLR). Estimation of a model ARDL yields the following QLR statistic with a trimming of 15%:

$$QLR = 97.6729 \text{ at } 2021 - 08 - 17$$

The associated p-value is lower than 0.00001.
Interpret and explain these results.

6.7.5 Let's now look at the pandemic2.docx file, in which the new variable "intensive care hospital admissions", denoted by INT, has been added and which, like the other three variables already presented, now covers the period from 1 February to 31 December 2021.

After estimating two ARDL models, we obtained the two cointegration relationships:

$$\begin{cases} LHOP = -2.231 - 0.306\ LVAC + 0.882\ LCONT \\ LINT = 5.732 - 0.390\ LVAC + 0.780\ LCONT. \end{cases}$$

1. Interpret each of these relationships.
2. Compare the results obtained in this way.

References

Bismans, F. (2016). *Probabilités et statistique inférentielle. Prélude à l'économétrie*. Ellipses.

Hendry, D. F., & Doornik, J. A. (2014). *Empirical model discovery and theory evaluation. Automatic selection methods in econometrics*. The MIT Press.

Pesaran, M. H. (2015). *Time series and panel data econometrics*. Oxford University Press.

Chapter 7
On Vector Autoregressions

The dynamic linear models developed in the previous chapter are univariate and mainly useful for forecasting. In this chapter, we highlight multivariate models and dynamic systems of the VAR (Vector Autoregressive) type. We discuss their formalisation and estimation, and show how such models can be used to evaluate dynamic responses to random shocks by calculating impulse response functions. In addition to forecasting techniques, these methods also provide a framework for the empirical evaluation of economic policies and the analysis of causality .

We will complement VAR modelling with an introduction to alternative methods known as Local Projections (LP), which have been increasingly used in recent years to measure the dynamic effect of economic policy variables.

7.1 The VAR Model

To understand the fundamentals of VAR models, we first need to look at their origins and philosophy.

Supplementary Information The online version contains supplementary material available at https://doi.org/10.1007/978-3-031-72910-2_7.

F. J. Bismans and O. Damette, *Dynamic Econometrics*,
https://doi.org/10.1007/978-3-031-72910-2_7

A Brief History of VAR Models

The development of macroeconometrics, particularly dynamic macroeconometrics, has paralleled the evolution of the measurement of economic fluctuations. The interwar period saw the emergence of "impulse-propagation" models aimed at determining the origins of the cycle [e.g. those of Slutsky (1927) and Frisch (1933)]. The search was then on for the sources of random fluctuations, be they technological shocks, on which the developers of real business cycles (*RBC*) focused, monetary shocks, banking shocks, agricultural shocks and so on.

As we showed in the preface to this book, until the 1970s, and particularly in the wake of the work of the *Cowles Foundation*, dynamic macroeconometric models were very large. They included numerous structural equations, each of which described the impact of one or more exogenous variables **X** on an endogenous variable **y**.

More precisely, these were generally systems of simultaneous equations, in which each equation linked endogenous variables to a set of so-called **predetermined** variables—i.e. literally, "determined beforehand"—the latter made up of exogenous variables and delayed endogenous variables.

The choice of variables considered to be predetermined and therefore truly exogenous is a matter of subjectivity. Based on this observation, Christopher Sims (1980) made several criticisms of this form of modelling. Firstly, the selection of the explanatory variables used in each of the structural equations of a macroeconomic model is not, in general, deduced from considerations of the overall theoretical structure of the system of equations, but is instead the result of considerations contingent on each equation examined individually. Such an approach then leads to the introduction of numerous a priori restrictions on the structural form of the general model, even though the theory would suggest very few. It follows—and this is Sims' second criticism—that other structural forms, based on other sets of restrictions, can give equally good simulations and forecasts.

Finally, the distinction between exogenous variables (which are supposed to incorporate economic policy variables such as interest rates, taxes, etc.) and endogenous variables (such as the money supply or GDP) is largely artificial and is never subject to causality tests to assess the effect of a variable **x** on a variable **y**.

In the end, based on his critique of the large-scale structural models developed in the 1940s–1970s, Sims provided a standard empirical framework, the VAR (Vector Autoregressive) model.

In other words, this analytical framework makes it possible to analyse the reactions of the economy—or of some of its variables—to exogenous or natural shocks. Local projection (LP) methods will be developed by Jorda later, i.e. from 2005 onwards, and will provide a new perspective on how to analyse dynamic responses to shocks.

Formalisation of VAR Models in Reduced Form

As we saw earlier, the dynamics of a time series y_t can be represented by a simple additive model of the form $y_t = \mu_t + x_t$, where the first term of the right-hand side refers to a deterministic process and the second to a stochastic process. Generally, the first term is simply a constant and the model then becomes $y_t = \mu_0 + x_t$ with μ_0 the constant.

The x_t process can be represented by an AR process, for example, where the dynamics of the variable of interest y_t is a function of a combination of its own *lags*.

VAR modelling is a natural extension of the class of univariate **autoregressive** (AR) models. Now, it is no longer a question of modelling the dynamics of a single time series over time, as in the previous example (y_t), but that of a vector made up of a set of several series ($\mathbf{y}$).

A VAR(p) model, i.e. vector autoregressions of order p, can be used to model the dynamic interactions between several variables. To this extent, it is referred to as a system of dynamic equations. This type of representation is very useful in macroeconomics, for example in the study of economic policy. The most commonly used application concerns the dynamics of monetary policy, when we seek to analyse the links between GDP, inflation, the money supply and the interest rate, i.e. between four interdependent variables, within a small system of equations , the relationships between these variables being multiple and potentially complex.

The special feature of this system is that each variable is expressed as a function detailing its linear dependence on its own delays and the offsets of the other variables making up the system. As we shall see, the structure of the shifts is the same for all the variables, and therefore for each of the equations making up the VAR system, which gives it a homogeneous form.

The VAR model therefore represents a system of simultaneous equations, in which the variables are linked to their own past values, but also to the lags of the other variables making up the system. We talk about vector autoregressions because the VAR model—as we have already emphasised—is a multidimensional extension of the autoregressive (AR) models seen previously. The latter included several variables defined on the same plane—the endogenous/exogenous distinction no longer really existed—which made it possible to represent them in the form of vectors.

We define the **order** of a VAR model of k variables, generally noted p, as the number of delays or lags in the system. In order to understand the mechanics of such systems, let's start with a model of order one, noted, for this reason, VAR(1).

VAR(1)

Let $\mathbf{y}_t$ be a vector containing k variables of the same system, three for example, which we will then define as $y_{1,t} y_{2,t}$ and $y_{3,t}$. These variables can be respectively GDP, inflation and the interest rate. Let's consider the specific case of a model where only one lag $p = 1$ is taken into account and where a constant is the only deterministic term. The system is therefore determined by three dynamic equations, each with a single lag for all the AR terms.

It is therefore written as

$$\begin{cases} y_{1,t} = \mu_1 + a_{1,1}y_{1,t-1} + a_{1,2}y_{2,t-1} + a_{1,3}y_{3,t-1} + u_{1,t} \\ y_{2,t} = \mu_2 + a_{2,1}y_{1,t-1} + a_{2,2}y_{2,t-1} + a_{2,3}y_{3,t-1} + u_{2,t} \\ y_{3,t} = \mu_3 + a_{3,1}y_{1,t-1} + a_{3,2}y_{2,t-1} + a_{3,3}y_{3,t-1} + u_{3,t}. \end{cases} \tag{7.1}$$

The terms $a_{i,j}$ constitute the elements of a matrix $\boldsymbol{A}_1$ representing the coefficients of the model with the exception of those of the constant, which are compacted into a vector $\boldsymbol{\mu}$. By way of illustration, the coefficient $a_{1,1}$ describes the linear dependence of the variable $y_{1,t}$ on its lag of order one, i.e. $y_{1,t-1}$—on the inertia of GDP for one year for example—and the coefficient $a_{1,2}$ is associated with the relationship between $y_{1,t}$ and the lag of order one of the variable y_2, i.e. $y_{2,t-1}$, is the effect of inflation on GDP if we continue with the previous example. Note, and this will have important consequences later in terms of estimation, that $u_{1,t}$, $u_{2,t}$, $u_{3,t}$ are innovations defined as uncorrelated white noises .

Rewriting the VAR(1) model in matrix form gives:

$$\begin{pmatrix} y_{1,t} \\ y_{2,t} \\ y_{3,t} \end{pmatrix} = \begin{pmatrix} \mu_1 \\ \mu_2 \\ \mu_3 \end{pmatrix} + \begin{pmatrix} a_{11} & a_{12} & a_{13} \\ a_{21} & a_{22} & a_{23} \\ a_{31} & a_{32} & a_{33} \end{pmatrix} \begin{pmatrix} y_{1,t-1} \\ y_{2,t-1} \\ y_{3,t-1} \end{pmatrix} + \begin{pmatrix} u_{1t} \\ u_{2t} \\ u_{3t} \end{pmatrix}. \tag{7.2}$$

This model can be represented in a more compact form by

$$\boldsymbol{y}_t = \boldsymbol{\mu} + \boldsymbol{A}_1\boldsymbol{y}_{t-1} + \boldsymbol{u}_t. \tag{7.3}$$

For ease of notation, it can even be rewritten as:

$$\boldsymbol{y}_t = \boldsymbol{a}_0 + \boldsymbol{A}_1\boldsymbol{y}_{t-1} + \boldsymbol{u}_t, \tag{7.4}$$

where $\boldsymbol{a}_0$ is a vector of constants of dimension $k = 3$, $\boldsymbol{y}_t = (y_{1t}, \ldots, y_{3t})'$ and $\boldsymbol{u}_t = (u_{1t}, \ldots, u_{3t})'$.

VAR(p)

By extension of the previous VAR(1) model, the expanded form of the general VAR(p) model incorporating a constant, p lags and k variables is given by

$$\begin{cases} y_{1t} = a_{10} + a_{11}y_{1,t-1} + \cdots + a_{1p}y_{k,t-p} + u_{1t} \\ y_{2t} = a_{20} + a_{21}y_{1,t-1} + \cdots + a_{2p}y_{k,t-p} + u_{2t} \\ \vdots \\ y_{kt} = a_{k0} + a_{k1}y_{1,t-1} + \cdots + a_{kp}y_{k,t-p} + u_{kt}. \end{cases} \tag{7.5}$$

The reduced form of the VAR (p) model is generally written in the following compact form:

$$\boldsymbol{y}_t = \boldsymbol{a}_0 + \boldsymbol{A}_1\boldsymbol{y}_{t-1} + \boldsymbol{A}_2\boldsymbol{y}_{t-2} + \cdots + \boldsymbol{A}_p\boldsymbol{y}_{t-p} + \boldsymbol{u}_t. \tag{7.6}$$

In the above, t is the time index with $t = p + 1, ..., T$, $\boldsymbol{y}_t$ is a column vector containing the k variables of the model and defined by $\boldsymbol{y}_t \equiv (y_{1t}, y_{2t}, ..., y_{kt})'$, $\boldsymbol{A}_i$ designates the matrices $k \times k$ of the model parameters, $\boldsymbol{a}_0$ is a vector $(k \times 1)$ of constants, while $\boldsymbol{u}_t$ groups the disturbances independently and identically distributed according to white noise processes with $\boldsymbol{u}_t \sim IID(\boldsymbol{0}, \boldsymbol{\Sigma}_u)$, where $\boldsymbol{\Sigma}_u$ is the variance–covariance matrix of the errors.

In other words, the economic variables included in **y** can be modelled as a sequence $\{\boldsymbol{y}_t\}$ of vectors $(k \times 1)$, generated by the VAR(p) model.

This model is also represented in its polynomial form

$$\left(\boldsymbol{I}_k - \boldsymbol{A}_1L - \boldsymbol{A}_2L^2 - \cdots - \boldsymbol{A}_pL^p\right)\boldsymbol{y}_t = \boldsymbol{a}_0 + \boldsymbol{u}_t, \tag{7.7}$$

where $\boldsymbol{I}_k$ is the k-squared identity matrix and L is the lag operator defined in general terms by $L^j x_t = x_{t-j}, \quad j = 0, 1, \ldots$.

Alternatively, (7.7) can be written as

$$\boldsymbol{A}(L)\,\boldsymbol{y}_t = \boldsymbol{a}_0 + \boldsymbol{u}_t, \tag{7.8}$$

with $\boldsymbol{A}(L) = \boldsymbol{I}_k - \sum_{i=1}^{p} \boldsymbol{A}_i L^i$.

We have already introduced stationary VAR(p) models in "Non-stationarity and Cointegration" chapter of this book (see "Non-stationary VAR Models" and "The VAR(p) Model" sections). It will therefore suffice to recall that the stationarity of such models depends on the values of λ which are solutions of the determinant equation

$$\det(\lambda^p \boldsymbol{I}_k - \lambda^{p-1}\boldsymbol{A}_1 - \cdots - \lambda\boldsymbol{A}_{p-1} - \boldsymbol{A}_p) = 0. \tag{7.9}$$

We will therefore have the following definition:

Definition 7.1 The process generated by a VAR(p) model is said to be **stationary** if the roots of (7.9), also called **eigenvalues**, are all **less than** unity.

Following the example of the ARMA and MA models designed for a univariate framework, the VAR(p) can be extended to take into account VARMA(p, q) processes, i.e. autoregressive moving averages.

Definition 7.2 The VAR(p) model is a special case of the VARMA(p, q) model, for which the q lags of the moving average are zero ($q = 0$).

In contrast, a VMA model can be written in its compact form

$$\boldsymbol{y}_t = \boldsymbol{\mu} + \boldsymbol{u}_t + \boldsymbol{\psi}_1 \boldsymbol{u}_{t-1} + \boldsymbol{\psi}_2 \boldsymbol{u}_{t-2}, \tag{7.10}$$

where the coefficients $\boldsymbol{\psi}_i$ are associated with delays in innovation.

Finally, we also have the following definition:

Definition 7.3 The system of equations represented by (7.7) or (7.8) is always **invertible**, provided that it is **stationary**, i.e. all associated **eigenvalues** lie in the open interval]0, 1[.

7.2 Estimating and Selecting the Number of Lags

There are several possible strategies for estimating a VAR(p) model: ordinary least squares (OLS) and generalised least squares (GLS), maximum likelihood and Bayesian methods. The parameter p is always assumed to be known or determined upstream. In the following, we will assume that the VAR models used are **stationary**.

Methods based on least squares (OLS and GLS) produce identical results, as demonstrated by Zellner (1962). Similarly, maximum likelihood (ML) estimation is asymptotically equivalent to least squares estimation, provided that the vector of innovations follows a multivariate normal distribution.

Bayesian methods are outside the scope of this book but can be found in the *Handbook of Bayesian Econometrics* by Geweke et al. (2013).

We will therefore restrict ourselves here to estimating VAR models using ordinary least squares.

Estimation by OLS

Let's start from (7.6), with the variance–covariance matrix of innovations $\boldsymbol{u}_t$ defined by $\boldsymbol{\Sigma}_u$. These innovations are assumed to be independently and identically distributed white noise , while the matrix $\boldsymbol{\Sigma}_u$ is non-singular, so that $\boldsymbol{u}_t \sim IID(0, \boldsymbol{\Sigma}_u)$.

If T is the size of the data sample, then the estimate will be based on $T - p$ observations. The number of parameters to be estimated grows very quickly: it is equal to pk^2, with p the number of lags and k the number of variables. In a parsimonious VAR(1) model with two variables, the number of coefficients to be estimated is already six, i.e. the constants $(a_{01}, a_{02}, a_{11}, a_{12}, a_{21}, a_{22})$. To this must be added the values to be determined for the variance of the two innovations $\text{Var}(u_{1t})$, $\text{Var}(u_{2t})$ and the covariance Cov (u_{1t}, u_{2t}).

It has been demonstrated in the econometric literature that estimation by ordinary least squares (OLS), separately, equation by equation, yields *efficient estimators*. The procedure is greatly facilitated by the fact that all the terms on the right-hand side of the equation to be estimated are lagged and therefore *predetermined*.

Therefore, if we take the example of a VAR(p) with four variables, namely GDP, inflation, money supply and the interest rate, then the OLS estimator should be applied to the GDP equation, the inflation equation and so on.

More formally, consider a VAR(p) model defined by (7.6), for which $\boldsymbol{Z}_{t-1} = (\boldsymbol{1}, \boldsymbol{y}_{t-1}, \ldots, \boldsymbol{y}_{t-p})$ and $\hat{\boldsymbol{A}}$ is a matrix of dimension $k \times (kp + 1)$.

The ordinary least squares estimator is then given by:

$$\hat{\boldsymbol{A}}_{k \times (kp+1)} = \left[\hat{\boldsymbol{\mu}}, \hat{\boldsymbol{A}}_1, \ldots, \hat{\boldsymbol{A}}_p\right] = \left(\sum_{t=1}^{T} \boldsymbol{Z}_{t-1} \boldsymbol{Z}'_{t-1}\right)^{-1}_{k \times k} \left(\sum_{t=1}^{T} \boldsymbol{Y}_t \boldsymbol{Z}'_{t-1}\right)_{k \times (kp+1)}. \tag{7.11}$$

By simplifying the notations, we can rewrite (7.11) in the equivalent form

$$\hat{\boldsymbol{A}} = \boldsymbol{YZ}'(\boldsymbol{ZZ}')^{-1}, \tag{7.12}$$

with $\boldsymbol{Y} \equiv \left[\boldsymbol{y}_1, \ldots, \boldsymbol{y}_T\right]$ of dimension $k \times T$ and $\boldsymbol{Z} \equiv \left[\mathbf{Z}_0, \ldots, \boldsymbol{Z}_{T-1}\right]$ of dimension $(kp + 1) \times T$.

Using the operator vec (.) and stacking, accordingly, the columns of $\boldsymbol{A}$ in a vector $\boldsymbol{\alpha} = \text{vec}(\boldsymbol{A})$ of dimension $(pk^2 + k) \times 1$, gives:

$$\sqrt{T}(\hat{\boldsymbol{\alpha}} - \boldsymbol{\alpha}) \sim N(0, \boldsymbol{\Sigma}_{\hat{\boldsymbol{\alpha}}}).$$

Proposition 7.1 *A sufficient condition for the* ***convergence*** *and* ***asymptomatic normality*** *of the estimator* $\hat{\mathbf{A}}$ *is that the* ***innovations*** *constitute a sequence of* ***absolutely continuous*** *IID random variables whose first four moments are* ***finite*** (see Lütkepohl, 2005, Chapter 3).

Maximum Likelihood Estimation

The maximum likelihood method—which we have used on many occasions—is based on maximising the likelihood of the sample of observations, which is assumed to admit a density function, generally noted $f(y_1, \ldots, y_T)$, with $y_1, \ldots, y_T$ observations.

Under conditions similar to those given for the OLS estimation, therefore assuming that the vector of innovations $\boldsymbol{u}_t$ follows a multivariate normal distribution, i.e.

$$\boldsymbol{u} = \text{vec}(\boldsymbol{U}) = (\boldsymbol{u}_1, \ldots, \boldsymbol{u}_T)' \sim N(\boldsymbol{0}, \boldsymbol{I}_T \otimes \boldsymbol{\Sigma}_u)$$

and consequently, its density function—see Bismans, 2016, Theorem 4.27—is given by

$$f(\boldsymbol{u}) = \frac{1}{(2\pi)^{kT/2}} \det(\boldsymbol{I}_T \otimes \boldsymbol{\Sigma}_u)\, e^{-(1/2)[\boldsymbol{u}'(\boldsymbol{I}_T \otimes\ \boldsymbol{\Sigma}_u)\boldsymbol{u}]},$$

it is then possible to obtain the vector $\tilde{\boldsymbol{A}}$ by maximising the likelihood function.

However, writing out the likelihood function is quite tedious, as is the resulting optimisation calculation, which allows us to determine the ML estimators. Readers interested in a full demonstration and the properties of these estimators should refer again to Lütkepohl (2005, Sect. 3.4). Note that the method yields the same estimators as those obtained by least squares, and with identical optimality properties.

7.3 Model Specification

We will essentially develop two techniques for selecting the optimal number of delays: firstly, the use of sequential tests; and secondly, the use of information criteria.

Sequential Test Procedures

Methods of the "*general to specific*" (also said *top down*) or conversely "specific to general" (or *bottom-up*) type can be used to select the optimum number of delays, i.e. the maximum ($p_{\max}$).

The first method involves applying a sequential testing strategy, based on a Wald or likelihood ratio (LR) test, to assess the significance of a VAR(m) against a VAR(m + *1*).

More precisely, it involves testing a series of null hypotheses in which the number of delays $p_{\max}$ is constrained to be zero:

$H_0 : A_{p\max} = 0$ vs $H_1 : A_{p\max} \neq 0$, $H_0 : A_{p\max-1} = 0$ vs $H_1 : A_{p\max-1} \neq 0$, $\ldots$, $H_0 : A_1 = 0$ vs $H_1 : A_1 \neq 0$.

The sequential procedure consists of a succession of tests conducted until the null hypothesis is rejected. If no null hypothesis can be rejected, then the number of lags p *is* zero.

Bottom-up methods, on the other hand, start by estimating a VAR(0) parsimonious model, then estimate a VAR(1) and so on. A sequential testing procedure is then applied, based on the quality of the residuals (in terms of autocorrelation, for example). Only the VAR(p) model with the least autocorrelated residuals is retained.

Information Criteria

An alternative strategy, which is widely used in practice, is to use information criteria. Minimising these statistics will help to identify the optimum number of delays.

We have introduced—see "The Akaike Criterion" and "Other Selection Criteria" sections—the information criteria. As a reminder, there are three of them:

- Akaike information criterion (AIC);
- Bayesian Information Criterion (BIC);
- the Hannan-Quinn information criterion (HQC).

Generally speaking, these criteria are based on maximum likelihood and involve a trade-off between two conflicting objectives: on the one hand, minimising the criterion, i.e. maximising likelihood; on the other, reducing the number of parameters to be estimated and adopting the most parsimonious model.

Our first task will be to adapt these criteria—initially applied to a single equation—to systems of equations such as VARs. In this case, they will take the following form:

$$\begin{aligned} AIC(m) &= \ln\big(\det\big(\widetilde{\Sigma}_u(m)\big)\big) + \frac{2}{T}\big(mk^2 + k\big) \\ BIC(m) &= \ln\big(\det\big(\widetilde{\Sigma}_u(m)\big)\big) + \frac{\ln(T)}{T}\big(mk^2 + k\big) \\ HQC(m) &= \ln\big(\det\big(\widetilde{\Sigma}_u(m)\big)\big) + \frac{2\ln(\ln(T))}{T}\big(mk^2 + k\big). \end{aligned}$$

with $\widetilde{\boldsymbol{\Sigma}}_u(m) = T^{-1}\sum_{t=1}^{T} \hat{\boldsymbol{u}}_t \hat{\boldsymbol{u}}'_t$, the estimator of the covariance matrix of the residuals for an m-order VAR model.

All in all, following Kilian and Lütkepohl (2017), the criteria in question can be summarised as:

$$C(m) = \ln\big(\det\big(\widetilde{\boldsymbol{\Sigma}}_u(m)\big)\big) + \text{Weight}\,(T)\,\varphi(m),$$

where $C(m)$ is the information criterion and $\varphi(m)$ is a function that negatively weights high numbers of lags m and corresponds to the total number of regressors in the VAR system.

As for Weight (T), it represents a series of weightings which may depend on the size of the sample (T). It has the following values for each of the criteria:

$$AIC{:}2/T;$$
$$BIC{:}\ln(T)/T;$$
$$HQC{:}2\ln[\ln(T)]/T.$$

Ultimately, the optimal VAR will be the one that minimises the criterion(s) chosen from the set of $m = \{0, ..., p_{\max}\}$.

7.4 Granger Causality Analysis

Any VAR model is a system of equations that describes the dynamic relationships between several variables over time. From such a system, it is possible to explain and analyse any causal relationships between these variables.

The Concept of Causality

Granger (1969) proposed an approach to causality based on an almost philosophical observation: an effect cannot precede its cause. In other words—econometrically speaking—the predictability of variables, i.e. the prediction of these variables on the basis of their past, is the basis of causality .

Definition 7.4 Let be the three sets of information $\boldsymbol{X}_t = \{X_{t-i},\ i \geq 0\}$, $\boldsymbol{Y}_t = \{Y_{t-i},\ i \geq 0\}$ and $(\boldsymbol{X}_t, \boldsymbol{Y}_t) = \{X_{t-i}, Y_{t-i},\ i \geq 0\}$. So

(i) Y **causes** X at t if and only if

$$E(X_t|\, E(X_t|\, \boldsymbol{X}_{t-1}, \boldsymbol{Y}_{t-1}) \neq E(X_t|\, \boldsymbol{X}_{t-1}).$$

(ii) Y **instantaneously causes** X at t if and only if

$$E(X_t|\, \boldsymbol{X}_{t-1}, \boldsymbol{Y}_t) \neq E(X_t|\, \boldsymbol{X}_{t-1}, \boldsymbol{Y}_{t-1}).$$

Expectations conditionally on sets of information, which appear in this definition, represent the best linear forecasts of a variable based on past information. These expectations are optimal because they minimise the *Mean Square Error* (MSE) of the forecast.

If, in the general case, we denote the error corresponding to the optimal forecast of X by $\varepsilon(X \mid I) = (X - E(X \mid I)$, then the MSE is calculated from the variance–covariance matrix **V** of the forecast errors, i.e.

$$V = \text{Var}(\varepsilon(X \mid I)).$$

We now need to particularise this approach to causality, in Granger's sense, by adapting it to systems of equations. Let's consider the VAR(1) model with two variables and no constants:

$$\begin{cases} y_{1t} = a_{11}y_{1,t-1} + a_{12}y_{2,t-1} + u_{1t} \\ y_{2t} = a_{21}y_{1,t-1} + a_{22}y_{2,t-1} + u_{2t}. \end{cases} \tag{7.13}$$

and let's suppose that we want to examine whether the variable y_2 causes in t the variable y_1.

To apply Definition 7.4, we first need to see what the corresponding information set in t is. Here it is reduced to two elements, namely $I_t = \{y_{1,t-1}, y_{2,t-1}\}$. The conditional expectations are easy to calculate, given that they depend only on lagged variables.

It follows that, in order to test the causality of y_2 on y_1, it is sufficient to see whether the two coefficients a_1 and a_2 are not equal to zero. Consequently, we will adopt as the null hypothesis the **non-causality** of y_2 on y_1, i.e.

$$H_0 : a_{12} = a_{21} = 0,$$

against the **alternative of causality**, i.e.

$$H_1 : a_{12} \neq 0 \text{ ou } a_{21} \neq 0 \text{ ou } a_{12} = a_{21} \neq 0.$$

If the result of a test—as we shall see, the Wald test—informs us that $a_{12} = 0$ with $a_{21} \neq 0$, then y_1 helps to predict the dynamics of y_2 and therefore y_2 depends on y_1, but, on the other hand, the variable y_1 is only a function of its own past values and is not affected by the dynamics of y_2. As a result, the variable y_1 causes the variable y_2, but the opposite relationship is not verified: in this case, we speak of **unidirectional causality** from y_1 to y_2.

On the other hand, if each of the two variables in the system causes the other variable, this is referred to as a **bidirectional causality** (*feedback effect*).

It should also be noted that causality in Granger's sense is linear, since it is based on systems of linear equations. However, some authors—see in particular Hiemstra and Jones (1994)—refer to a framework in which the relationship between variables can be non-linear.

Using Wald Statistics

The H_0 hypothesis of non-causality requires us to test the joint nullity of several coefficients, in our case just two: a_1 and a_2.. It is therefore possible to use a Wald test, the purpose of which is precisely to test the nullity of a set of estimated coefficients.

We presented the Wald test in "Presence of Autocorrelated Disturbances" and "Lagrange Multiplier Test" sections. It is based on maximum likelihood and is an asymptotic test just like the likelihood ratio or Lagrange multiplier test . It should also be noted that it uses the estimators of the free model.

Before formulating the corresponding statistic, we need to introduce the column vector $\boldsymbol{r}(\boldsymbol{\theta}) = \mathbf{0}$ of the m restrictions imposed by H_0 on the k-vector $\boldsymbol{\theta}$, with $m < k$, and the $(m \times k)$-matrix $\mathbf{R}$ of the partial first derivatives of the log-likelihood function , evaluated in $\hat{\boldsymbol{\theta}}$. We also need to take into account the Fisher information matrix, $\boldsymbol{I}(\boldsymbol{\theta})$, which is equal to minus the expectation of the matrix of partial second derivatives of the log-likelihood function . It is therefore a matrix of dimension $k \times k$.

This statistic is then given by

$$W = \boldsymbol{r}(\hat{\boldsymbol{\theta}})'\left[\boldsymbol{R}\boldsymbol{I}(\hat{\boldsymbol{\theta}})^{-1}\boldsymbol{R}'\right]^{-1}\boldsymbol{r}(\hat{\boldsymbol{\theta}}). \tag{7.14}$$

In (7.14), W represents a quadratic form in $\boldsymbol{r}(\hat{\boldsymbol{\theta}})$, which is the product of a row vector $(1 \times m)$ and the inverse of the matrix $\boldsymbol{R}\boldsymbol{I}(\hat{\boldsymbol{\theta}})^{-1}\boldsymbol{R}'$ of dimension $(m \times m)$, post-multiplied by the column vector $\boldsymbol{r}(\hat{\boldsymbol{\theta}})$ with m components.

Moreover—see Proposition 4.2, "Lagrange Multiplier Test" section—under H_0, the Wald statistic asymptotically follows a **Chi-square distribution** with m degrees of freedom.

Adapted to the VAR(1) of Eq. (7.13), the set of constraints is determined by

$$\boldsymbol{R}\hat{\boldsymbol{a}} = \begin{pmatrix} 0\,1\,0\,0 \\ 1\,0\,0\,0 \end{pmatrix} \begin{pmatrix} \hat{a}_{11} \\ \hat{a}_{12} \\ \hat{a}_{21} \\ \hat{a}_{22} \end{pmatrix} = \begin{pmatrix} 0 \\ 0 \end{pmatrix} = \mathbf{0},$$

where $\hat{\mathbf{a}}$ is the vector of free, unconstrained model estimators, the latter being CAN (Convergent and Asymptotically Normal). The inverse of the Fisher information matrix is equal to

$$\boldsymbol{I}^{-1}(\hat{\boldsymbol{a}}) = -E\left(\frac{\partial^2 \ln L(\hat{\boldsymbol{a}})}{\partial \boldsymbol{a}\, \partial \boldsymbol{a}'}\right).$$

The Wald statistic is then asymptotically distributed according to a Chi-square with 2 degrees of freedom:

$$W \sim \chi^2(2).$$

It should be noted that we have made the assumption—see Sect. 7.2—that all the VAR models we use are **stationary**. This assumption is essential, because if this were not the case, W would no longer follow a standard distribution. Consequently, non-causality can only be tested if all the variables are stationary or stationarised, for example by differentiation in the case where they are integrated of order one.

As we have seen, causality in Granger's sense is fundamentally bivariate. However, Triacca (1998, 2001) has shown that the omission of variables is likely to bias the results of a classic test based on only two variables.

Let us consider the two variables Y and X whose potential causal links we are seeking to establish. The link between Y and X must be studied conditional on the information obtained for Y and X, but also conditional on the information in the subgame Y, X and Z, where Z is a potentially omitted variable.

This differentiated conditioning leads to the following proposition:

Proposition 7.2 *If Y does* ***not cause*** *X* ***conditionally on*** *the available information set based on the observations of X and Y, but Y* ***causes*** *X* ***conditionally*** *on the information set based on X, Y and Z, then Z* ***causes*** *X* ***conditionally on*** *the set X, Y and Z.*

7.5 Structural Form of VAR Models

Until now, we have reasoned within the framework of a reduced-form VAR model in which all the terms of the right-hand side were lagged and therefore predetermined, which facilitated estimation. However, it is common practice to write the VAR(p) model in its so-called structural form.

Presentation of the Model

The structural model is written as follows:

$$\boldsymbol{B}_0 \boldsymbol{y}_t = \boldsymbol{\gamma}_0 + \boldsymbol{\Gamma}_1 \boldsymbol{y}_{t-1} + \boldsymbol{v}_t, \tag{7.15}$$

with $\boldsymbol{\gamma}_0$ the vector of constants, $\boldsymbol{B}_0$ and $\boldsymbol{\Gamma}_1$ the parameter matrices.

Analysis of the disturbance vector is fundamental here. This vector has the following properties: $E(\boldsymbol{v}_t) = \boldsymbol{0}$ and $E(\boldsymbol{v}_t \boldsymbol{v}')_t = \boldsymbol{D}$, where **D** represents a diagonal matrix in which only the variances are shown. There is therefore no correlation between the innovations.

The variance–covariance matrix $\boldsymbol{\Sigma}$ is positive semi-definite, while the matrix $\boldsymbol{B}_0$ of dimension $(K \times K)$ expresses the contemporaneous effects—i.e. dated to the same period t—of each endogenous variable in the system on the other endogenous variable.

In matrix form, the structural model corresponding to a VAR(1) with two variables but no constants is written as:

$$\begin{pmatrix} 1 & b_{12} \\ b_{21} & 1 \end{pmatrix}\begin{pmatrix} y_{1t} \\ y_{2t} \end{pmatrix} = \begin{pmatrix} \gamma_{11} & \gamma_{12} \\ \gamma_{21} & \gamma_{22} \end{pmatrix}\begin{pmatrix} y_{1,\,t-1} \\ y_{2,\,t-1} \end{pmatrix} + \begin{pmatrix} v_{1t} \\ v_{2t} \end{pmatrix}. \tag{7.16}$$

This so-called **structural** model has a well-established correspondence with the **reduced-form** model.

Thus, neglecting the constants, as we pointed out, and assuming the matrix $\boldsymbol{B}_0$ is invertible, we get:

$$\boldsymbol{y}_t = \boldsymbol{B}_0^{-1}\boldsymbol{\Gamma}_1\boldsymbol{y}_{t-1} + \boldsymbol{B}_0^{-1}\boldsymbol{v}_t = \boldsymbol{A}_1\boldsymbol{y}_{t-1} + \boldsymbol{u}_t. \tag{7.17}$$

This is the single offset version of the reduced-form model (7.6).

The correspondence between the reduced-form and structural VAR models is therefore given by the following relationship:

$$\boldsymbol{A}_1 = \boldsymbol{B}_0^{-1}\,\boldsymbol{\Gamma}_1. \tag{7.18}$$

By means of this inverse matrix, it is therefore possible to link the innovations of the structural model to those of the reduced model.

In other words, the error terms of the reduced-form model can be written as combinations of the errors—more commonly known as **shocks**—of the structural form.

To do this, with reference to (7.15) and (7.16), we start by inverting the (2×2) matrix $\boldsymbol{B}_0$, which can be done using the following formula (see Appendix B, definition 24):

$$\boldsymbol{B}_0^{-1} = (1/\det(\boldsymbol{B}_0))(\text{Adj } \boldsymbol{B}_0)',$$

where the adjoint (Adj) of $\boldsymbol{B}_0$ is the matrix of cofactors for each element of this matrix.

The result is:

$$\boldsymbol{B}_0^{-1} = \begin{pmatrix} 1 & b_{12} \\ b_{21} & 1 \end{pmatrix}^{-1} = \frac{1}{1 - b_{12}b_{21}}\begin{pmatrix} 1 & -b_{12} \\ -b_{21} & 1 \end{pmatrix}. \tag{7.19}$$

It immediately follows that the relationship between $\boldsymbol{u}_t$ and $\boldsymbol{v}_t$ is given by $\boldsymbol{u}_t = \boldsymbol{B}_0^{-1}\boldsymbol{v}_t$, or

$$\begin{cases} u_{1t} = (v_{1t} - b_{12}v_{2t})/(1 - b_{12}b_{21}) \\ u_{2t} = (v_{2t} - b_{21}v_{1t})/(1 - b_{12}b_{21}). \end{cases} \tag{7.20}$$

As the innovations in the reduced-form model are white noise , the errors u_{1t} and u_{2t} have zero mean and constant variance, and are not individually autocorrelated.

However, still referring to (7.18), the covariance between the shocks u_{1t} and u_{2t} is not zero, since

$$E(u_{1t}u_{2t}) = E[(v_{1t} - b_{12}v_{2t})(v_{2t} - b_{21}v_{1t})]/(1 - b_{12}b_{21})^2. \tag{7.21}$$

It follows that, in general, the two shocks will be correlated.

The Problem of Identification

The structural representation makes it possible to express what is at the heart of the interest of VAR models in the study of economic policy, namely shocks and their identification. It is through these shocks that we can evaluate economic policy scenarios and test theoretical forecasts.

Nevertheless, the structural form is not neutral for identification and estimation. We have seen that there is no problem in estimating the VAR model in reduced form [Eqs. (7.7) and in compact form (7.8)] and thus providing the estimated coefficients of $\boldsymbol{A_0}$ and $\boldsymbol{A_1}$, since the variables in the right-hand side are predetermined.

The identification of the structural model, on the other hand, is more problematic. Indeed, if we reason from a VAR(1) with two variables, given (7.21), there is an additional parameter to estimate. The system is therefore *under-identified*.

It is therefore necessary to impose nullity constraints on certain parameters, in this case at least one, so that the VAR system can be correctly identified.

A popular method of solving this problem is to use the **Choleski decomposition**.

The basic idea behind this decomposition is based on a result from matrix algebra.

Definition 7.5 Let $\boldsymbol{A}_{n\times n}$ be a positive definite matrix. Then there exists an **upper triangular** matrix **L**, all the elements of whose main diagonal are real numbers, such that

$$\boldsymbol{A} = \boldsymbol{LL'}.$$

Suppose we triangularise (7.20) by imposing $b_{21} = 0$, then we get

$$\begin{cases} u_{1t} = v_{1t} + b_{12}y_{2t} \\ u_{2t} = v_{2t}. \end{cases} \tag{7.22}$$

The matrix of coefficients in (7.22) is now upper triangular and equal to

$$L = \begin{pmatrix} 1 & b_{12} \\ 0 & 1 \end{pmatrix}.$$

As a result, the system is correctly identified and there is no longer any problem of under-identification. Furthermore, using the inverse of **L**, the structural VAR model in its expanded form, including the restriction introduced, is then written as:

$$\begin{pmatrix} y_{1t} \\ y_{2t} \end{pmatrix} = \begin{pmatrix} 1 & -b_{12} \\ 0 & 1 \end{pmatrix} \begin{pmatrix} \gamma_{11} & \gamma_{12} \\ \gamma_{21} & \gamma_{22} \end{pmatrix} \begin{pmatrix} y_{1,t-1} \\ y_{2,t-1} \end{pmatrix} + \begin{pmatrix} 1 & -b_{12} \\ 0 & 1 \end{pmatrix} \begin{pmatrix} v_{1t} \\ v_{2t} \end{pmatrix}$$

or alternatively,

$$\begin{pmatrix} y_{1t} \\ y_{2t} \end{pmatrix} = \begin{pmatrix} \gamma_{11} - b_{12}\gamma_{21} & \gamma_{12} - b_{12}\gamma_{22} \\ \gamma_{21} & \gamma_{22} \end{pmatrix} \begin{pmatrix} y_{1,t-1} \\ y_{2,t-1} \end{pmatrix} + \begin{pmatrix} v_{1t} - b_{12}v_{2t} \\ v_{2t} \end{pmatrix}. \tag{7.23}$$

The following system can then be estimated by OLS:

$$\begin{pmatrix} y_{1t} \\ y_{2t} \end{pmatrix} = \begin{pmatrix} a_{11} & a_{12} \\ a_{21} & a_{22} \end{pmatrix} \begin{pmatrix} y_{1,t-1} \\ y_{2,t-1} \end{pmatrix} + \begin{pmatrix} u_{1t} \\ u_{2t} \end{pmatrix},$$

which gives the estimators $a_{11} = \gamma_{11} - b_{12}\gamma_{21}, a_{12} = \gamma_{12} - b_{12}\gamma_{22}$, $a_{21} = \gamma_{21}$ and $a_{22} = \gamma_{22}$. With $b_{21} = 0$, the model (7.16) becomes

$$\begin{cases} y_{1t} = -b_{12}y_{2t} + \gamma_{11}y_{1,t-1} + \gamma_{12}y_{2,t-1} + v_{1t} \\ y_{2t} = \gamma_{21}y_{1,t-1} + \gamma_{22}y_{2,t-1} + v_{2t}. \end{cases} \tag{7.24}$$

Estimation of (7.24) also gives us $\hat{b}_{12}$, and also the residue sequences $(v_{1t})_t$ and $(v_{2t})_t$. Knowing that—see (7.22)—$u_{1t} = v_{1t} - b_{12}v_{2t}$ and $u_{2t} = v_{2t}$, we can then evaluate

$$\begin{aligned} \text{Var}(u_1) &= \sigma_{1v}^2 + b_{12}^2\sigma_{2v}^2 \\ \text{Var}(u_2) &= \sigma_{2v}^2 \\ \text{Cov}(u_1, u_2) &= -b_{12}\sigma_{2v}^2. \end{aligned}$$

The interpretation of these results is obvious. In (7.23), the fact that b_{21} is equal to zero indicates that y_{1t} has no effect on y_{2t} during the current period. Moreover, this restriction also implies that the shocks v_{1t} and v_{2t} together affect the values, in t, of y_{t-1}, even though only the shocks v_{2t} influence the current values of y_{2t}. In other words, the effect of these shocks is entirely attributed to y_{2t}. This is the result of the Choleski decomposition adopted.

Identification and Economic Interpretation

The VAR as such can be considered as a purely statistical and atheoretical model which would be used for pure forecasting purposes and which would therefore have a limited theoretical content.

However, as we saw in (7.22), the model identification procedure requires us to introduce nullity restrictions on some of the parameters to be estimated.

We can, of course, generalise the decomposition procedure developed in the previous section to a matrix **B** of dimension $(k \times k)$. We are then dealing with a VAR with k variables and no longer just two, which will provide k regression residuals and k structural shocks.

Let's take the compact-form structural model formulated in (7.17), i.e.

$$\boldsymbol{B}_0\boldsymbol{y}_t = \boldsymbol{\Gamma}_{t-1}\boldsymbol{y}_{t-1} + \boldsymbol{v}_t.$$

Let's premultiply it by the inverse of the parameter matrix $\boldsymbol{B}_0$, which gives

$$\boldsymbol{y}_t = \boldsymbol{B}_0^{-1}\boldsymbol{\Gamma}_{t-1}\boldsymbol{y}_{t-1} + \boldsymbol{B}_0^{-1}\boldsymbol{v}_t = \boldsymbol{B}_0^{-1}\boldsymbol{\Gamma}_{t-1}\boldsymbol{y}_{t-1} + \boldsymbol{u}_t.$$

with $\boldsymbol{u}_t = \boldsymbol{B}_0^{-1}\boldsymbol{v}_t$.

The variance of $\boldsymbol{u}_t$ can be obtained as follows:

$$\boldsymbol{\Sigma}_u = E\left(\boldsymbol{u}_t\boldsymbol{u}'_t\right) = (\boldsymbol{B}_0^{-1})E\left(\boldsymbol{v}_t\boldsymbol{v}'_t\right)(\boldsymbol{B}_0^{-1})' = (\boldsymbol{B}_0^{-1})(\boldsymbol{B}_0^{-1})'$$

where $E(\boldsymbol{v}_t\boldsymbol{v}'_t) = \boldsymbol{I}_k$.

To solve the identification problem, we simply need to count the equations and unknowns. Since the matrix $\boldsymbol{\Sigma}_u$ is symmetric, it contains $k(k-1)/2$ distinct elements and there are also $\left(k^2 + k\right)/2$ independent equations. Using the OLS estimator, we can therefore obtain the variance–covariance matrix of the reduced form $\boldsymbol{\Sigma}_u$.

Since the diagonal elements of the matrix $\boldsymbol{B}_0$ are all equal to unity, $\boldsymbol{B}_0$ contains $\left(k^2 - k\right)$ unknown values. There are also k unknown values for a total of k^2 unknown values in the structural model. The identification problem then comes down to identifying the k^2 unknowns among the $\left(k^2 + k\right)/2$ distinct and independent elements of the variance–covariance matrix . It is therefore necessary to impose $k^2 - \left[\left(k^2 + k\right)/2\right]$, i.e. $\left(k^2 - k\right)/2$ restrictions on the system.

The Choleski decomposition then requires us to triangularise $\boldsymbol{B}_0$, which means imposing $k(k-1)/2$ nullity restrictions on this matrix. These restrictions will in turn affect the relationships between residuals and shocks.

We can therefore see that the number of restrictions required to identify the system has been reached and that triangularisation is an efficient method of solving the identification problem.

However, although this decomposition procedure is widely used in practice, it has no economic interpretation in itself. It requires the econometric practitioner to reason in terms of the **order** of the variables that is the most plausible economically, i.e. by moving from the variable at the origin of a sequence to the other variables that are affected by the shock, after one or more indirect effects.

Furthermore, we have only considered restrictions on *short-run* effects using the recursive method, which is the most popular, even though other methods exist.

It is also possible to impose *long-run* restrictions in the case of vector error correction models (VECMs). The interested reader can continue the analysis in Kilian and Lütkepohl (2017, Chapter 11).

7.6 Impulse Response Functions

There are several ways of using dynamic models to estimate the dynamic effects of shocks. We begin by explaining the principle of impulse response functions in VAR models, and then present the concept of local projections developed by Jorda (2005).

Impulse Response Functions in VAR Models

Let's start with a reduced-form model with two variables y_{1t} and y_{2t} :

$$\begin{pmatrix} y_{1t} \\ y_{2t} \end{pmatrix} = \begin{pmatrix} a_{11} & a_{22} \\ a_{21} & a_{22} \end{pmatrix} \begin{pmatrix} y_{1,t-1} \\ y_{2,t-1} \end{pmatrix} + \begin{pmatrix} u_{1t} \\ u_{2t} \end{pmatrix}. \tag{7.25}$$

However, Zellner and Palm (1974) pointed out that there was no reason why such a model should be appropriate for the needs of macroeconomic practitioners. They also suggested using a VARMA (*Vector Autoregressive Moving Average*) type model instead. (We introduced moving average processes in "Autoregressive Moving Average Processes" section of the second chapter).

On a one-dimensional level, Wold (1938, Theorem 7) demonstrated that any stationary process could be written as the sum of two linear processes, one deterministic, the other stochastic, the latter being an infinite sum of moving average (MA) processes. (The convergence of this sum is on quadratic average).

We record this result in Proposition 7.3:

Proposition 7.3 *Any* ***stationary*** *process* $\{y_t\}$ *can be* ***decomposed*** *into two uncorrelated processes* $\{z_t\}$ *and* $\{x_t\}$*, such that*

$$\{y_t\} = \{z_t\} + \{x_t\},$$

where $\{z_t\}$ *represents the* ***deterministic*** *component and* $\{x_t\}$ *the* ***linear stochastic*** *component. Furthermore, if* $\{z_t\} = 0$*, then* $\{z_t\}$ *is a* ***purely stochastic*** *process such that*

$$y_t = \sum_{j=0}^{+\infty} \phi_j \varepsilon_{t-j},$$

where ε_t ***is white noise*** , $\phi_0 = 1$ and $\sum_j \phi_j^2 < +\infty$.

We now need to generalise these results to multidimensional linear systems. As shown, for example, by Enders (2010), any VAR model can, in fact, be rewritten in terms of a *Vector Moving Average* (VMA) representation, whereby the variables in the model are expressed as a function of the current and past values of present and previous shocks (in this case, the two shocks $u_{1,t}$ and $u_{2,t}$). In other words, these variables are expressed as sequences of innovations.

Consider, from this point of view, the compact matrix equivalent of (7.25), $\boldsymbol{y}_t$ being, of course, stationary.

$$\boldsymbol{y}_t = \boldsymbol{A}_j \boldsymbol{y}_{t-j} + \boldsymbol{u}_t. \tag{7.26}$$

By repeated substitution in (7.26), we get

$$\begin{aligned}
\boldsymbol{y}_t &= \boldsymbol{A}_1(\boldsymbol{A}_1\boldsymbol{y}_{t-2}) + \boldsymbol{u}_1 + \boldsymbol{A}_1\boldsymbol{u}_{t-1} \\
&= \boldsymbol{A}_1^2\boldsymbol{y}_{t-2} + \boldsymbol{u}_1 + \boldsymbol{A}_1\boldsymbol{u}_{t-1} \\
&= \boldsymbol{A}_1^3\boldsymbol{y}_{t-2} + \boldsymbol{u}_1 + \boldsymbol{A}_1\boldsymbol{u}_{t-1} + \boldsymbol{A}_1^2\boldsymbol{u}_{t-2} \\
&\vdots \\
&= \sum_{j=0}^{+\infty} \boldsymbol{A}_1^j \boldsymbol{u}_{t-j}.
\end{aligned}$$

The VMA form of $\boldsymbol{y}_t$ is therefore written as follows

$$\boldsymbol{y}_t = \sum_{j=0}^{+\infty} \boldsymbol{A}_1^j\, \boldsymbol{u}_{t-j}. \tag{7.27}$$

It should be added that this sum converges rapidly.

This result, known as the Wold representation theorem, is extremely powerful since it allows us to write that any stationary VAR model admits a moving average representation. What's more, given that infinite summation is performed as a quadratic mean, the coefficient matrices converge rapidly to zero. It follows that any VAR process can be correctly approximated by a finite-order VAR process.

Applying the above to the **reduced** two-variable model (7.25) while using (7.19) and (7.27), we obtain

$$\begin{pmatrix} y_{1t} \\ y_{2t} \end{pmatrix} = \frac{1}{1 - b_{12}b_{21}} \sum_{i=0}^{+\infty} \begin{pmatrix} a_{11} & a_{12} \\ a_{21} & a_{22} \end{pmatrix}^i \begin{pmatrix} 1 & -b_{12} \\ -b_{21} & 1 \end{pmatrix} \begin{pmatrix} u_{1,\,t-i} \\ u_{2,\,t-i} \end{pmatrix}, \tag{7.28}$$

If we introduce the matrix $\boldsymbol{\Phi}_i$ of dimension (2×2) and of typical element $\phi_{jk}(i)$, either

$$\boldsymbol{\Phi}_i = \frac{\boldsymbol{A}_1^i}{1 - b_{12}b_{21}} \begin{pmatrix} 1 & -b_{12} \\ -b_{21} & 1 \end{pmatrix},$$

(7.28) then becomes

$$\begin{pmatrix} y_{1t} \\ y_{2t} \end{pmatrix} = \sum_{i=0}^{+\infty} \begin{pmatrix} \phi_{11}(i) & \phi_{12}(i) \\ \phi_{21}(i) & \phi_{22}(i) \end{pmatrix} \begin{pmatrix} u_{1,\ t-i} \\ u_{2,\ t-i} \end{pmatrix}. \tag{7.29}$$

The four sets of coefficients $\phi_{11}(i)$, $\phi_{12}(i)$, $\phi_{21}(i)$ and $\phi_{22}(i)$ are referred to as **impulse response functions**. A graph representing the values of these coefficients as a function of i is very useful for visualising the time trajectory of y_{1t} and y_{2t} in response to differentiated exogenous shocks.

The moving average representation of a VAR model is therefore an effective tool for evaluating the interactions between the sequences (y_{1t}) and (y_{2t}), while using the coefficients of $\boldsymbol{\Phi}_i$.

So, for example, $\phi_{12}(2)$ will designate the response of the variation of one unit of $u_{2,\ t-2}$ on y_{1t}. From this point of view, if we refer to $i = 0$, the corresponding coefficients $\phi_{jk}(0)$ are called **impact multipliers** .

The cumulative effects of a unit impulse, for example from u_{1t} to y_{2t}, are easily obtained by summing the coefficients of the response functions. So, after n periods, the cumulative effects of u_{1t} on y_{2t} will be given by

$$\sum_{i=1}^{n} \phi_{21}(i). \tag{7.30}$$

Since our VAR(1) model has been assumed to be stationary, we can be sure that the sum (7.30) will converge to a finite number when $n \to +\infty$.

Impulse Response Functions and Local Projections

Let's start by introducing the concept of mean square error , abbreviated as MSE.

Definition 7.6 Let $\theta \in \Theta$ be an unknown parameter and $\hat{\theta}$ an estimator of θ.. The **mean square error** of $\hat{\theta}$ is the real number given by

$$EQM\,(\hat{\theta}) = E\big[(\theta - \hat{\theta})^2\big].$$

The central problem in forecasting is to make the forecast error as small as possible, in other words to minimise the predictive mean square error .

Suppose, from this point of view, that we wish to forecast the value of the variable y_{t+1} on the basis of a set of variables $\boldsymbol{X}_t$ observed at date t. If we denote by $\hat{y}_{t+1|t}$ a forecast of y_{t+1} made on the basis of the information available at t, we must select the forecast which minimises the mean square error, i.e.

$$EQM\,(\hat{y}_{t+1|t}) = E\big[(y_{t+1} - \hat{y}_{t+1|t})^2\big]. \tag{7.31}$$

Such a forecast minimising (7.31) turns out to be the **expectation** of y_{t+1} **conditional** on $\boldsymbol{X}_t$:

$$\hat{y}_{t+1|t} = E(y_{t+1}\,|\,\boldsymbol{X}_t). \tag{7.32}$$

We can now take a further step by restricting the class of forecasting procedures to those for which $\hat{y}_{t+1|t}$ is a linear function of $\boldsymbol{X}_t$:

$$\hat{y}_{t+1|t} = \boldsymbol{\alpha}'\boldsymbol{X}_t.$$

Assuming that there is a vector of values for $\boldsymbol{\alpha}$ such that the forecast error $(y_{t+1} - \boldsymbol{\alpha}'\boldsymbol{X}_t)$ is not correlated with $\boldsymbol{X}_t$, we get

$$E\big[(y_{t+1} - \boldsymbol{\alpha}'\boldsymbol{X}_t)\boldsymbol{X}'_t\big] = \mathbf{0}'. \tag{7.33}$$

A forecast $\boldsymbol{\alpha}'\boldsymbol{X}_t$ which verifies (7.33) is called a **linear projection** of y_{t+1} onto $\boldsymbol{X}_t$. This projection therefore minimises the associated mean square error in the class of all linear predictors.

Following Jorda (2005), an impulse response (IR) function can be defined as the difference between two **optimal** forecasts, thus minimising the mean square error :

$$RI(t, s, \boldsymbol{d}_i) = E\big(\boldsymbol{y}_{t+s}|\,\boldsymbol{u}_t = \boldsymbol{d}_i\ ;\ \boldsymbol{X}_t\big) - E\big(\boldsymbol{y}_{t+s}|\,\boldsymbol{u}_t = \mathbf{0}\ ;\ \boldsymbol{X}_t\big),\quad s = 0, 1, \ldots \tag{7.34}$$

where $\boldsymbol{X}_t \equiv (\boldsymbol{y}_{t-1}, \boldsymbol{y}_{t-2}, \ldots)$, $\boldsymbol{u}_t$ is the vector of disturbances of the VAR model in reduced form and, finally, $\mathbf{D}$ is a matrix whose components in each column $\boldsymbol{d}_i$ are the predetermined structural shocks.

More generally, consider the **projection** of $\boldsymbol{y}_{t+s}$ onto the linear space generated by the p shifted values of $\boldsymbol{y}_t$, i.e. $(\boldsymbol{y}_{t-1}, \ldots, \boldsymbol{y}_{t-p})$. Ignoring the vector of constants, we obtain

$$\boldsymbol{y}_{t+s} = \boldsymbol{B}_1^{s+1}\boldsymbol{y}_{t-1} + \boldsymbol{B}_2^{s+1}\boldsymbol{y}_{t-2} + \cdots + \boldsymbol{B}_p^{s+1}\boldsymbol{y}_{t-p} + \boldsymbol{e}_{t+s}^{s},\quad s = 0, 1, \ldots, h, \tag{7.35}$$

where $\boldsymbol{B}_i^{s+1}$ is the coefficient matrix for each i and each forecast horizon h.

We will refer to the set of h regressions in (7.35) as the linear **local projection.** Consequently, the response functions associated with these local projections will be given, in view of (7.34), by

$$\widehat{\mathrm{RI}}(t, s, \boldsymbol{d}_i) = \hat{\boldsymbol{B}}_1^s \boldsymbol{d}_i, \quad s = 0, 1, \ldots, h, \tag{7.36}$$

with $\hat{\boldsymbol{B}}_1^0 = \boldsymbol{I}$, **I** being the identity matrix.

It remains to be verified that condition (7.33) is satisfied. The residuals $\boldsymbol{e}_{t+s}^s$ in (7.35) are a moving average of the forecast errors from period t to period $t+s$. They are therefore not correlated with the regressors, which cover the period $t-1$ to $t-p$.

7.7 Application: The Dynamics of the Foreign Exchange Market

We begin by giving some details of the problem addressed, in this case the analysis of the functioning of the foreign exchange market, as well as the data used. We then discuss the estimation of the VAR model and its dynamics, using impulse response functions. We will also analyse causality in the Granger sense.

The Data

The foreign exchange market (or *Forex*, short for foreign exchange) has two main compartments: the *spot* market and the forward market, which is subdivided into *forward outright* and *swap* contracts. In the first of these compartments, a distinction is made between the official market (*fixing*) and the continuous market.

The term "continuous" should be taken in its strictest sense. Foreign exchange transactions take place around the clock, via banks, financial institutions, hedge funds, asset managers and so on.

The main characteristic of "continuous" trading is its opaque and decentralised nature, with no exchange or central authority. A spot foreign exchange transaction is the purchase of an amount of currency Y against an amount of currency Z. For example, the USD/EUR rate represents the number of units of the dollar currency expressed in one unit of the euro currency.

Since the spot market has no central authority, prices are quoted on the basis of supply and demand. As a result, every exchange rate has two components: the bid price, more commonly known as the *bid*, and the *ask* price, corresponding to the offer price.

The *bid* price of a currency is obviously always lower than the *ask* price. The *bid-ask spread* is therefore the difference between the two prices, which is necessarily negative.

Finally, the last technical term in the field, a *tick* represents the difference between two consecutive exchange rates for a currency. It is not a unit of time, but a variation in price due to past transactions in the exchange rate of the currency in question.

Because a tick is not a unit of time, it can be used to take into account all the transactions made. This is known as *tick by tick*.

Our data, acquired from *Olsen Financial*, relates exclusively to the euro-dollar exchange rate. The data is daily and covers a fairly long period during the *subprime* crisis, from 1 October 2008 to the end of October 2010—not including closed days, of course—i.e. 516 observations.

We will use three series to describe the dynamics of the foreign exchange market, namely

- transaction volume (VOL);
- price volatility (VOLA);
- the *spreads* (SPREAD) recorded on the spot market.

The construction of the first of these series poses a number of problems, given the completely decentralised and continuous nature of the spot foreign exchange market. We will therefore be forced to use proxies.

The choice, generally adopted in the literature, is to approximate trading volume by quotation frequency, which is obtained by means of the number, defined above, of transactions tick by tick, without reference to time.

Here is a graph of this VOL series, expressed in logarithms to base *e* (Fig. 7.1):

Fig. 7.1 Change in transaction volume

The graph shows that until May 2009, volumes were on a downward trend, with two drastic decreases in December 2008 and, above all, in April 2009. Thereafter, the trend was towards a slight increase in volumes traded on the spot market.

I would now like to explain how the other two series are constructed. The VOLA series, which measures price volatility, is obtained by using relative returns, so that VOLA is given as the sum of the squares of the differences in the logarithms of the relative returns.

Finally, the spread series is based on the median *Bid* and *Ask* series. Obviously, they incorporate the transaction costs associated with foreign exchange transactions.

The SPREAD is then calculated by taking the *Bid-Ask* spread, i.e. the difference, as we have explained, between the bid and ask prices of a foreign exchange transaction.

Figure 7.2 shows a graphical representation of the SPREAD series, again expressed in logarithms:

As can be seen, the SPREAD is almost always higher than 9.2 in absolute terms, while in very rare cases it is lower.

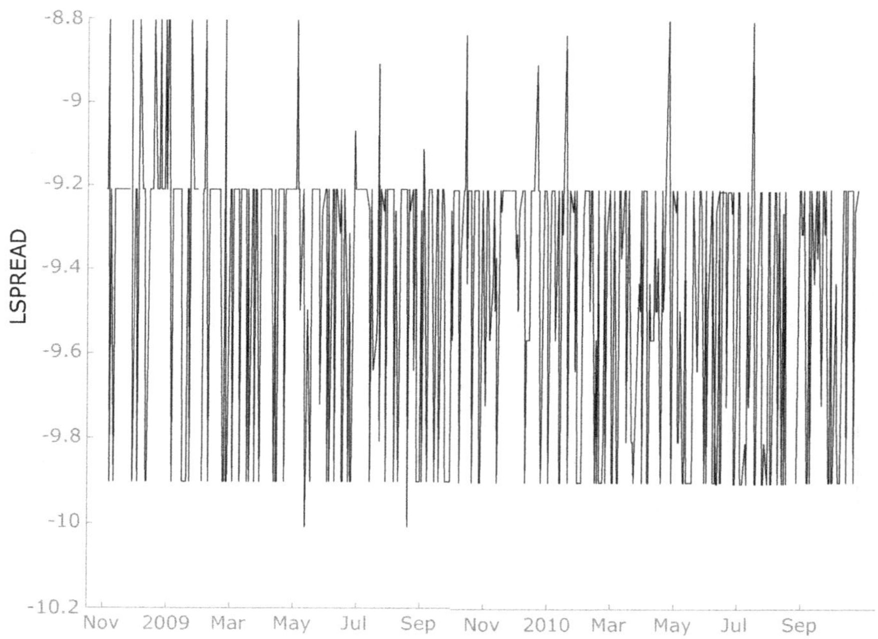

Fig. 7.2 Graph of the SPREAD series

The Equation to Be Estimated

Initially, we could try to represent the dynamics at work on the foreign exchange market by a dynamic equation linking the volume of transactions to the two other exogenous variables, namely volatility and the spread.

This would lead to the following univariate model, in which L represents the logarithm, while p_v et p_s denotes the number of volatility-specific and spread-specific lags, respectively:

$$\begin{aligned} LVOL_t &= m + \sum_{r=1}^{p} \alpha_r \, LVOL_{t-r} + \sum_{j=0}^{p_v} \beta_{1j} \, LVOLA_{t-j} \\ &\quad + \sum_{j=1}^{p_s} \beta_{2j} \, LSPREAD_{t-j} + \varepsilon_t . \end{aligned}$$

However, such a modelling strategy is not the right one for at least two reasons. First, volatility, as Shiryaev (1999) explains, is itself volatile, which means that it is a random variable. Moreover, its probability distribution is a mixture of statistical distributions.

Secondly, we can conjecture that the *spread* variable is not truly exogenous; that, in any case, it contains at least a certain amount of endogeneity. In other words, if the spread determines trading volume, the reverse is also true.

From a methodological point of view, it follows that it is necessary to estimate a system of simultaneous dynamic equations rather than a single equation. Consequently, in the first instance and for the sake of parsimony, we will adopt a VAR(1) model of the form

$$\begin{cases} LVOL_t = a_{10} + a_{11} LVOL_{t-1} + a_{12} LVOLA_{t-1} + a_{13} LSPREAD_{t-1} + u_{1t} \\ LVOLA_t = a_{20} + a_{21} LVOL_{t-1} + a_{22} LVOLA_{t-1} + a_{23} LSPREAD_{t-1} + u_{2t} \\ LSPREAD_t = a_{30} + a_{31} LVOL_{t-1} + a_{32} LVOLA_{t-1} + a_{33} LSPREAD_{t-1} + u_{3t} \end{cases} \tag{7.37}$$

The first equation describes the dynamics of FX volume as a function of its own lags and those of volatility and spread. The second equation in (7.37) models volatility and, in particular, its relationship with volume and spread.

The third of these equations aims to capture the effect of trading volume on the spread. The explanation for this relationship lies in the fact that higher volumes are associated with economies of scale and therefore also with reductions in the exchange spread.

Estimating the Model and Impulse Responses

We now estimate model (7.37). To do this, we first need to ensure that the variables are stationary, which we have already done.

Next, the exact order of the VAR(p) must be determined to check whether such a parsimonious VAR(1) model is appropriate or whether, on the contrary, a higher order VAR model should be estimated.

As explained above, we calculated the values of the usual information criteria: AIC , BIC and HQC (see "Estimating the Model" chapter, "The Akaike Criterion" and "Other Selection Criteria" sections). The results, in which the star indicates the minimum and therefore optimum value for the criterion in question, are as follows (Table 7.1):

Table 7.1 Comparison of models

Lags (p)	AIC	BIC	HQC
0	5.442122	5.468209	5.452376
1	4.503831	4.608176*	4.544847*
2	4.528212	4.710815	4.599989
3	4.494912	4.755774	4.597451
4	4.493760	4.832880	4.627061
5	4.487816*	4.905194	4.651878
6	4.503885	4.999522	4.698709
7	4.514156	5.088052	4.739742
8	4.548241	5.200395	4.804589

This table shows that the VAR(1) is selected on the basis of two of the three criteria. The AIC criterion leads us to consider a higher number of lags, five in this case. This discrepancy is not surprising. We said ourselves—see "Estimating the Model" chapter, "The Akaike Criterion" section—that the Akaike criterion tended to select the least parsimonious model.

We will therefore consider a first-order VAR model and estimate it by appropriate procedures. As we have seen, the least squares and maximum likelihood methods provide convergent results under certain conditions of errors normality.

The results in Table 7.2 provide some interesting insights. The dynamics of trading volume seems to depend essentially on its own inertia. The relationship between volume and volatility as described in the *Mixture Distribution Hypothesis* is not significant, even if the coefficient associated with volatility is positive as expected (in fact, an increase in volume should be linked to an increase in volatility and vice versa).

The volatility equation does not produce any significant determinant and volatility does not seem to be affected by volume or spread, as some of the literature suggests.

Table 7.2 Estimation of the VAR(1) model

Equation 1 (volume)	Coefficients	p-values
Volume (−1)	0.779	28.10
Volatility (−1)	0.003	0.69
Spread (−1)	−0.031	−1.17
Equation 2 (Volatility)	Coefficients	t-statistics
Volume (−1)	−0.081	−0.24
Volatility (−1)	0.037	0.83
Spread (−1)	−0.151	−0.47
Equation 3 (Spread)	Coefficients	p-values
Volume (−1)	−0.094	−2.05
Volatility (−1)	−0.001	−0.21
Spread (−1)	0.110	2.50

On the other hand, the spread is indeed determined by the volume of foreign exchange transactions, which confirms our forecast of a possible two-way relationship between these two variables.

We have tried to illustrate the dynamics of variables over time using the impulse response functions described above. The idea behind such functions is to present the response of a variable to a shock. If we assume, for example, that the spread is at its average level and that it is affected by a shock of one unit during the current period, how will its future trajectory evolve? How long will it take before it returns to its average value?

The graph below shows, by way of example, the impulse responses (IR) of the spread to changes in the volume of foreign exchange and volatility.

Figure 7.3 shows that the spread response function, following a one standard deviation change in volatility, is not significant, as evidenced by the flat shape of the response function. The value of the coefficients is close to zero for all periods up to the tenth, while the confidence intervals are distributed on both sides of the x-axis.

On the other hand, the spread's response to volume is significantly negative: a volume shock today will have a negative impact on the spread, but the effect gradually fades after ten periods.

However, our model has a number of limitations. Firstly, it is based on the assumption of linearity, even though we might expect the relationships between our three variables to be dependent on the phases or regimes of the foreign exchange market.

Secondly, heteroscedastic effects, which are common in financial data series, may also exist and bias the estimate produced.

Finally, the reduced-form VAR estimated in this way is based on the assumption of zero covariance between the disturbances in the three equations, which may be highly debatable in reality.

More detailed estimates based on non-linear VAR models, which have been attracting growing interest in recent years, particularly structural VARs that allow the theoretical links between series to be broken down, would be a natural extension

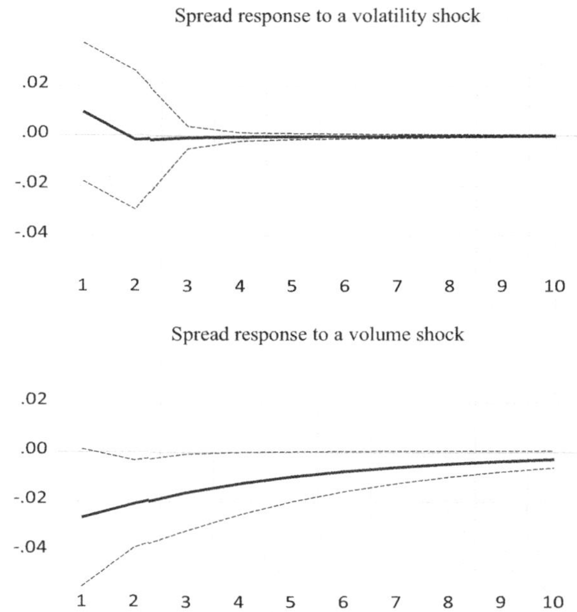

Fig. 7.3 Impulse responses of the SPREAD

for researchers in applied econometrics. However, such an approach is beyond the scope of this book.

Causality Analysis

Let's now apply the Granger causality analysis to highlight any causal links between the three variables.

Table 7.3 lists the results of the causality tests, in which the null hypothesis of non-causality is tested six times since we have three variables. In accordance with the information criteria used and the VAR(1) chosen previously, we use a single lag to carry out the various tests.

Table 7.3 Non-causality tests (1 lag)

Null hypothesis	Statistics	Probability
Volatility $\nrightarrow$ volume	0.458	0.50
Volume $\nrightarrow$ volatility	0.024	0.88
Spread $\nrightarrow$ volume	1.241	0.266
Volume $\nrightarrow$ spread	4.065	0.044
Spread $\nrightarrow$ volatility	0.198	0.657
Volatility $\nrightarrow$ spread	0.172	0.679

The table shows that the null hypothesis of non-causality between the selected variables cannot be rejected in five out of six cases. Trading volume is indeed the cause of volatility.

On the other hand, the probability associated with the Wald statistic for the relationship between volume and spread is less than 5%. We therefore reject, at an acceptable threshold, the null hypothesis of non-causality, which amounts to concluding that the volume of foreign exchange transactions causes, in the Granger sense, the spread.

This result is fully in line with historical trends in the foreign exchange market and the steady decline in transaction costs resulting from the increase in trading volume over time. It also confirms the negative relationship between market liquidity and spreads.

It should also be noted that the degree of parsimony of the model, i.e. the choice of the number of lags, can have a significant influence on the results of the Granger test and therefore on its conclusions in terms of causality . This test, it should be remembered, was based on a comparison of forecasts conditional on different sets of information.

We also decided to determine the order of integration of the series using the usual unit root tests. Our variables were all clearly stationary and integrated to the same order, so we had no difficulty estimating a VAR(1) model.

Nevertheless, when the order of integration of the series is not obvious and they may be non-stationary, the method implemented by Toda and Yamamoto (1995) is recommended, in order to avoid a spurious causality bias linked to unit root tests.

The method consists first of all in selecting the lag p of the VAR model using the usual criteria, and then determining the maximum order of integration of the series: for example, $m = 1$, if we suspect that one variable can be integrated of order one, while the others are $I(0)$.

On this basis, we simply estimate the VAR model of order $(p+m)$, after which we carry out the non-causality analysis in the Granger sense by assessing the significance of the lagged parameters using a standard Wald test.

7.8 Exercises

7.8.1 Using your own words, explain what is the difference between the "reduced form" and the “structural form” of the VAR model and how they can be used for policy applications.

7.8.2 Consider a structural VAR(1) model given by

$$\boldsymbol{A}_0\boldsymbol{y}_t = \boldsymbol{A}_1\boldsymbol{y}_{t-1} + \boldsymbol{u}_t.$$

(i) Show that this structural model can be transformed into the reduced-form model:

$$\boldsymbol{y}_t = \boldsymbol{\Phi}_1 \boldsymbol{y}_{t-1} + \boldsymbol{v}_t.$$

(ii) Identify the relationship between the parameters of the structural and reduced-form models.

(iii) Explain the identification problem in the context of structural VAR models. Why is it necessary to impose constraints on the parameters to achieve identification?

(iv) Describe the Choleski decomposition method for identifying a structural VAR model.

(v) How does triangularising the variance–covariance matrix Σ help in solving the identification problem?

7.8.3 Consider a VAR(1) model with two variables and no constants. Write the structural model as:

$$\begin{pmatrix} 1 & -\alpha_{12} \\ -\alpha_{21} & 1 \end{pmatrix} \begin{pmatrix} y_{1t} \\ y_{2t} \end{pmatrix} = \begin{pmatrix} \beta_{11} & \beta_{12} \\ \beta_{21} & \beta_{22} \end{pmatrix} \begin{pmatrix} y_{1,t-1} \\ y_{2,t-1} \end{pmatrix} + \begin{pmatrix} u_{1t} \\ u_{2t} \end{pmatrix}.$$

Transform this into the reduced-form model and express in terms of α and β.

7.8.4 Given the structural model form derived in the previous question, use the Choleski decomposition to identify the parameters. Write the corresponding system of equations and show how to estimate the set of these parameters. How do the imposed constraints (e.g. $\alpha_{12} = 0$) affect the interpretation of the results?

7.8.5 Verify that the residuals $\boldsymbol{e}_{t+s}^{s}$ in the local projection equations are not correlated with the regressors.

7.8.6 The file `cds.xls` contains two series: the daily observations of CDS spreads (denoted CDS) and sovereign bond spreads (denoted OAT). Sovereign bond spreads are computed as the difference between the rate of 10-year sovereign bonds and the rate of asset swaps (a proxy for a risk-free sovereign rate) for the same period. These data, which concern France, are daily and cover the period from September 1, 2008, to January 29, 2010.

You are required, using VAR methodology, to:

(i) Plot the series and adequately specify (using either levels or first-difference variables) the system VAR model with CDS and OAT as the two endogenous variables.

(ii) Find the optimal lag length for the VAR model composed of CDS and OAT as the two endogenous variables using AIC, BIC and HQC criteria.

(iii) Adequately estimate (using either levels or first-difference variables) the system VAR model with CDS and OAT as the two endogenous variables and interpret the results in an economic context.

(iv) Determine the Granger causality relationship between the two variables and provide an economic interpretation, particularly regarding the existence of potential CDS speculation behaviours during the 2008 crisis period.

References

Bismans, F. (2016). *Probabilités et statistique inférentielle. Prélude à l'économétrie*. Ellipses.

Enders, W. (2010). *Applied econometric time series* (3rd ed.). Wiley.

Frisch, R. (1933). Propagation problems and impulse problems in dynamic economics. In *Economic essays in honour of Gustav Cassel*. George Allen & Unwin (pp. 171–205).

Geweke, J., Koop, G., & Van Dijck, H. (Eds.). (2013). *The Oxford handbook of Bayesian econometrics*. Oxford University Press.

Granger, C. W. J. (1969). Investigating causal relations by econometric models and cross-spectral methods. *Econometrica, 37*, 424–438.

Hiemstra, C., & Jones, J. D. (1994). Testing for linear and nonlinear Granger causality in the price-volume relation. *Journal of Finance, 49*, 1639–1664.

Jorda, O. (2005). Estimation and inference of impulse responses by local projections. *The American Economic Review, 95*, 161–182.

Kilian, L., & Lütkepohl, H. (2017). *Structural vector autoregressive analysis*. Cambridge University Press.

Lütkepohl, H. (2005). *New introduction to multiple time series analysis*. Springer.

Sims, C. (1980). Macroeconomics and reality. *Econometrica, 48*, 1–48.

Slutsky, E. (1927, 1937). The summation of random causes as the source of cyclic processes. *Econometrica, 5*, 105–146.

Shiryaev, A. N. (1999). *Essentials of stochastic finance, models, theory*. World Scientific.

Toda, H. Y., & Yamamoto, T. (1995). Statistical inference in vector autoregressions with possibly integrated processes. *Journal of Econometrics, 66*, 225–250.

Triacca, U. (1998). Non-causality: The role of the omitted variables. *Economic Letters, 60*, 317–320.

Triacca, U. (2001). On the use of Granger causality to investigate the human influence on climate. *Theoretical and Applied Climatology, 69*, 137–138.

Wold, H. (1954). *A study in the analysis of stationary time series* (2nd ed.), Almqvist & Wicksell.

Zellner, A. (1962). An efficient method of estimating seemingly unrelated regressions and tests of aggregation bias. *Journal of the American Statistical Association, 57*, 348–368.

Zellner, A., & Palm, F. (1974). Time series analysis and simultaneous equation econometric models. *Journal of Econometrics, 2*, 17–54.

Chapter 8
Panel Data Models

A panel data series, also known as a longitudinal series, is characterised by a double dimension: a set of individual units such as households, companies, countries, regions, etc., on the one hand, and a set of temporal observations on the other. In other words, such a series is a mixture of a cross-sectional series and a time series.

8.1 A Range of Models

At this stage, we will distinguish three main categories of models specific to longitudinal series: the fixed-effects model, the random effects model and the compound-error model, the latter of which may be single or double.

The Fixed-Effects Model

Let's assume that we have the time observations y_t and $\boldsymbol{x}_{it}$ in number T on a set of individual units in number N. By analogy with the multiple regression model, we get

$$y_{it} = \boldsymbol{x'}_{it}\boldsymbol{\beta} + u_{it}, \qquad i = 1, \ldots, N \ ; \ \ t = 1, \ldots, T, \tag{8.1}$$

where $\boldsymbol{x'}_{it}$ is a row vector with NT components and $\boldsymbol{\beta}$ is a column vector with the same number of components.

Supplementary Information The online version contains supplementary material available at https://doi.org/10.1007/978-3-031-72910-2_8.

F. J. Bismans and O. Damette, *Dynamic Econometrics*,
https://doi.org/10.1007/978-3-031-72910-2_8

Apparently, taking account of the cross-sectional dimension does not raise any new econometric problems. However, the advantage of panel data is that it allows greater flexibility in modelling differences in the behaviour of individual units.

From this point of view, the first possibility is to introduce a unit-specific effect in the form of a constant α_i, so that we obtain the **fixed-effects model**

$$y_{it} = \alpha_i + \boldsymbol{x}'_{it}\boldsymbol{\beta} + u_{it}, \qquad i = 1, \ldots, N \; ; \; t = 1, \ldots, T. \tag{8.2}$$

In matrix terms, the *NT* equations above can be written as:

$$\underset{NT \times 1}{\begin{pmatrix} \boldsymbol{y}_1 \\ \boldsymbol{y}_2 \\ \vdots \\ \boldsymbol{y}_N \end{pmatrix}} = \underset{NT \times 1}{\begin{pmatrix} \boldsymbol{e}_T \\ \boldsymbol{0}_T \\ \vdots \\ \boldsymbol{0}_T \end{pmatrix}} \alpha_1 + \cdots + \underset{NT \times 1}{\begin{pmatrix} \boldsymbol{0}_T \\ \boldsymbol{0}_T \\ \vdots \\ \boldsymbol{e}_T \end{pmatrix}} \alpha_N + \underset{NT \times K}{\begin{pmatrix} \boldsymbol{x}_1 \\ \boldsymbol{x}_2 \\ \vdots \\ \boldsymbol{x}_N \end{pmatrix}} \underset{K \times 1}{\boldsymbol{\beta}} + \underset{NT \times 1}{\begin{pmatrix} \boldsymbol{u}_1 \\ \boldsymbol{u}_2 \\ \vdots \\ \boldsymbol{u}_N \end{pmatrix}}, \tag{8.3}$$

where all the vectors in this system are of dimension $T \times 1$, with the exception of $\boldsymbol{x}_i$ which are $1 \times T$. In addition, $\boldsymbol{e}_T = (1, 1, \ldots, 1)'$ is a vector made up of T units. Finally, the perturbations $\boldsymbol{u}_i$ are assumed to satisfy the following usual assumptions:

$$E(\boldsymbol{u}_i) = \boldsymbol{0}, \quad E(\boldsymbol{u}_i\boldsymbol{u}'_i) = \sigma^2\boldsymbol{I}_T, \quad E(\boldsymbol{u}_i\boldsymbol{u}'_j) = \boldsymbol{0}, \quad \forall i \neq j,$$

with $\boldsymbol{I}_T$ the identity matrix of dimension $T \times T$. Furthermore, they are independently and identically distributed.

We can give an even more compact version of the system (8.3):

$$\boldsymbol{y} = \boldsymbol{E}\boldsymbol{\alpha} + \boldsymbol{X}\boldsymbol{\beta} + \boldsymbol{u}, \tag{8.4}$$

where $\boldsymbol{y} = (\boldsymbol{y}_1, \ldots, \boldsymbol{y}_N)'$,$\boldsymbol{E}$ is a matrix of dummy variables of dimension $NT \times N$, which is composed of the N vectors $(\boldsymbol{e}_T, \boldsymbol{0}_T, \ldots, \boldsymbol{0}_T)', \ldots, (\boldsymbol{0}_T, \boldsymbol{0}_T, \ldots, \boldsymbol{e}_T)'$, $\boldsymbol{\alpha} = (\alpha_1, \ldots, \alpha_N)$ and $\boldsymbol{X} = (\boldsymbol{x}_1, \ldots, \boldsymbol{x}_N)'$ is a matrix of dimension $NT \times K$.

Random Effects and Error Component Model

Until now, the effects of variables specific to an individual unit have been considered as fixed constants. However, there is nothing to prevent these constants from being seen as random.

The equivalent of model (8.4) when random effects are introduced is therefore written as:

$$\boldsymbol{y} = \alpha\, \boldsymbol{e}_{NT} + \boldsymbol{X}\boldsymbol{\beta} + \boldsymbol{v}, \tag{8.5}$$

where $\boldsymbol{e}_{NT}$ is a vector of units of dimension NT and α is a constant. The essential difference now lies in the random term **v**, which is such that

$$v_{it} = \varepsilon_i + u_{it}, \quad i = 1, \ldots, N \; ; \quad t = 1, \ldots, T. \tag{8.6}$$

Each disturbance is therefore the sum of two stochastic variables, one representing the random effect specific to the individual unit and the other the traditional error term.

In view of (8.6), it is easy to understand why the model is called a simple **error component model.** (We use the qualifier "simple" because, as we shall soon see, we can add a temporal random effect, in which case we obtain the double or general error component model).

In any case, we will introduce the following restrictions on perturbations v_{it} : $\varepsilon_i \sim IID(0, \sigma_\varepsilon^2)$, $u_{it} \sim IID(0, \sigma_u^2)$ and the disturbance ε_i is independent of u_{it} for all i. Hence:

$$(i) \quad E(\varepsilon_i) = E(u_{it}) = 0 \; ;$$

$$(ii) \quad E(\varepsilon_i . u_{it}) = 0 \; ;$$

$$(iii) \quad E(\varepsilon_i \varepsilon_j) = \begin{cases} \sigma_\varepsilon^2 & \text{if } i = j \\ 0 & \text{if } i \neq j \end{cases}$$

$$(iv) \quad E(u_{it} u_{js}) = \begin{cases} \sigma_u^2 & \text{if } i = j, \; t = s \\ 0 & \text{otherwise} \end{cases}$$

$$(v) \quad E(v_{it} v_{ts}) = \sigma_\varepsilon^2, \quad t = s.$$

Consequently, for each observation on the individual unit i, the disturbance vector $\boldsymbol{v}_i$, of size $T \times 1$, admits the variance–covariance matrix of dimension $T \times T$

$$\boldsymbol{V} = E(\boldsymbol{v}_i \boldsymbol{v'}_i) = \begin{pmatrix} \sigma_\varepsilon^2 + \sigma_u^2 & \sigma_\varepsilon^2 & \cdots & \sigma_\varepsilon^2 \\ \sigma_\varepsilon^2 & \sigma_\varepsilon^2 + \sigma_u^2 & \cdots & \sigma_\varepsilon^2 \\ \vdots & \vdots & \ddots & \vdots \\ \sigma_\varepsilon^2 & \sigma_\varepsilon^2 & \cdots & \sigma_\varepsilon^2 + \sigma_u^2 \end{pmatrix}. \tag{8.7}$$

This matrix **V** can also be written as

$$\boldsymbol{V} = \sigma_u^2 \boldsymbol{I}_T + \sigma_\varepsilon^2 \boldsymbol{e}_T \boldsymbol{e'}_T,$$

where $\boldsymbol{e}_T = (1, \ldots, 1)'$ is, as before, a column vector made up of T units.

Taking into account all the individual units i, $i = 1, \ldots, N$, the overall variance–covariance matrix, $E(\boldsymbol{v}\boldsymbol{v'}) = \boldsymbol{\Omega}$, is given by

$$\boldsymbol{\Omega} = E(\boldsymbol{v}\boldsymbol{v}') = \begin{pmatrix} \boldsymbol{V} & \boldsymbol{0} & \cdots & \boldsymbol{0} \\ \boldsymbol{0} & \boldsymbol{V} & \cdots & \boldsymbol{0} \\ \vdots & \vdots & \ddots & \vdots \\ \boldsymbol{0} & \boldsymbol{0} & \ldots & \boldsymbol{V} \end{pmatrix}. \tag{8.8}$$

This matrix $\boldsymbol{\Omega}$ is of dimension $NT \times NT$ and therefore has the structure of a diagonal block matrix, each of these blocks being, for its part, of dimension $T \times T$.

Possible Extensions

An immediate extension of the fixed-effects model (8.2) is to add a specific time effect, which leads to a set of equations of the form

$$y_{it} = \alpha_i + \gamma_t + \boldsymbol{x}'_{it}\boldsymbol{\beta} + u_{it}, \qquad i = 1, \ldots, N \ ; \ \ t = 1, \ldots, T. \tag{8.9}$$

In other words, we have included $(T - 1)$ additional dummy variables among the regressors. (The base period is discarded to avoid the problem of multicollinearity). This gives us the **double fixed-effects model.**

In a similar way, a set of time-specific random effects can be introduced into the simple error component model represented by (8.5), so that the disturbances then take the following form:

$$v_{it} = \varepsilon_i + \lambda_t + u_{it}. \tag{8.10}$$

If the errors defined by $\varepsilon_i \sim IID(0, \sigma_\varepsilon^2)$, $\lambda_t \sim IID(0, \sigma_\lambda^2)$, $u_{it} \sim IID(0, \sigma_u^2)$ are independent of each other, then we have the **double or general error component model**. (We will not study this, as the addition of a temporal random effect is not a priority).

8.2 Estimation of the Fixed-Effects Model

We will simply give two methods for estimating the parameters of the (simple) fixed-effects model: one which consists of the pure and simple application of ordinary least squares, and the other which transforms the observations before applying least squares.

Least Squares Estimators

Let's start with the fixed-effects model represented by (8.4):

$$y = E\alpha + X\beta + u.$$

The sum of the squares **u'u** is given by

$$\begin{aligned} u'u &= (y - E\alpha - X\beta)'(y - E\alpha - X\beta) \\ &= y'y - 2y'E\alpha - 2y'X\beta + 2\beta'X'E\alpha + \alpha'EE\alpha + \beta'X'X\beta. \end{aligned} \tag{8.11}$$

Deriving (8.11) with respect to α and β, gives the two systems of normal equations

$$\begin{cases} E'E\hat{\alpha} + E'X\hat{\beta} = E'y \\ X'E\hat{\alpha} + X'X\hat{\beta} = X'y. \end{cases}$$

From the first of these equations, we can immediately derive

$$\hat{\alpha} = (E'E)^{-1}(E'y - E'X\hat{\beta}). \tag{8.12}$$

Transferring (8.12) to the second set of normal equations gives

$$\begin{aligned} & X'E(E'E)^{-1}(E'y - E'X\hat{\beta}) + X'X\hat{\beta} = X'y \\ & \Leftrightarrow X'X\hat{\beta} - X'E(E'E)^{-1}E'X\hat{\beta} = X'y - X'E(E'E)^{-1}E'y \\ & \Leftrightarrow X'(I_{NT} - E(E'E)^{-1}E')X\hat{\beta} = X'(I_{NT} - E(E'E)^{-1}E')y, \end{aligned}$$

from which we deduce, given $M_e = (I_{NT} - E(E'E)^{-1}E')$, that

$$\hat{\beta} = (X'M_eX)^{-1}X'M_e y. \tag{8.13}$$

All that remains is to substitute (8.13) into (8.12) to obtain the parameters characterising the fixed effects we are looking for.

The estimation method we have just used is sometimes referred to as Least Squares Dummy Variables (LSDV).

Let us add that, given the assumptions made on **u**, which are the same as those of the classical linear regression model, the Gauss-Markov theorem—see "The Gauss–Markov Theorem" section—allows us to state that estimators (8.12) and (8.13) are optimal and constitute the best linear unbiased estimators (BLUES). In other words, they have the smallest variance in the class of unbiased linear estimators.

From a practical point of view, to obtain $\hat{\beta}$, we have to use (8.12) and therefore invert the matrix $E'E$, which is of dimension $N \times N$. When the number of individual units is large—for example several thousand—inversion can pose certain computational problems.

This is why another estimator of the fixed-effects model, which is easier to calculate, has been developed. It is called the *Within* estimator.

The Within Estimator

The principle behind this method is to apply least squares to the data obtained by expressing the observations of each group i ($i = 1, \ldots, N$) as a deviation from their own mean.

Note that if the vector $z = (z_1, \ldots, z_T)$ is multiplied by the matrix $\boldsymbol{Q} = \boldsymbol{I}_T - (1/T)\,\boldsymbol{e}_T \boldsymbol{e}'_T$, where $\boldsymbol{e}_T$ is a vector made up of T units, it is expressed as a deviation from its mean.

Example 8.1 Let $z = (2, 4)'$ and $\boldsymbol{Q} = \boldsymbol{I}_2 - (1/2)(1, 1)'(1, 1)$. Then

$$\begin{aligned}\boldsymbol{Q}\,z &= \begin{pmatrix}1 & 0\\ 0 & 1\end{pmatrix}\begin{pmatrix}2\\ 4\end{pmatrix} - \frac{1}{2}\begin{pmatrix}1 & 1\\ 1 & 1\end{pmatrix}\begin{pmatrix}2\\ 4\end{pmatrix} = \begin{pmatrix}2\\ 4\end{pmatrix} - \frac{1}{2}\begin{pmatrix}6\\ 6\end{pmatrix}\\ &= \begin{pmatrix}2-3\\ 4-3\end{pmatrix}.\end{aligned}$$

What's more, the **Q** matrix is idempotent, i.e. $\boldsymbol{QQ} = \boldsymbol{Q}$, as is easy to check.

Consider system (8.3), whose typical equation is

$$\boldsymbol{y}_i = \boldsymbol{e}_T\alpha_i + \boldsymbol{X}_i\boldsymbol{\beta} + \boldsymbol{u}_i, \qquad i = 1, \ldots, N. \tag{8.14}$$

Let's then premultiply (8.14) by $\boldsymbol{Q} = \boldsymbol{I}_T - (1/T)\,\boldsymbol{e}_T\boldsymbol{e}'_T$ of dimension $T \times T$

$$\boldsymbol{Qy}_i = \boldsymbol{Q}\,\boldsymbol{e}_T\alpha_i + \boldsymbol{QX}_i\,\boldsymbol{\beta} + \boldsymbol{Qu}_i \tag{8.15}$$

After a few calculations, (8.15) becomes

$$\begin{aligned}\boldsymbol{Qy}_i &= (\boldsymbol{I}_T - (1/T)\boldsymbol{e}_T\boldsymbol{e}'_T)\boldsymbol{e}_T\alpha_i + \boldsymbol{QX}_i\boldsymbol{\beta} + \boldsymbol{Qu}_i\\ &= \boldsymbol{e}_T\alpha_i - (1/T)\,\boldsymbol{e}_T(T)\alpha_i + \boldsymbol{QX}_i\boldsymbol{\beta} + \boldsymbol{Qu}_i\\ &= \boldsymbol{QX}_i\boldsymbol{\beta} + \boldsymbol{Qu}_i, \qquad i = 1, \ldots, N.\end{aligned} \tag{8.16}$$

It is then sufficient to minimise the sum of the squares of the disturbances over all i, the matrix **Q** being idempotent, symmetrical and therefore equal to its transpose, to obtain an estimator of $\boldsymbol{\beta}$:

$$\sum_{i=1}^{N} \boldsymbol{u}'_i\boldsymbol{QQu}_i = \sum_{i=1}^{N} (\boldsymbol{Qy}_i - \boldsymbol{QX}_i\,\boldsymbol{\beta})'\,(\boldsymbol{Qy}_i - \boldsymbol{QX}_i\,\boldsymbol{\beta})$$

$$= \sum_{i=1}^{N} (y'_i QQ y_i - 2y'_i QQ y'_i X_i \beta + \beta' X'_i QQ X_i \beta)$$

$$= \sum_{i=1}^{N} (y'_i Q y_i - 2y'_i Q y'_i X_i \beta + \beta' X'_i Q X_i \beta).$$

The estimator in question, $\hat{\beta}_W$ (where w stands for "*Within*"), is then equal to

$$\hat{\beta}_W = \left(\sum_{i=1}^{N} X'_i Q X_i \right)^{-1} \left(\sum_{i=1}^{N} X'_i Q y_i \right). \tag{8.17}$$

Given the definition of **Q**, (8.17) can still be written as

$$\hat{\beta}_W = \left(\sum_{i=1}^{N} \sum_{t=1}^{T} (x_{it} - \overline{x}_i)(x_{it} - \overline{x}_i)' \right)^{-1} \left(\sum_{i=1}^{N} \sum_{t=1}^{T} (x_{it} - \overline{x}_i)(y_{it} - \overline{y}_i) \right), \tag{8.18}$$

where $\overline{x}_i = (1/T) \sum_{t=1}^{T} x_{it}$ and $\overline{y}_i = (1/T) \sum_{t=1}^{T} y_{it}$.

The operational advantage of the Within estimator over the least squares estimator with binary or dummy variables is immediately apparent: the calculation of the former, once the observations have been transformed and expressed as deviations from the group time averages, requires only the inversion of the matrix $X'_i Q X_i$, of dimension $K \times K$.

Moreover, the Within estimator is unbiased and its variance–covariance matrix

$$\text{Var}(\hat{\beta}_W) = \sigma_u^2 \left(\sum_{i=1}^{N} X'_i Q X_i \right)^{-1}$$

is minimal in the class of unbiased linear estimators of β. It is also convergent when T and/or N tend towards infinity.

However, since the parameters α_i have been eliminated by the observation transformation procedure, we will use the following formula to obtain them:

$$\hat{\alpha}_i = \overline{y}_i - \hat{\beta}_W \overline{x}_i, \qquad i = 1, \ldots, N. \tag{8.19}$$

The estimators in (8.19) are unbiased and convergent when $T \to +\infty$. However, they are not convergent when $N \to +\infty$, since, in this case, the number of constants to be estimated increases for a fixed T.

8.3 Estimation of the Error Component Model

There are several possible estimators of the random effects in the simple error component model. The difficulty lies in estimating the variance–covariance matrix of the random effects specific to this type of model.

Preliminaries

Before getting to the heart of the matter, a few details on the mathematical and statistical tools to be used will be welcome. They will focus on three points: the Kronecker product (also known as the outer product) of two matrices; the Within- and Between-individual operators; and finally, generalised and feasible generalised least squares.

Start with the first point, which we'll introduce by means of a definition.

Product of Kronecker

Definition 8.1 Let be $\boldsymbol{A}_{m\times n} = (a_{ij})$ and $\boldsymbol{B}_{p\times q} = (b_{ij})$. **The Kronecker product** of matrices **A** and **B**, denoted by $\boldsymbol{A}\otimes\boldsymbol{B}$, is the matrix of dimension $mp \times nq$ given by.

$$\boldsymbol{A}\otimes\boldsymbol{B} = \begin{pmatrix} a_{11}\boldsymbol{B} & \cdots & a_{1n}\boldsymbol{B} \\ \vdots & \ddots & \vdots \\ a_{m1}\boldsymbol{B} & \cdots & a_{mn}\boldsymbol{B} \end{pmatrix}.$$

Example 8.2 Let be the matrices

$$\boldsymbol{A} = \begin{pmatrix} 3 & 4 \\ 1 & 0 \end{pmatrix} \text{ and } \boldsymbol{B} = \begin{pmatrix} 4 & -1 \\ 3 & 3 \end{pmatrix}.$$

Their Kronecker product is

$$\boldsymbol{A}\otimes\boldsymbol{B} = \begin{pmatrix} 8 & -2 & 16 & -4 \\ 6 & 6 & 12 & 12 \\ 4 & -1 & 0 & 0 \\ 3 & 3 & 0 & 0 \end{pmatrix}.$$

Here are some of the properties of the Kronecker product:

(i) $(\boldsymbol{A} \otimes \boldsymbol{B})' = \boldsymbol{A}' \otimes \boldsymbol{B}'$

(ii) $\boldsymbol{A} \otimes (\boldsymbol{B} + \boldsymbol{C}) = (\boldsymbol{A} \otimes \boldsymbol{B}) + (\boldsymbol{A} \otimes \boldsymbol{C})$

(iii) $(\boldsymbol{A} \otimes \boldsymbol{B})(\boldsymbol{C} \otimes \boldsymbol{D}) = \boldsymbol{AC} \otimes \boldsymbol{BD}$

(iv) if $\boldsymbol{A}$ and $\boldsymbol{B}$ are invertible, $(\boldsymbol{A} \otimes \boldsymbol{B})^{-1} = \boldsymbol{A}^{-1} \otimes \boldsymbol{B}^{-1}$.

Within and Between Operators

When we derived the Within estimator, we used the **Q** matrix defined by $\boldsymbol{Q} = \boldsymbol{I}_T - (1/T)\, \boldsymbol{e}_T \boldsymbol{e'}_T$, where $\boldsymbol{e}_T$ is a column vector composed of T units, to express the observations in deviation from their group mean [see Eq. (8.15)].

To apply this same transformation to each of the components $\boldsymbol{y}_i$ of the vector $\boldsymbol{y} = (y_1, \ldots, y_N)$, we will use the **Within operator** defined by

$$\boldsymbol{W} = \boldsymbol{I}_N \otimes \left(\boldsymbol{I}_T - (1/T)\, \boldsymbol{e}_T \boldsymbol{e'}_T\right). \tag{8.20}$$

Clearly, this operator is **symmetric**, since the matrices of which it is composed are symmetric. It is also **idempotent** because, by virtue of property (iii) of the Kronecker product and the fact that the matrix $\boldsymbol{Q} = \boldsymbol{I}_T - (1/T)\, \boldsymbol{e}_T \boldsymbol{e'}_T$ is itself idempotent, it follows that

$$\begin{aligned} \boldsymbol{W}^2 &= \left[\left(\boldsymbol{I}_N \otimes \left(\boldsymbol{I}_T - (1/T)\, \boldsymbol{e}_T \boldsymbol{e'}_T\right)\right)\left(\boldsymbol{I}_N \otimes \left(\boldsymbol{I}_T - (1/T)\, \boldsymbol{e}_T \boldsymbol{e'}_T\right)\right)\right] \\ &= \boldsymbol{I}_N \otimes \left(\boldsymbol{I}_T - (1/T)\, \boldsymbol{e}_T \boldsymbol{e'}_T\right)\left(\boldsymbol{I}_T - (1/T)\, \boldsymbol{e}_T \boldsymbol{e'}_T\right) \\ &= \boldsymbol{I}_N \otimes \left(\boldsymbol{I}_T - (1/T)\, \boldsymbol{e}_T \boldsymbol{e'}_T\right). \end{aligned}$$

Similarly, we introduce the **Between operator** defined by

$$\boldsymbol{B} = \boldsymbol{I}_N \otimes (1/T)\, \boldsymbol{e}_T \boldsymbol{e'}_T. \tag{8.21}$$

Applying **B** to **y**, i.e. the vector of dimension $NT \times 1$ of time observations on each of the individual units $i = 1, \ldots, N$, then yields

$$\boldsymbol{By} = (\boldsymbol{I}_N \otimes (1/T)\, \boldsymbol{e}_T \boldsymbol{e'}_T)\boldsymbol{y} = \begin{pmatrix} (1/T)\, \boldsymbol{e}_T \boldsymbol{e'}_T & \ldots & \boldsymbol{0} \\ \vdots & \ddots & \vdots \\ \boldsymbol{0} & \ldots & (1/T)\, \boldsymbol{e}_T \boldsymbol{e'}_T \end{pmatrix} \begin{pmatrix} y_{11} \\ \vdots \\ y_{1T} \\ \vdots \\ y_{N1} \\ \vdots \\ y_{NT} \end{pmatrix} = \begin{pmatrix} \overline{y}_1 \\ \vdots \\ \overline{y}_1 \\ \vdots \\ \overline{y}_N \\ \vdots \\ \overline{y}_N \end{pmatrix}.$$

Consequently, the vector resulting from the **By** product is the vector $NT \times 1$ of the time averages of the groups of individual units, each of which is repeated T times.

It is easy to check, in the same way as the Within operator, that the Between operator is also **symmetrical** and **idempotent**.

Generalised and Feasible Generalised Least Squares

Either the classic linear model, but, restriction, in which the variance can be heteroscedastic and/or the errors autocorrelated

$$\boldsymbol{y} = \boldsymbol{X\beta} + \boldsymbol{u}, \qquad E(\boldsymbol{uu}') = \boldsymbol{\Omega}, \tag{8.22}$$

where $\boldsymbol{\Omega}$ therefore differs from the classical variance–covariance matrix $\sigma^2 \boldsymbol{I}_T$ for the reasons given. However, this matrix is assumed to be *known* and to satisfy $\boldsymbol{u} | \boldsymbol{X} \sim MN(\boldsymbol{0}, \sigma^2 \boldsymbol{\Omega})$, $\sigma^2 < +\infty$, with $\boldsymbol{\Omega}$ symmetric and positive definite.

On the basis of these assumptions, we can obtain an initial result.

Proposition 8.1 *There exists a non-singular matrix* **P** *such that*

$$\boldsymbol{\Omega}^{-1} = \boldsymbol{P}'\boldsymbol{P}.$$

Premultiplying (8.22) by **P** gives

$$\boldsymbol{Py} = \boldsymbol{PX\beta} + \boldsymbol{Pu}. \tag{8.23}$$

Let's then formulate two hypotheses that we would like the model transformed in this way to satisfy.

(H1) $\boldsymbol{u} \,|\, \boldsymbol{X} \sim MN(\boldsymbol{0}, \sigma^2 \boldsymbol{I}_T)$.

(H2) **X** is a non-stochastic matrix of rank $K < T$ such that

$$\boldsymbol{L} = \operatorname*{plim}_{T \to +\infty} \left(\frac{\boldsymbol{X}'\boldsymbol{X}}{T}\right)$$

exists and is **non-singular**

As it happens, (8.23) satisfies both these assumptions, which is recorded by the

Proposition 8.2 *The transformed model* $\boldsymbol{Py} = \boldsymbol{PX\beta} + \boldsymbol{Pu}$, *where* **P** *is non-singular verifies conditions* (H1) *and* (H2).

Since the transformed model satisfies the assumptions necessary for the application of ordinary least squares, a proposition known as Aitken's theorem arises.

Proposition 8.3 *Let the transformed model* $\boldsymbol{Py} = \boldsymbol{PX\beta} + \boldsymbol{Pu}$ *satisfy conditions* (H1) *and* (H2). *Then the best linear unbiased estimator* (BLUES) *of* $\boldsymbol{\beta}$ *is the* ***generalised least squares*** (GLS) *estimator*:

$$\hat{\beta}_{GLS} = (X'\Omega^{-1}X)^{-1}X'\Omega^{-1}y. \tag{8.24}$$

Let us add that the variance–covariance matrix of the generalised least squares estimator is equal to $\sigma^2(X'\Omega^{-1}X)^{-1}$.

Until now, it has been assumed that the matrix Ω was **known**. In practice, this is not the case and it is therefore necessary to estimate it. Of course, this is not always possible: for example, if we had to estimate all the distinct elements of Ω—of which there are $T(T+1)/2$—the task would obviously be hopeless, since we would only have T observations.

As long as we have an estimator from Ω with the "right" properties, we have the following definition:

Definition 8.2 Let be the model $y = X\beta + u, \quad u \sim MN(0, \Omega)$, and $\widehat{\Omega}$ an estimator of Ω. Then the **Feasible generalized Least Squares** (F.G.L.S.) estimator of β is given by

$$\hat{\beta}_{MCQG} = \left(X'\widehat{\Omega}^{-1}X\right)^{-1}X'\widehat{\Omega}^{-1}y.$$

Estimation Using Generalised Least Squares

Consider the error component model (8.5) rewritten as

$$y = X\beta + v, \tag{8.25}$$

with $v_{it} = \varepsilon_i + u_{it}, \ i = 1, \ldots, N$ and $t = 1, \ldots, T$. (The matrix $\mathbf{X}$ therefore incorporates the unit vector e_{NT} and the vector β the constant α.)

Furthermore, in "Random Effects and Error Component Model" section, we saw that the variance–covariance matrix of each $\mathbf{v}_i$ was given by

$$V_{T\times T} = \sigma_u^2 I_T + \sigma_\varepsilon^2 e_T e'_T. \tag{8.26}$$

We also know—see Eq. (8.8)—that $\Omega = E(vv')$ is a diagonal block matrix of dimension $NT \times NT$, $\mathbf{V}$ being repeated N times. Moreover, each block has the structure given by Eq. (8.26).

It follows that Ω can still be written using the Kronecker product and especially its property (ii):

$$\begin{aligned}\Omega &= \sigma_u^2 I_{NT} + (\sigma_\varepsilon^2 I_N \otimes e_T e'_T)\\ &= \sigma_u^2 (I_N \otimes e_T e'_T) + (\sigma_\varepsilon^2 I_N \otimes e_T e'_T)\end{aligned}$$

$$= \sigma_u^2 \boldsymbol{I}_N \otimes \left(\boldsymbol{I}_T + \frac{\sigma_\varepsilon^2}{\sigma_u^2} \boldsymbol{e}_T \boldsymbol{e'}_T \right).$$

We then transform this last expression of the matrix $\boldsymbol{\Omega}$, to show its relationship with the operators Within and Between:

$$\begin{aligned}
\boldsymbol{\Omega} &= \sigma_u^2 \boldsymbol{I}_N \otimes \left(\boldsymbol{I}_T - (1/T)\boldsymbol{e}_T \boldsymbol{e'}_T + (1/T)\,\boldsymbol{e}_T \boldsymbol{e'}_T + \frac{\sigma_\varepsilon^2}{\sigma_u^2} \boldsymbol{e}_T \boldsymbol{e'}_T \right) \\
&= \sigma_u^2 \boldsymbol{I}_N \otimes \left((\boldsymbol{I}_T - (1/T)\boldsymbol{e}_T \boldsymbol{e'}_T) + (1/T)\,\boldsymbol{e}_T \boldsymbol{e'}_T \left(1 + T \frac{\sigma_\varepsilon^2}{\sigma_u^2} \right) \right) \\
&= \sigma_u^2 \boldsymbol{I}_N \otimes \left((\boldsymbol{I}_T - (1/T)\boldsymbol{e}_T \boldsymbol{e'}_T) + \left(\frac{\sigma_u^2 + T\sigma_\varepsilon^2}{\sigma_u^2} \right) (1/T)\,\boldsymbol{e}_T \boldsymbol{e'}_T \right) \\
&= \sigma_u^2 \left[(\boldsymbol{I}_N \otimes (\boldsymbol{I}_T - (1/T)\boldsymbol{e}_T \boldsymbol{e'}_T)) + \left(\frac{\sigma_u^2 + T\sigma_\varepsilon^2}{\sigma_u^2} \right) (\boldsymbol{I}_N \otimes (1/T)\boldsymbol{e}_T \boldsymbol{e'}_T) \right] \\
&= \sigma_u^2 \left(\boldsymbol{W} + \left(\frac{\sigma_u^2 + T\sigma_\varepsilon^2}{\sigma_u^2} \right) \boldsymbol{B} \right) \\
&= \sigma_u^2 \boldsymbol{W} + (\sigma_u^2 + T\sigma_\varepsilon^2)\boldsymbol{B}.
\end{aligned}$$

These developments show that the variance–covariance matrix of the disturbances of the error component model, i.e.

$$\boldsymbol{\Omega} = \sigma_u^2 \mathbf{W} + (\sigma_u^2 + T\sigma_\varepsilon^2)\boldsymbol{B}, \tag{8.27}$$

combines the Within (**W**) and Between (**B**) matrices in a given proportion.

By posing $\theta = \sigma_u^2/(\sigma_u^2 + T\sigma_\varepsilon^2)$, (8.27) can still be written as

$$\boldsymbol{\Omega} = \mathbf{W} + \theta^{-1}\,\boldsymbol{B}. \tag{8.28}$$

If we assume that this matrix $\boldsymbol{\Omega}$ is **known,** then, by virtue of Proposition 8.3 and using (8.28), the best linear unbiased estimator of $\boldsymbol{\beta}$ in (8.25) is the **generalised least squares** estimator given by

$$\begin{aligned}
\hat{\boldsymbol{\beta}}_{GLS} &= (\boldsymbol{X'\Omega}^{-1}\boldsymbol{X})^{-1}\boldsymbol{X'\Omega}^{-1}\boldsymbol{y} \\
&= \left(\boldsymbol{X}'(\boldsymbol{W} + \theta^{-1}\,\boldsymbol{B})^{-1}\boldsymbol{X}\right)^{-1}\boldsymbol{X}'(\boldsymbol{W} + \theta^{-1}\,\boldsymbol{B})^{-1}\boldsymbol{X} \\
&= \left(\boldsymbol{X'WX} + \theta\,\boldsymbol{X'BX}\right)^{-1}\left(\boldsymbol{X'Wy} + \theta\,\boldsymbol{X'By}\right),
\end{aligned}$$

given that **W** and **B** are their own inverse.

The GLS estimator therefore combines the Within and Between estimators in a proportion given by θ. In practice, however, it is of little use, quite simply because we do not know the matrix $\boldsymbol{\Omega}$.. We will therefore turn to its estimable version, the feasible generalised least squares estimator.

Feasible Generalised Least Squares Estimation

Given definition (8.2), the estimator for FGLSs is identical to that for GLSs, except that the former requires us to obtain a convergent estimator of the variance–covariance matrix of the disturbances.

We therefore have the following formulation:

$$\begin{aligned}\hat{\boldsymbol{\beta}}_{QGLS} &= (\boldsymbol{X}'\hat{\boldsymbol{\Omega}}^{-1}\boldsymbol{X})^{-1}\boldsymbol{X}'\hat{\boldsymbol{\Omega}}^{-1}\boldsymbol{y} \\ &= \left(\boldsymbol{X}'\boldsymbol{W}\boldsymbol{X} + \hat{\theta}\,\boldsymbol{X}'\boldsymbol{B}\boldsymbol{X}\right)^{-1}\left(\boldsymbol{X}'\boldsymbol{W}\boldsymbol{y} + \hat{\theta}\,\boldsymbol{X}'\boldsymbol{B}\boldsymbol{y}\right),\end{aligned}$$

where

$$\hat{\theta} = \frac{\hat{\sigma}_u^2 + T\hat{\sigma}_\varepsilon^2}{\hat{\sigma}_u^2}.$$

The whole problem therefore boils down to obtaining estimators of the unknown variances σ_u^2 and σ_ε^2, which is usually done using the procedure devised by Arora and Swamy.

This essentially involves deriving estimators of the variances of the residuals resulting from the Within and Between regressions.

Within (or Within-individual) regression consists, as we have seen, in estimating by OLS the transformed model, in which the variables are expressed as deviations from the individual means, thus the group, i.e. model

$$\boldsymbol{W}\boldsymbol{y} = \boldsymbol{W}\boldsymbol{X}\boldsymbol{\beta}_W + \boldsymbol{W}\boldsymbol{v}.$$

This gives the estimator $\hat{\boldsymbol{\beta}}_W$ and the corresponding residuals

$$\hat{\boldsymbol{v}}_W = \boldsymbol{W}\boldsymbol{y} - \boldsymbol{W}\boldsymbol{X}\hat{\boldsymbol{\beta}}_W,$$

and the variance of these residuals

$$\hat{\sigma}_W^2 = \frac{\hat{\boldsymbol{v}}'_W\hat{\boldsymbol{v}}_W}{NT - N - K}. \tag{8.29}$$

Similarly, applying least squares to the transformed model in which **B** is the matrix Between

$$\boldsymbol{B}\boldsymbol{y} = \boldsymbol{B}\boldsymbol{X}\boldsymbol{\beta}_B + \boldsymbol{B}\boldsymbol{v},$$

we obtain the estimator $\hat{\boldsymbol{\beta}}_B$ and the residuals

$$\hat{\boldsymbol{v}}_B = \boldsymbol{B}\boldsymbol{y} - \boldsymbol{B}\boldsymbol{X}\hat{\boldsymbol{\beta}}_B,$$

and the variance of these residuals

$$\hat{\sigma}_B^2 = \frac{\hat{\boldsymbol{v}}'_B \hat{\boldsymbol{v}}_B}{N-K}. \tag{8.30}$$

We then record that (8.29) and (8.30) are unbiased estimators of σ_u^2 and $(\sigma_u^2 + T\sigma_\varepsilon^2)$. respectively. More precisely, we have a new proposition:

Proposition 8.4 *The estimators $\hat{\sigma}_W^2$ and $\hat{\sigma}_B^2$ are such that*

$$E(\hat{\sigma}_W^2) = E\left(\frac{\hat{\boldsymbol{v}}'_W \hat{\boldsymbol{v}}_W}{NT-N-K}\right) = \sigma_u^2$$
$$E(\hat{\sigma}_B^2) = E\left(\frac{\hat{\boldsymbol{v}}'_B \hat{\boldsymbol{v}}_B}{N-K}\right) = \sigma_u^2 + T\sigma_\varepsilon^2.$$

From this proposition, we can easily obtain an unbiased and convergent estimator of θ:

$$\hat{\theta} = \frac{\hat{\sigma}_W^2}{\hat{\sigma}_B^2},$$

from which we derive, on the basis of (8.27), that

$$\hat{\boldsymbol{\Omega}} = \hat{\sigma}_W^2 \left(\boldsymbol{W} + \frac{\hat{\sigma}_B^2}{\hat{\sigma}_W^2} \boldsymbol{B} \right).$$

The inverse of this matrix is given by

$$\hat{\boldsymbol{\Omega}}^{-1} = (\hat{\sigma}_W^2)^{-1} \left(\boldsymbol{W} + \frac{\hat{\sigma}_W^2}{\hat{\sigma}_B^2} \boldsymbol{B} \right). \tag{8.31}$$

The feasible generalised least squares estimator is then obtained using (8.31), such that

$$\hat{\boldsymbol{\beta}}_{QGLS} = \left(\boldsymbol{X}' \widehat{\boldsymbol{\Omega}}^{-1} \boldsymbol{X} \right)^{-1} \boldsymbol{X}' \widehat{\boldsymbol{\Omega}}^{-1} \boldsymbol{y}$$
$$= \left[\boldsymbol{X}' \left((\hat{\sigma}_W^2)^{-1} \left(\boldsymbol{W} + \frac{\hat{\sigma}_W^2}{\hat{\sigma}_B^2} \boldsymbol{B} \right) \right) \boldsymbol{X} \right]^{-1} \left[\boldsymbol{X}' \left((\hat{\sigma}_W^2)^{-1} \left(\boldsymbol{W} + \frac{\hat{\sigma}_W^2}{\hat{\sigma}_B^2} \boldsymbol{B} \right) \right) \boldsymbol{y} \right].$$

Its variance–covariance matrix is obviously equal to

$$\operatorname{Var}(\hat{\boldsymbol{\beta}}_{QGLS}) = \sigma_W^2 \left(\boldsymbol{X}' (\boldsymbol{W} + \hat{\theta}\, \boldsymbol{B}) \boldsymbol{X} \right)^{-1}$$
$$= \sigma_W^2 \left(\boldsymbol{X}' \left(\boldsymbol{W} + \frac{\sigma_W^2}{\sigma_B^2} \boldsymbol{B} \right) \boldsymbol{X} \right)^{-1}.$$

There are methods or procedures other than that of Arora and Swamy (1972) for estimating $\hat{\theta}$ and thus obtaining the feasible generalised least squares estimator. Without wishing to be exhaustive, let us mention the procedures recommended by Wansbeek and Kapteyn (1982), Fuller and Battese (1973, 1974), Wallace and Hussain (1969), Amemiya (1971) and Nerlove (1971).

Nevertheless, the Arora–Swamy method, based on the calculation of the two regressions Within and Between, has a great computational advantage.

Maximum Likelihood Estimation

Let's start by recalling the density of a random vector distributed according to a multidimensional normal distribution.

Definition 8.3 If $X = (X_1, \ldots, X_n)$ follows a **multinormal distribution,** then its **density** is

$$f_X(x_1, \ldots, x_n) = \frac{\left|\det \boldsymbol{\Sigma}^{-1}\right|^{1/2}}{(2\pi)^{n/2}} e^{-\frac{1}{2}(\boldsymbol{x}-\boldsymbol{\mu})'\boldsymbol{\Sigma}^{-1}(\boldsymbol{x}-\boldsymbol{\mu})}, \quad \boldsymbol{x} \in \mathbb{R}^n,$$

where $\boldsymbol{\mu} = E(\boldsymbol{X})$ and $\boldsymbol{\Sigma} = \mathrm{Var}(\boldsymbol{X})$.

Since $\boldsymbol{v} \sim NM\,(\boldsymbol{0}, \boldsymbol{\Omega})$ in the model

$$\boldsymbol{y} = \alpha\, \boldsymbol{e}_{NT} + \boldsymbol{X\beta} + \boldsymbol{v},$$

the likelihood function of the perturbations is written as

$$L(\alpha, \boldsymbol{\beta}, \boldsymbol{\Omega} \,|\, \boldsymbol{y}) = (2\pi)^{-NT/2} (\det \boldsymbol{\Omega})^{-1/2} e^{(-\,1/2)\ (\boldsymbol{y}-\boldsymbol{e}_{NT}\alpha-\boldsymbol{X\beta})'\boldsymbol{\Omega}^{-1}(\boldsymbol{y}-\boldsymbol{e}_{NT}\alpha-\boldsymbol{X\beta})}.$$

The log-likelihood function is therefore equal to

$$\begin{aligned} \ln L(\alpha, \boldsymbol{\beta}, \boldsymbol{\Omega} \,|\, \boldsymbol{y}) &= -\frac{NT}{2}\ln(2\pi) - \frac{1}{2}\ln(\det \boldsymbol{\Omega}) \\ &\quad - \frac{1}{2}(\boldsymbol{y} - \boldsymbol{e}_{NT}\alpha - \boldsymbol{X\beta})'\boldsymbol{\Omega}^{-1}(\boldsymbol{y} - \boldsymbol{e}_{NT}\alpha - \boldsymbol{X\beta}). \end{aligned} \tag{8.32}$$

We now need to determine the values of $\boldsymbol{\Omega}^{-1}$ and the determinant of this same matrix. We already know that

$$\boldsymbol{\Omega}^{-1} = (\sigma_u^2)^{-1}(\boldsymbol{W} + \theta\, \boldsymbol{B}), \tag{8.33}$$

where **W** and **B** are the Within and Between operators respectively, and θ is given by

$$\theta = \frac{\sigma_u^2}{\sigma_u^2 + T\sigma_\varepsilon^2}. \tag{8.34}$$

The calculation of the determinant of $\boldsymbol{\Omega}$ is long and complicated. It involves a theorem which states that the determinant of a matrix is equal to the product of its eigenvalues. For this reason, we will simply give the final result, i.e.

$$\det(\boldsymbol{\Omega}) = (\sigma_u^2)^{NT}\,\theta^{\,-N}. \tag{8.35}$$

Using these results (Eqs. 8.33–8.35), the log-likelihood function becomes:

$$\begin{aligned}\ln L(\alpha, \beta, \sigma_u^2, \sigma_\varepsilon^2) = k &- \frac{NT}{2}\ln(\sigma_u^2) + \frac{N}{2}\ln(\theta)\\ &- \frac{1}{2\sigma_u^2}(\boldsymbol{y} - \boldsymbol{e}_{NT}\alpha - \boldsymbol{X\beta})'(\boldsymbol{W} + \theta\,\boldsymbol{B})(\boldsymbol{y} - \boldsymbol{e}_{NT}\alpha - \boldsymbol{X\beta}).\end{aligned} \tag{8.36}$$

Maximising such a function is particularly difficult, as the first-order conditions for an extremum lead to a system of non-linear equations that are not easy to solve. Consequently, it is better to use an algorithm—due to Breusch (1978)—which, moreover, leads to some very useful practical recommendations.

The essence of the approach consists in concentrating the log-likelihood (8.36) by substituting the maximants $\hat{\alpha}$ and $\hat{\sigma}_u^2$, while taking into account a set of values for $\boldsymbol{\beta}$ and θ.

The maximum likelihood estimator of α is easily obtained:

$$\hat{\alpha}_{MV} = \frac{1}{NT}\,\boldsymbol{e}'_{NT}\left(\boldsymbol{y} - \boldsymbol{X}\hat{\boldsymbol{\beta}}_{MV}\right). \tag{8.37}$$

On the other hand, the procedure for determining the estimator $\hat{\sigma}_u^2$ is much more delicate.

To do this, let's look at several matrices that share the common characteristic of being both symmetric and idempotent. First, there are the well-known matrices Within (**W**) and Between (**B**) defined by $\boldsymbol{W} = \boldsymbol{I}_N \otimes (\boldsymbol{I}_T - (1/T)\,\boldsymbol{e}_T\boldsymbol{e}'_T)$ and $\boldsymbol{B} = \boldsymbol{I}_N \otimes (1/T)\,\boldsymbol{e}_T\boldsymbol{e}'_T$. Then there are two other matrices:

$$\begin{aligned}\boldsymbol{W}_0 &= \boldsymbol{I}_{NT} - (1/NT)\,\boldsymbol{e}_{NT}\boldsymbol{e}'_{NT}\\ \boldsymbol{Q}_B &= \boldsymbol{W}_0 - \boldsymbol{W}.\end{aligned}$$

Deriving the log-likelihood (8.36) with respect to σ_u^2 and cancelling this derivative, while adopting the convention that $\boldsymbol{v} = \boldsymbol{y} - \boldsymbol{e}_{NT}\alpha - \boldsymbol{X\beta}$, gives:

$$\frac{\partial \ln L}{\partial \sigma_u^2} = -\frac{NT}{2\sigma_u^2} + \frac{1}{2\sigma_u^4} \boldsymbol{v}'(\boldsymbol{W} + \theta\, \boldsymbol{B})^{-1}\boldsymbol{v} = 0,$$

which leads to the ML estimator

$$\hat{\sigma}_u^2 = (1/NT)\boldsymbol{v}'(\boldsymbol{W} + \theta\, \boldsymbol{B})^{-1}\boldsymbol{v}. \tag{8.38}$$

However, (8.38) can be rewritten as

$$\hat{\sigma}_u^2 = (1/NT)\boldsymbol{d}'(\boldsymbol{W} + \theta\, \boldsymbol{Q}_B)\boldsymbol{d}, \tag{8.39}$$

with $\boldsymbol{d} = \boldsymbol{y} - \boldsymbol{X}\boldsymbol{\beta}$.

By transferring the two previous equations into (8.36), we obtain the concentrated likelihood

$$\begin{aligned}\ln L_c(\boldsymbol{\beta}, \theta) = k &- \frac{NT}{2} \ln\left(\frac{1}{NT}\boldsymbol{d}'(\boldsymbol{W} + \theta\, \boldsymbol{Q}_C)^{-1}\boldsymbol{d}\right) + \frac{N}{2}\ln(\theta)\\ &- \frac{NT}{2}\frac{\boldsymbol{v}'(\boldsymbol{W} + \theta\, \boldsymbol{B})^{-1}\boldsymbol{v}}{\boldsymbol{v}'(\boldsymbol{W} + \theta\, \boldsymbol{B})^{-1}\boldsymbol{v}}.\end{aligned}$$

After simplification, this function becomes

$$\ln L_c(\boldsymbol{\beta}, \theta) = k * -\frac{NT}{2}\ln\left(\frac{1}{T}\right) - \frac{NT}{2}\ln\big(\boldsymbol{d}'(\boldsymbol{W} + \theta\, \boldsymbol{Q}_B)\boldsymbol{d}\big) + \frac{N}{2}\ln(\theta). \tag{8.40}$$

It is now a question of maximising the concentrated log-likelihood with respect to $\boldsymbol{\beta}$ and θ. The first-order conditions are:

$$\left\{\begin{aligned}\frac{\partial \ln L_c(\boldsymbol{\beta},\theta)}{\partial \boldsymbol{\beta}} &= -\frac{NT}{2}\cdot\frac{-2\boldsymbol{X}'(\boldsymbol{W}+\theta\,\boldsymbol{Q}_B)\boldsymbol{y}+2\boldsymbol{X}'(\boldsymbol{W}+\theta\,\boldsymbol{Q}_B)\boldsymbol{X}\boldsymbol{\beta}}{(\boldsymbol{y}-\boldsymbol{X}\boldsymbol{\beta})'(\boldsymbol{W}+\theta\,\boldsymbol{Q}_B)(\boldsymbol{y}-\boldsymbol{X}\boldsymbol{\beta})} = 0\\ \frac{\partial \ln L_c(\boldsymbol{\beta},\theta)}{\partial \theta} &= -\frac{NT}{2}\cdot\frac{1}{\boldsymbol{d}'(\boldsymbol{W}+\theta\,\boldsymbol{Q}_B)\boldsymbol{d}}\cdot \boldsymbol{d}'\boldsymbol{Q}_B\boldsymbol{d} - \frac{N}{2\theta}\\ &= -\theta NT\,\boldsymbol{d}'\boldsymbol{Q}_B\boldsymbol{d} + N\,\boldsymbol{d}'\boldsymbol{W} + N\theta\,\boldsymbol{d}'\boldsymbol{Q}_B\boldsymbol{d} = 0.\end{aligned}\right.$$

From the first of these conditions, we derive that

$$\hat{\boldsymbol{\beta}}_{ML} = \big(\boldsymbol{X}'(\boldsymbol{W} + \theta\, \boldsymbol{Q}_B)\boldsymbol{X}\big)^{-1}\boldsymbol{X}'(\boldsymbol{W} + \theta\, \boldsymbol{Q}_B)\boldsymbol{y} \tag{8.41}$$

and the second that

$$\hat{\theta}_{ML} = \frac{\boldsymbol{d}'\boldsymbol{W}\boldsymbol{d}}{(T-1)\,\boldsymbol{d}'\boldsymbol{Q}_B\boldsymbol{d}}. \tag{8.42}$$

Note that in (8.41), if $\theta = 1$, knowing that $\boldsymbol{W} + \boldsymbol{Q}_B = \boldsymbol{W}_0$, we obtain the OLS estimator

$$\hat{\boldsymbol{\beta}}_{OLS} = (\boldsymbol{X}'\boldsymbol{W}_0\boldsymbol{X})^{-1}\boldsymbol{X}'\boldsymbol{W}_0\,\boldsymbol{y},$$

whereas if $\theta = 0$, the Within estimator is obtained:

$$\hat{\boldsymbol{\beta}}_W = (\boldsymbol{X}'\boldsymbol{W}\boldsymbol{X})^{-1}\boldsymbol{X}'\boldsymbol{W}\,\boldsymbol{y}.$$

For all other values of θ, the maximum likelihood estimator of $\boldsymbol{\beta}$ corresponds to the feasible generalised least squares estimator.

Let us now consider the sequence of operators resulting from the successive application of formulae (8.41) and (8.42):

$$\theta^i_{ML} \to \boldsymbol{\beta}^i_{ML} \to \theta^{i+1}_{ML} \to \boldsymbol{\beta}^{i+1}_{ML} \to \cdots$$

This sequence has two remarkable properties which we will state in the form of a proposition:

Proposition 8.5

(i) *If* $T > 1$, *any* $\boldsymbol{\beta}^i$ *induces* $0 < \theta^{\,i+1} < +\infty$.
(ii) *If* $0 \leq \theta^{\,i} < \theta^j$, *then* $0 < \theta^{\,i+1} < \theta^{j+1}$.

This last sub-proposition (ii) implies that for all $\theta^{\,i} \in\to [0, +\infty[$,

1. If $\theta^{\,i} < \theta^{\,i+1}$, then $\theta^i < \theta^{\,i+1} < \theta^{\,i+2} < \cdots$;
2. If $\theta^{\,i} > \theta^{\,i+1}$, then $\theta^i > \theta^{\,i+1} > \theta^{\,i+2} > \cdots$.

This leads to a set of practical recommendations based on the increasing or decreasing monotonicity of the sequences $(\theta^{\,i})_{i\in\mathbb{N}^*}$, which we will summarise in four points.

1. From the Within estimator obtained for $\theta^{\,1} = 0$, we construct a sequence $(\theta^{\,i})$ which is strictly increasing. Similarly, from the Between estimator obtained when $\theta^{\,1} \to +\infty$, we construct a sequence $(\theta^{\,i})$ which is monotonically decreasing.
2. There is a single local maximum of the concentrated likelihood—which is therefore also a global maximum—if and only if the sequences initialised with $\theta^{\,1} = 0$ (Within estimator) and $\theta^{\,1} \to +\infty$ (Between estimator) converge towards the same estimator.
3. If the iterations applied on the basis of the Within estimator give a value of θ which violates the non-negativity requirement of the composite variance, then there is no local maximum inside the interval $]0, 1]$ and the estimators subject to the non-negativity constraint are $\hat{\theta}_{ML} = 1$ and $\hat{\boldsymbol{\beta}}_{ML} = \hat{\boldsymbol{\beta}}_{OLS}$.
4. Maddala (1971) showed that there are at most two maxima of the concentrated likelihood on the interval $0 < \theta \leq 1$ and that a maximum of $\ln L_c(\theta)$ is reached at the boundary $\theta = 1$ if and only if

$$\frac{\partial}{\partial\theta}\left(T \ln \sigma_u^2 - \ln\theta\right)_{\theta=1} < \; 0.$$

This last requirement is easily met by starting iterations from $\theta^{\,0} = 1$,, i.e. by using the least squares estimator $\hat{\boldsymbol{\beta}}_{OLS}$, and observing whether the next iteration provides a $\theta^{\,1} > 1$: local maximum at the boundary $\theta = 1$ is then reached.

8.4 Hypothesis Testing

Logically, having shown which estimators are appropriate for the various panel data models, we now need to consider the corresponding tests which are, of course, specific to longitudinal series.

The Fixed-Effects Model

What interests us here is essentially to develop one or other test which is, let us repeat, specific to this type of model. The best way to do this is to start from Eqs. (8.1) and (8.2), either on one side,

$$y_{it} = \alpha_0 + \boldsymbol{x}'_{it}\boldsymbol{\beta} + u_{it}, \qquad i = 1, \ldots, N \ ; \quad t = 1, \ldots, T \tag{8.1}$$

and the other,

$$y_{it} = \alpha_i + \boldsymbol{x}'_{it}\boldsymbol{\beta} + u_{it}, \qquad i = 1, \ldots, N \ ; \quad t = 1, \ldots, T. \tag{8.2}$$

The first of these models—we have slightly modified Eq. (8.1) for ease of comparison—is simply a set of "stacked" linear least squares regressions, while the second is the fixed-effects model, estimated by least squares with dummy variables (LSDV) or Within regression. (The latter two regressions give computationally equivalent results, at least when N is large).

We can of course test the joint significance of the dummy (or binary) variables, i.e. the null hypothesis

$$H_0: \ \alpha_1 = \alpha_2 = \cdots = \alpha_{N-1} = 0,$$

by performing an F-test.

In this case, the test statistic—see "Lagrange Multiplier Test" section—will be given, under the null hypothesis, using the sum of the squares of the constrained residuals obtained from the "stacked" regressions in (8.1 and the sum of the squares of the free residuals resulting from the application of the LSDV or Within regression:

$$S_0 = \frac{(\hat{\boldsymbol{u}}'_C\hat{\boldsymbol{u}}_C - \hat{\boldsymbol{u}}'_L\hat{\boldsymbol{u}}_L)/N - 1}{\hat{\boldsymbol{u}}'_L\hat{\boldsymbol{u}}_L/(NT - N - K)} \sim F(N - 1, N(T - 1) - K). \tag{8.43}$$

The conclusion of the test is then obvious. If the statistic S_0 is greater than the fractile F of order $(1 - \alpha)$ of the Fisher-Snedecor distribution, then we reject the null hypothesis and deduce that the model must incorporate fixed effects, i.e. the N dummy variables α_i. On the other hand, if this same statistic is less than the fractile in question, we accept the null hypothesis and conclude that there are no fixed effects.

However, an alternative approach is possible. It consists of examining whether the individual or group units all incorporate the same constant so that the null hypothesis in this case is as follows:

$$H_0: \quad \alpha_1 = \alpha_2 = \cdots = \alpha_{N-1} = \alpha_0,$$

whereas the alternative hypothesis is that the different units incorporate a constant that is specific to each of them.

The corresponding statistic, noted here $\overline{S}_0$, is obviously a statistic that also follows a Fisher-Snedecor distribution with the same number of degrees of freedom as in the previous case:

$$\overline{S}_0 \sim F(N-1, N(T-1)-K).$$

The same conclusions apply as above, depending on whether the latter statistic $\overline{S}_0$ is above or below the order fractile $(1-\alpha)$ of the Fisher-Snedecor distribution.

Instrumental Variables and General Hausman Test

Let's return to the linear model as formulated in Eq. (3.2), i.e.

$$\boldsymbol{y} = \boldsymbol{X\beta} + \boldsymbol{u}, \tag{8.44}$$

where **y** and **u** are vectors $T \times 1$, $\boldsymbol{\beta}$ a column vector with K components and **X** a matrix of dimension $T \times K$.

By assumption, the matrix **X** and the vector **u** are assumed to be uncorrelated, i.e. $E(\boldsymbol{u}|\boldsymbol{X}) = \boldsymbol{0}$. Alternatively, **u** is also said to be **orthogonal** to **X.** What happens when this assumption is not verified and, as a result, the least squares estimator of $\boldsymbol{\beta}$ is no longer the best (BLUES)?

One solution to this problem is to use the **instrumental variables estimator** (IV). To do this, let's assume that there is a matrix **Z** made up of K **instruments**, i.e. of format $T \times K$, such that

(i) the instruments in **Z** are correlated with the corresponding variables in **X**;
(ii) the instruments are asymptotically uncorrelated with the error vector **u**, i.e. $\text{plim}((1/T)\boldsymbol{Z}'\boldsymbol{u}) = \boldsymbol{0}$.

Asymptotically, the non-correlation of **u** with the instrument matrix results in a set of conditions on the moments

$$\boldsymbol{Z}'\boldsymbol{u} = \boldsymbol{Z}'(\boldsymbol{y} - \boldsymbol{X\beta}) = \boldsymbol{0},$$

from which, given that $\boldsymbol{Z}'\boldsymbol{X}$ has the dimension $K \times K$ and is assumed to be invertible, we derive the estimator of the instrumental variables:

$$\hat{\boldsymbol{\beta}}_{IV} = (\boldsymbol{Z}'\boldsymbol{X})^{-1}\boldsymbol{Z}'\boldsymbol{y}. \tag{8.45}$$

This IV estimator is therefore asymptotically **convergent**, but its uniqueness depends crucially on the assumption that there are as many instruments as explanatory variables and that the $\boldsymbol{Z}'\boldsymbol{X}$ matrix is non-singular.

Hausman developed a test whose null hypothesis for model (8.44) is

$$H_0 \; : \; E(\boldsymbol{X}'\boldsymbol{u}) = \boldsymbol{0}.$$

This assumption is an orthogonality assumption between the regressors and the errors; its violation indicates that the former are neither exogenous nor deterministic.

The principle on which this test is based is easy to understand: under H_0, the least squares estimator is convergent and that of the instrumental variables is also convergent; if we denote by $\hat{\boldsymbol{\beta}}_{OLS}$ the least squares estimator and by $\hat{\boldsymbol{\beta}}_{IV}$ that of the instrumental variables, then under the null hypothesis of orthogonality, the difference between these two estimators, i.e. $\hat{\boldsymbol{q}} = \hat{\boldsymbol{\beta}}_{IV} - \hat{\boldsymbol{\beta}}_{OLS}$, must converge towards $\boldsymbol{0}$.

However, a test based on $\hat{\boldsymbol{q}}$ requires knowledge of both the probability limit of $\hat{\boldsymbol{q}}$, and the asymptotic variance of the distribution of $\sqrt{T}\hat{\boldsymbol{q}}$, i.e. Vas$(\hat{\boldsymbol{q}})$. A priori, this would seem to raise a problem, since $\hat{\boldsymbol{\beta}}_{IV}$ and $\hat{\boldsymbol{\beta}}_{OLS}$ use, at least in part, the same data and are therefore correlated. Fortunately, it can be shown that this is not the case and that under the null hypothesis of orthogonality, the variance of the difference $\sqrt{T}\hat{\boldsymbol{q}} = \sqrt{T}(\hat{\boldsymbol{\beta}}_{IV} - \hat{\boldsymbol{\beta}})$ is equal to the difference between the variances of the two estimators.

The test procedure can now be formulated in appropriate terms. It is similar to a Wald test since only the model defined by the null hypothesis of orthogonality is used. Specifically, we consider the statistic

$$m = T\hat{\boldsymbol{q}}' \; \hat{\boldsymbol{V}}(\hat{\boldsymbol{q}})^{-1}\hat{\boldsymbol{q}}, \tag{8.46}$$

where $\hat{\boldsymbol{V}}(\hat{\boldsymbol{q}})^{-1}$ is a **convergent** estimator of $\boldsymbol{V}(\hat{\boldsymbol{q}})^{-1}$, and we show that this statistic follows a Chi-square distribution with K degrees of freedom, K being the rank of the (non-singular) matrix involved in the quadratic form (8.46). (Note that the notation Vas$(\hat{\boldsymbol{q}})^{-1}$ for the asymptotic variance has been abbreviated to $\boldsymbol{V}(\hat{\boldsymbol{q}})^{-1}$; and the same simplification has been used for the convergent estimator of the variance–covariance matrix).

Using these different results, we arrive at the final proposal:

Proposition 8.6 Under H_0, *the test statistic m converges asymptotically to a Chi-square distribution with K degrees of freedom*

$$m = T\hat{\boldsymbol{q}}'\hat{\boldsymbol{V}}(\hat{\boldsymbol{q}})^{-1}\hat{\boldsymbol{q}} \sim As\chi^2(K),$$

where $\hat{\boldsymbol{V}}(\hat{\boldsymbol{q}})^{-1}$ *is a convergent estimator* of $\boldsymbol{V}(\hat{\boldsymbol{q}})^{-1}$.

The Hausman Test and the Error Component Model

The task now is to adapt the Hausman test described in the previous section to the specific case of the error component model, characterised by the presence of an error term of the form

$$v_{it} = \varepsilon_i + u_{it}, \quad i = 1, \ldots, N \; ; \quad t = 1, \ldots, T,$$

with ε_i representing the random effect specific to unit i.

One of the important assumptions underlying the estimation of the error component model is that $E(v_{it}|X_{it}) = 0$. This is all the more important as the error term includes the unobservable component ε_i, which may be correlated with the X_{it}.

When $E(v_{it}|X_{it}) \neq 0$, the feasible generalised least squares estimator $\hat{\boldsymbol{\beta}}_{QGLS}$ is biased and non-convergent. Using the Within operator, however, eliminates ε_i. In addition, this operator leads to the Within estimator $\hat{\boldsymbol{\beta}}_W$, which is unbiased and asymptotically convergent.

The Hausman test therefore amounts to adopting the null hypothesis H_0 : $E(v_{it}|X_{it}) = 0$. Under H_0, both $\hat{\boldsymbol{\beta}}_{QGLS}$ and $\hat{\boldsymbol{\beta}}_W$ are unbiased and convergent. However, they have different probability limits if the null hypothesis is invalidated.

In reality, $\hat{\boldsymbol{\beta}}_W$ is convergent whether the null hypothesis is validated or invalidated. On the other hand, $\hat{\boldsymbol{\beta}}_{QGLS}$ is BLUES and convergent when H_0 is verified, but it is non-convergent when the null hypothesis is invalidated.

It is therefore logical to use the following test statistic

$$\hat{\boldsymbol{q}}_1 = \hat{\boldsymbol{\beta}}_{QGLS} - \hat{\boldsymbol{\beta}}_W.$$

Under H_0, plim$\hat{\boldsymbol{q}}_1 = \mathbf{0}$ and Cov$(\hat{\boldsymbol{q}}_1, \hat{\boldsymbol{\beta}}_{QGLS}) = \mathbf{0}$.

Therefore, using Proposition 8.6, it follows that the Hausman statistic asymptotically follows, under H_0, a Chi-square distribution with a number of degrees of freedom equal to the dimension of the vector $\boldsymbol{\beta}$:

$$m_1 = \hat{\boldsymbol{q}}'_1 \widehat{\text{Var}(\hat{\boldsymbol{q}}_1)}^{-1} \hat{\boldsymbol{q}}_1 \sim As\chi^2(K).$$

8.5 Stationary Dynamic Models

The "stationary" restriction is far from negligible. It basically corresponds to the case where the number of time observations T is small or very small. We begin by giving some characteristics of dynamic panel models, after which we provide some information on an estimation method known as the Generalized Method *of* Moments (GMM). Finally, we will see how to apply this method to the estimation procedures for dynamic panel models.

Dynamic Models

As a first step, we can specify a staggered delay panel model of the form

$$y_{it} = \sum_{j=0}^{r} \boldsymbol{x}'_{i,t-j}\boldsymbol{\beta}_j + u_{it}.$$

Estimating such a model does not pose any genuinely new problems, provided that a sufficient number of time observations are available.

However, this is not the case when we adopt an autoregressive specification in which the lagged dependent variable is one of the regressors. As an example, consider the dynamic error component model

$$y_{it} = \rho y_{i,t-1} + \boldsymbol{x}'_{it}\boldsymbol{\beta} + \varepsilon_i + u_{it}, \quad i = 1, \ldots, N\ ; \quad t = 1, \ldots, T, \tag{8.47}$$

where ρ is a scalar, $\varepsilon_i \sim IID(0, \sigma_\varepsilon^2)$ and $u_{it} \sim IID(0, \sigma_u^2)$, while $\boldsymbol{x}'_{it}$ and $\boldsymbol{\beta}$ are vectors respectively $(1 \times K)$ and $(K \times 1)$.

Given that, in (8.47), y_{it} depends on ε_i, it follows that $y_{i,\, t-1}$ also depends on ε_i. Consequently, $y_{i,\, t-1}$ is correlated with the error term and the least squares estimator is biased and non-convergent, even if the components u_{it} are not autocorrelated.

It is therefore necessary to use estimation methods other than least squares. These other methods rely on the use of instrumental variables and, above all, on a technique based on generalised moments.

Instruments and Generalised Moments

We will again use the linear model $\boldsymbol{y} = \boldsymbol{X\beta} + \boldsymbol{u}$. No new assumptions will be made about the distribution of $\mathbf{u}$ and we will continue to assume, by simplifying, that $E(\boldsymbol{u}) = 0$, $\mathrm{Var}(\boldsymbol{u}) = \sigma^2 \boldsymbol{I}_T$. Above all, we will admit that there exists an instrument matrix $\mathbf{Z}$ of dimension $T \times Q$, which satisfies $E(z'_t u_t) = \mathbf{0}'$ for all t, where z'_t represents the t-th row of $\mathbf{Z}$. (This is therefore an orthogonality condition).

We then introduce the **symmetrical positive definite** matrix $\boldsymbol{R}_T$ (known as the **weighting** matrix, in the dimension $Q \times Q$),) which converges in probability to a matrix of constants $\mathbf{R}$.

Let's now define the quadratic form $Q_T(\boldsymbol{\beta})$ by

$$\begin{aligned} Q_T(\boldsymbol{\beta}) &= \left[T^{-1}\boldsymbol{u}'\boldsymbol{Z}\right]\boldsymbol{R}_T\left[T^{-1}\boldsymbol{Z}'\boldsymbol{u}\right] \\ &= \left[T^{-1}(\boldsymbol{y} - \boldsymbol{X\beta})'\boldsymbol{Z}\right]\boldsymbol{R}_T\left[T^{-1}\boldsymbol{Z}'(\boldsymbol{y} - \boldsymbol{X\beta})\right] \\ &= (1/T^2)\left(\boldsymbol{y}'\boldsymbol{Z}\boldsymbol{R}_T\boldsymbol{Z}'\boldsymbol{y} - 2\boldsymbol{y}'\boldsymbol{Z}\boldsymbol{R}_T\boldsymbol{Z}'\boldsymbol{X\beta} + \boldsymbol{\beta}'\boldsymbol{X}'\boldsymbol{Z}\boldsymbol{R}_T\boldsymbol{Z}'\boldsymbol{X\beta}\right). \end{aligned} \tag{8.48}$$

Clearly, the terms in square brackets in (8.48) represent a set of **empirical moments** (moreover, $\boldsymbol{y}'\boldsymbol{Z}\boldsymbol{W}_T\boldsymbol{Z}'\boldsymbol{X}\boldsymbol{\beta}$ occurs twice, because the transpose of $\boldsymbol{\beta}'\boldsymbol{X}'\boldsymbol{Z}\boldsymbol{W}_T\boldsymbol{Z}'\boldsymbol{y}$, a scalar, is equal to $\boldsymbol{y}'\boldsymbol{Z}\boldsymbol{W}_T\boldsymbol{Z}'\boldsymbol{X}\boldsymbol{\beta}$).

At this stage, two major cases should be highlighted:

1[er] **case**: the number of instruments Q is equal to the number of parameters to be estimated K $(Q = K)$; the model is said to be **exactly identified.**

2[e] **case**: the number of instruments Q is greater than the number of parameters to be estimated K $(Q > K)$; the model is said to be **over-identified**.

When $K = Q$, we can obtain the estimators of the generalised moments$\hat{\boldsymbol{\beta}}_{MG}$ by simply minimising (8.48). First, there are the first-order conditions:

$$\frac{\partial}{\partial \boldsymbol{\beta}} Q_T(\boldsymbol{\beta}) = \frac{\partial}{\partial \boldsymbol{\beta}}\left[(1/T^2)\left(\boldsymbol{y}'\boldsymbol{Z}\boldsymbol{R}_T\boldsymbol{Z}'\boldsymbol{y} - 2\boldsymbol{y}'\boldsymbol{Z}\boldsymbol{R}_T\boldsymbol{Z}'\boldsymbol{X}\boldsymbol{\beta} + \boldsymbol{\beta}'\boldsymbol{X}'\boldsymbol{Z}\boldsymbol{R}_T\boldsymbol{Z}'\boldsymbol{X}\boldsymbol{\beta}\right)\right]$$
$$= (1/T^2)(-2\boldsymbol{X}'\boldsymbol{Z}\boldsymbol{R}_T\boldsymbol{Z}'\boldsymbol{y} + 2\boldsymbol{X}'\boldsymbol{Z}\boldsymbol{R}_T\boldsymbol{Z}'\boldsymbol{X}\hat{\boldsymbol{\beta}}_{MG}) = \boldsymbol{0}.$$

We deduce from these K partial derivatives that

$$\boldsymbol{X}'\boldsymbol{Z}\boldsymbol{R}_T\boldsymbol{Z}'\boldsymbol{X}\hat{\boldsymbol{\beta}}_{GM} = \boldsymbol{X}'\boldsymbol{Z}\boldsymbol{R}_T\boldsymbol{Z}'\boldsymbol{y} \Rightarrow \boldsymbol{Z}'\boldsymbol{X}\hat{\boldsymbol{\beta}}_{GM} = \boldsymbol{Z}'\boldsymbol{y}, \tag{8.49}$$

which gives the desired estimator

$$\hat{\boldsymbol{\beta}}_{GM} = (\boldsymbol{Z}'\boldsymbol{X})^{-1}\boldsymbol{Z}'\boldsymbol{y}. \tag{8.50}$$

By comparing (8.45) and (8.50), we see that the estimator of the generalised moments and that of the instrumental variables coincide when the number of instruments is equal to the number of regressors.

Let us now assume that the model is over-identified, i.e. that the number of instruments Q is greater than the number of explanatory variables K. The solution to (8.49) is then given by

$$\hat{\boldsymbol{\beta}}_{GM} = (\boldsymbol{X}'\boldsymbol{Z}\boldsymbol{R}_T\boldsymbol{Z}'\boldsymbol{X})^{-1}\boldsymbol{X}'\boldsymbol{Z}\boldsymbol{R}_T\boldsymbol{Z}'\boldsymbol{y}. \tag{8.51}$$

The weighting matrix $\boldsymbol{R}_T$ now assigns the value of $\hat{\boldsymbol{\beta}}_{GM}$. To make the role of $\boldsymbol{R}_T$, more intuitive, let's rewrite (8.49) as

$$\boldsymbol{X}'\boldsymbol{Z}\boldsymbol{R}_T\boldsymbol{Z}'(\boldsymbol{y} - \boldsymbol{X}\hat{\boldsymbol{\beta}}_{GM}) = \boldsymbol{X}'\boldsymbol{Z}\boldsymbol{R}_T\boldsymbol{Z}'\hat{\boldsymbol{u}}_{GM} = \boldsymbol{0}.$$

The matrix $\boldsymbol{X}'\boldsymbol{Z}\boldsymbol{W}_T$ has the dimension $K \times Q$, while $\boldsymbol{Z}'\hat{\boldsymbol{u}}_{GM}$ represents a set of conditions on the empirical moments. Therefore, the value of $\hat{\boldsymbol{\beta}}_{GM}$ is such that it poses K **linear combinations** of the Q **conditions on the moments equal to zero**, the weights of each of these combinations being given by $\boldsymbol{X}'\boldsymbol{Z}\boldsymbol{R}_T$.

However, two problems remain. The first relates to the optimal choice of the weighting matrix $\boldsymbol{R}_T$. The second lies in the selection of the instruments making up the **Z** matrix, which must—see "Instrumental Variables and General Hausman Test" section—meet the dual condition of being highly correlated with the explanatory variables while being asymptotically decorrelated with the error term. These are the famous **orthogonality** conditions. However, we have said enough to address the specific case of panel models.

Estimation of the Dynamic Error Component Model

Let's return to model (8.47), simplifying it and taking into account no explanatory variable other than y lagged by one period:

$$y_{it} = \rho y_{i,t-1} + \varepsilon_i + u_{it}, \quad i = 1, \dots, N\,; \quad t = 1, \dots, T. \tag{8.52}$$

The basic principle of the method proposed by Arellano and Bond (1991) is to make use of the orthogonality conditions which link the shifted values of y_{it} to the perturbations u_{it}. In this way, to obtain a convergent estimator of ρ, when $N \to +\infty$ and T is fixed, we rewrite (8.52) in first differences, so as to eliminate the ε_i:

$$y_{it} - y_{i,\,t-1} = \rho(y_{i,\,t-1} - y_{i,\,t-2}) + (u_{it} - u_{i,\,t-1}). \tag{8.53}$$

Note that $(u_{it} - u_{i,\,t-1})$ is a first-order moving average process with a unit root.
If $t = 3$, (8.53) becomes

$$y_{i3} - y_{i2} = \rho(y_{i2} - y_{i1}) + (u_{i3} - u_{i2}).$$

It follows that y_{i1} is then a valid instrument, since it is strongly correlated with $(y_{i2} - y_{i1})$ while being decorrelated from $(u_{i3} - u_{i2})$, u_{it} being, by hypothesis, independent.

If $t = 4$, using (8.53) makes

$$y_{i4} - y_{i3} = \rho(y_{i3} - y_{i2}) + (u_{i4} - u_{i3}).$$

For the same reasons as in the previous case, y_{i2}, like y_{i1} for that matter, is a valid instrument.

This iterative procedure can be repeated to obtain the sequence of instruments $(y_{i1}, y_{i2}, \dots, y_{i,\,T-2})$. However, this procedure does not take into account the differentiated error terms in (8.53). From this point of view

$$E\big(\Delta \boldsymbol{u}_i \Delta \boldsymbol{u}'_i\big) = \sigma_u^2 \boldsymbol{G}, \tag{8.54}$$

where $\Delta \boldsymbol{u}'_i = (u_{i3} - u_{i2}, \dots, u_{iT} - u_{i,\,T-1})$ and

$$
G_{(T-2)\times(T-2)} = \begin{pmatrix} 2 & -1 & 0 & \dots & 0 & 0 & 0 \\ -1 & 2 & -1 & \dots & 0 & 0 & 0 \\ 0 & -1 & 2 & \dots & 0 & 0 & 0 \\ \vdots & \vdots & \vdots & \ddots & \vdots & \vdots & \vdots \\ 0 & 0 & 0 & \dots & 2 & -1 & 0 \\ 0 & 0 & 0 & \dots & -1 & 2 & -1 \\ 0 & 0 & 0 & \cdots & 0 & -1 & 2 \end{pmatrix}.
$$

Next, let's define the matrix $\boldsymbol{Z}_i$ such that

$$
\boldsymbol{Z}_i = \begin{pmatrix} (y_{i1}) & 0 & \dots & 0 \\ 0 & (y_{i1}, y_{i2}) & \dots & 0 \\ \vdots & \vdots & \ddots & \vdots \\ 0 & 0 & \dots & (y_{i1}, \dots, y_{i,\,T-2}) \end{pmatrix}.
$$

Using $\boldsymbol{Z}_i$, the instrument matrix is given by $\boldsymbol{Z} = (\boldsymbol{Z}'_1, \dots, \boldsymbol{Z}'_N)'$, while the moment equations are given by $E\big(\boldsymbol{Z}'_i \Delta \boldsymbol{u}_i\big) = \boldsymbol{0}$.

Let's premultiply Eq. (8.53) in vector form by $\boldsymbol{Z}'$:

$$
\boldsymbol{Z}' \Delta \boldsymbol{y} = \boldsymbol{Z}'(\Delta \boldsymbol{y}_{-1})\rho + \boldsymbol{Z}' \Delta \boldsymbol{u}. \tag{8.55}
$$

From this last set of equations, we can deduce the convergent **first-step** estimator from ρ, i.e.

$$
\begin{aligned} \hat{\rho}_1 = &\big((\Delta \boldsymbol{y}_{-1})'\boldsymbol{Z}(\boldsymbol{Z}'(\boldsymbol{I}_N \otimes \boldsymbol{G})\boldsymbol{Z})^{-1}\boldsymbol{Z}'(\Delta \boldsymbol{y}_{-1})\big)^{-1} \\ &\times \big((\Delta \boldsymbol{y}_{-1})'\boldsymbol{Z}(\boldsymbol{Z}'(\boldsymbol{I}_N \otimes \boldsymbol{G})\boldsymbol{Z})^{-1}\boldsymbol{Z}'(\Delta \boldsymbol{y})\big). \end{aligned} \tag{8.56}
$$

The estimator of the generalised method of moments of ρ for $N \to +\infty$ and T fixed, using exclusively the restrictions on the empirical moments, leads to the same formula as in (8.56), except that $\boldsymbol{Z}'(\boldsymbol{I}_N \otimes \boldsymbol{G})\boldsymbol{Z}$ is replaced by

$$
\boldsymbol{R}_N = \sum_{i=1}^{N} \boldsymbol{Z}'_i(\Delta \boldsymbol{u}_i)(\Delta \boldsymbol{u}_i)'\boldsymbol{Z}_i.
$$

The GM estimator does not require any information on the initial conditions, nor on the djstributions of $\boldsymbol{u}_i$ and ε_i. Making this estimator feasible involves substituting for $\Delta \boldsymbol{u}$ the differentiated residuals obtained from the convergent estimator $\hat{\rho}_1$.

In this way, the resulting estimator is the **second-step generalised moments** estimator

$$
\hat{\rho}_{2GM} = \Big((\Delta \boldsymbol{y}_{-1})'\boldsymbol{Z}\hat{\boldsymbol{R}}_N^{-1}\boldsymbol{Z}'(\Delta \boldsymbol{y}_{-1})\Big)^{-1}\Big((\Delta \boldsymbol{y}_{-1})'\boldsymbol{Z}\hat{\boldsymbol{R}}_N^{-1}\boldsymbol{Z}'(\Delta \boldsymbol{y})\Big). \tag{8.57}
$$

A convergent estimator of the asymptotic matrix $\text{Var}(\hat{\rho}_{2GM})$ is given by the first term of (8.57), i.e.

$$\widehat{\text{Var}}(\hat{\rho}_{2GM}) = \left((\Delta y_{-1})' Z \hat{R}_N^{-1} Z' (\Delta y_{-1})\right)^{-1}. \tag{8.58}$$

Asymptotically, the estimators $\hat{\rho}_1$ and $\hat{\rho}_{2GM}$ are equivalent, at least if the u_{it} are $IID(0, \sigma_u^2)$.

Now a word about the over-identification test associated with Sargan's name. Its purpose can be illustrated as follows:

Example 8.3 Assume that $T = 4$ in Eq. (8.52). There are then three conditions on the moments to identify a single parameter:

$$\begin{aligned} E(y_{i1}(v_{i3} - v_{i2})) &= 0 \\ E(y_{i1}(v_{i4} - v_{i3})) &= 0 \\ E(y_{i2}(v_{i4} - v_{i3})) &= 0 \end{aligned}$$

To estimate ρ, any of these conditions can in principle be used. The other two are over-identification restrictions.

Under the null hypothesis that the instruments are valid, the Sargan statistic asymptotically follows a Chi-square distribution with a number of degrees of freedom equal to the number of over-identifying restrictions.

We can also perform a Wald test of the joint significance of the different regressors (including dummy—also called binary—variables if applicable), and test for the presence of first- or second-order autocorrelation in the residuals of the difference equations.

8.6 Application: The Convergence Equation

We will begin by giving some details of the convergence equation and the data used, before moving on to the estimation of the fixed-effects model, the error component model and finally the dynamic model.

The Equation to Be Estimated and the Data

The starting point can be found in the articles by Solow (1956) and Swan (1956). Solow uses a constant returns to scale production function, so that the lower the initial level of real output per capita in an economy, the higher its rate of growth towards the *steady* state. The convergence of economies towards this state is in fact the result of the hypothesis of diminishing marginal returns to capital.

The Solow-Swan model involves a large number of variables, including output per capita (y_t), the capital-labour ratio (k_t), the savings rate (s), capital depreciation (δ), and the exogenous population growth rate (n).

In discussions on convergence, the starting point is a production function of the Cobb–Douglas type

$$y_t = A_t k_t^{\alpha},$$

where A_t groups together all the determinants of growth other than the traditional factors of production and α is the elasticity of output per capita in relation to the capital-labour ratio.

Based on this Cobb–Douglas, the equation most widely used in empirical studies is written as follows

$$\ln y_t = \frac{\alpha(1-\rho)}{1-\alpha}[\ln s - \ln(n+\delta)] + \frac{1-\rho}{1-\alpha}\ln A_t + \rho \ln y_{t-1}.$$

From an econometric point of view, this equation still needs to be simplified before it can be estimated. In particular, per capita output is replaced by real GDP *per capita*, while s is replaced by the average savings rate over 5 years and n by the average population growth rate over the same 5 years. In addition, the assumption of increasing returns to scale allows ln s and ln $(n+\delta)$. to be merged into a single variable.

The convergence equation then takes the simplified form

$$\ln y_{it} = a_i + \kappa \ln x_{it} + \rho \ln y_{i,t-1} + v_{it}, \tag{8.59}$$

where, it should be noted, x is equal to the difference between the savings rate and the population growth rate. (This variable is usually positive, except of course for countries with high population growth).

In these conditions, testing the hypothesis of convergence of the economies essentially comes down to examining the estimated value of the coefficient ρ. If it is less than unity and close to zero, then convergence is rapid. On the other hand, if the value is close to unity, convergence will take place at a snail's pace (for example, $\rho = 0.9$ implies that convergence will take 110 years).

Finally, our data are those used by Nerlove and come from the *Penn World Tables* (PWT). They cover the period 1960–1985.

Estimation of the Fixed-Effects Model

Using the GETS methodology—see "From General to Specific" section—the first step is to estimate the model

Table 8.1 Comparison of models

	AIC	BIC	HQC
Model 1 (−2)	−283.5	69.73	−141.87
Model 2 (−1)	−255.3	94.31	−115.8

$$y_{it} = \alpha_i + \beta_0 x_{it} + \beta_1 x_{i,t-1} + \beta_2 x_{i,t-2} + u_{it}, \quad i = 1, \ldots, 94\ ; \quad t = 1, \ldots, 6. \tag{8.60}$$

We introduced two time lags for the explanatory variable, corresponding to 10 years given the 5-year averages.

We then made an initial reduction, limiting ourselves to a single shift in x. Table 8.1 compares the two estimates on the basis of the three usual information criteria.

The first model as formulated in Eq. (8.60) improves all three information criteria. It should therefore be selected. In this way, the reduction operations—limited in this case—have led to the initial model being retained.

The following is a summary of the main results relating to the estimation of the fixed-effects model (Table 8.2).

Firstly, all the model's coefficients are significantly different from zero to one ten-thousandth. Secondly, the coefficient of determination is very high. Finally, the F-test for joint nullity of the estimated coefficients leads to the conclusion that the coefficients in question are significantly different from zero.

With regard to group fixed effects—in this case for the 94 countries considered—the test with the null hypothesis "a constant common to all units" shows that with the static and the p-value

$$F(93,\ 185) = 74.84, \qquad p\text{-value} < 0.0001,$$

this null hypothesis must be rejected. Consequently, the presence of fixed effects in the model is fully justified.

Table 8.2 Regression with binary (or dummy) variables

	Coefficients	p-values
Constant	7.565	< 0.0001
x	0.112	< 0.0001
x (−1)	0.130	< 0.0001
x (−2)	0.141	< 0.0001
$R^2 LSDV = 0.99$	$F(3,\ 185) = 12.419$	with a p-value < 0.0001

The Error Component Model

We are now going to estimate the following error component model, i.e. with random effects

$$y_{it} = \alpha + \sum_{j=0}^{2} \beta_j x_{i,\,t-j} + \varepsilon_i + u_{it}, \quad i = 1, \ldots, 94\ ; \quad t = 1, \ldots, 6. \tag{8.61}$$

(We take two lags, because this proved correct for estimating the fixed-effects model, but above all, because we have only six time series data).

Table 8.3 shows the results of estimating model (8.61) using feasible generalised least squares.

Once again, we see that all the estimated coefficients are very largely significantly different from zero. As is to be expected, they also all have a positive sign, meaning that an increase (in percentage terms) in one of the corresponding variables translates into an increase in real GDP *per capita*.

We can add that the (asymptotic) test statistic for the joint nullity of the three estimated coefficients and its associated p-value are respectively

$$\chi^2(3) = 140.74 \quad \text{with } p\text{-value } < 0.0001,$$

which implies that the coefficients in question are fully significant.

Let's move on to the Hausman test, for which the null hypothesis (see "The Hausman Test and the Error Component Model" section) is that the FGLS estimators are convergent. In this case, the corresponding statistic and p-value are as follows:

$$\chi^2(3) = 46.99 \quad \text{with } p\text{-value } < 0.0001.$$

It follows that H_0 is invalidated and that, consequently, the FGLS estimators are, in this specific case, non-convergent.

Table 8.3 Estimation of the error component model

	Coefficients	p-values
Constant	7.40	< 0.0001
x	0.214	< 0.0001
x (−1)	0.203	< 0.0001
x (−2)	0.227	< 0.0001
$AIC = 623.0$	$BIC = 637.58$	$HQC = 628.85$

Table 8.4 Estimation of the dynamic model

	Coefficients	*p*-values
y (−1)	0.578	< 0.0001
x	0/132	< 0.0001
x (−1)	0.093	0.012

The Stationary Dynamic Model

The model to be treated is similar to an **autoregressive distributed lag model** (ARDL) on panel data, since it includes, among the regressors, both the lagged dependent variable and the explanatory variable x lagged by one period:

$$y_{it} = \rho y_{i,t-1} + \sum_{j=0}^{1} \beta_j x_{i,\,t-j} + \varepsilon_i + u_{it}, \quad i = 1, \ldots, 94\ ; \quad t = 1, \ldots, 6. \qquad (8.62)$$

(Taking two lag periods is out of the question because, as we pointed out, the Arellano-Bond estimator works with the first differences of the variables concerned).

We will estimate (8.62) using the Arellano-Bond procedure, i.e. the two-stage generalised moments technique (see "Estimation of the Dynamic Error Component Model" section).

Table 8.4 details the results obtained.

All the estimated parameters are significantly different from zero. They are also all of the expected signs, i.e. positive. In particular, the elasticity of real GDP with respect to itself lagged by one period takes on a value close to 0.6.

With regard to the tests specific to this estimation technique, the Wald test of joint nullity of the three coefficients leads to the $\chi^2(3) = 208.96$ statistic with a p-value close to zero. The conclusion is that all three coefficients are significant.

However, the Sargan over-identifying restrictions test gives a statistic $\chi^2(8) = 22.25$, associated with a p-value equal to 0.0045, which leads us to reject the null hypothesis that the instruments used are all valid.

The value of ρ obtained here, even if the model used is not entirely statistically justified, bears witness, in any case, to a very slow convergence towards the steady—of the order of two or three decades.

8.7 Exercises

8.7.1 Explain what you see as the advantages and disadvantages of panel data series.

8.7.2 Consider the Within matrix applied to an individual unit of a longitudinal series and defined by

$$Q = I_T - (1/T)\, e_T e'_T,$$

where $\mathbf{e}_T$ is a vector composed of T units.

You are required to:

(i) carry out the product **Qz**, where **z'** = (5, 9, 16).
(ii) interpret the result obtained in terms of deviation from the mean for the individual unit in question.
(iii) show that **Q** is idempotent, i.e. **Q** = **QQ**.

8.7.3 Let be the two matrices **A** and **G** such that

$$A = \begin{pmatrix} a\ b \\ c\ d \end{pmatrix} \quad \text{and} \quad G = \begin{pmatrix} w\ x \\ y\ z \end{pmatrix}.$$

We ask

(i) carry out the Kronecker product of these two matrices.
(ii) determine the dimension of the resulting matrix.

8.7.4 Prove that the matrix Between defined by

$$B = I_N \otimes (1/T)\, e_T e'_T,$$

where $\mathbf{e}_T$ is a column vector composed of T units, is

(i) symmetric.
(ii) idempotent.

8.7.5 Take the series from the Penn World Tables used in "The Equation to Be Estimated and the Data" section. Thus, we have:

- y_{it} is the log of per capita real GDP;
- x_{it} is the log of the difference between the savings rate and the population growth rate.

Note also that $N = 94$, i.e. there are 94 individual units (countries) taken into account, but that T is equal to 6 only.

Finally, we will look at the model with fixed effects,

$$y_{it} = \alpha_i + \beta x_{it} + u_{it}, \quad i = 1, \ldots, 94\ ; \ t = 1, \ldots, 6, \tag{8.63}$$

and then the model incorporating random effects

$$y_{it} = \alpha + \beta x_{it} + \varepsilon_i + u_{it}, \quad i = 1, \ldots, 94\ ; \ t = 1, \ldots, 6. \tag{8.64}$$

You will be asked:

(i) to estimate (8.63), why is the Within method preferable to the LSDV method?

(ii) analyse the results of estimating the fixed effects model (8.63).
(iii) perform an F-test on the null hypothesis that the individual units have a common intercept.
(iv) compare your estimates with those that would be obtained by a Between regression.
(v) estimate the random effects model (8.64) by using the Generalyzed Least Squares (GLS) method and analyse the estimates you have obtained.
(vi) Use the Breusch-Pagan and Hausman tests to evaluate the estimates of the model (8.64).

8.7.6 We continue to use the data from the previous exercise, but we will now focus on the estimation of the dynamic model:

$$y_{it} = \rho y_{i,t-1} + \beta x_{it} + \varepsilon_i + u_{it}, \quad i = 1, \ldots, 94 \; ; \quad t = 1, \ldots, 6. \tag{8.65}$$

You are required to

(i) estimate the model (8.65) by using the Arellano-Bond method.
(ii) interpret the results thus obtained.
(iii) apply the Sargan test for fully evaluating over-identification restrictions.

References

Amemiya, H. (1971). The estimation of the variances in a variance-components model, *International Economic Review*, *12*, 1–13.

Arellano, M., & Bond, S. (1991). Some tests of specification for panel data: Monte Carlo evidence and an application to employment equations, *Review of Economic Studies*, *58*, 277–297.

Breusch, T. S. (1987). Maximum likelihood estimation of random effects models, *Journal of Econometrics*, *36*, 383–389.

Fuller, W. A., & Battese, G. E. (1973). Transformations for estimation of linear models with nested error structure, *Journal of the American Statistical Association*, *68*, 626–632.

Fuller, W. A., & Battese, G. E. (1974). Estimation of linear models with cross-error structure, *Journal of Econometrics*, *2*, 67–78.

Maddala, G. S. (1971). The use of variance components models in pooling cross section and time series data. *Econometrica*, *39*, 341–358.

Nerlove, M. (1971). A note on error components models, *Econometrica*, *39*, 383–396.

Solow, R. M. (1956). A contribution to the theory of economic growth, *The Quarterly Journal of Economics*, *70*, 65–94.

Swamy, P. A. V. B., & Arora, S. S. (1972). The exact finite sample properties of the estimators of coefficients in the error components regression models. *Econometrica*, *40*, 261–275.

Swan, T. (1956). Economic growth and capital accumulation, *Economic Record*, *32*, 334–361.

Wallace, T. D., & Hussain, A. (1969). The use of error components models in combining cross-section and time-series data, *Econometrica*, *37*, 55–72.

Wansbeek, T. J, & Kapteyn, A. (1982). A class of decompositions of the variance-covariance matrix of a generalized error components model, *Econometrica*, *50*, 713–724.

Chapter 9
Non-stationary Panels

We will consider, as in the previous chapter, N cross-sectional units and T temporal observations on each of these N units, that is the panel (y_{it}), $i = 1, \ldots, N$; $t = 1, \ldots, T$. However, we will now allow these series to be non-stationary.

Let's then assume that they follow an autoregressive process of the form

$$\Delta y_{it} = \phi_i y_{i,t-1} + \sum_{j=1}^{k_i} \beta_{ij} \Delta y_{i,t-j} + \kappa_i + \gamma_i\, t + u_{it}, \tag{9.1}$$

where $u_{it} \sim IID\big(0, \sigma_i^2\big)$ for $t = k_{i+2}, \ldots, T$, k_i being known, while κ_i and $\gamma_i t$ respectively represent a constant and a deterministic trend.

To test the non-stationarity of Eqs. (9.1), we will first add an additional assumption, namely that of the **independence** of the different cross-sectional series: $(y_{1t})_{t=k_{1+2}}^{T}, \ldots, (y_{1t})_{t=k_{N+2}}^{T}$.

9.1 First-Generation Tests

The common characteristic of these tests, except for one of them, is that they are based on the independence of the cross-sectional series. We will start by examining the most common one, associated with the names of Levin, Lin, and Chu (LLC).

Supplementary Information The online version contains supplementary material available at https://doi.org/10.1007/978-3-031-72910-2_9.

F. J. Bismans and O. Damette, *Dynamic Econometrics*,
https://doi.org/10.1007/978-3-031-72910-2_9

The LLC Test

Rather than performing a unit root test on each individual unit, Levin et al. (2002) consider that there is a unit root process common to all of these units, so that Eq. (9.1) can then be rewritten as follows:

$$\Delta y_{it} = \phi y_{i,t-1} + \sum_{j=1}^{k_i} \beta_{ij} \Delta y_{i,t-j} + \boldsymbol{\alpha}'_i \boldsymbol{d}_t + u_{it}, \tag{9.2}$$

where $\boldsymbol{\alpha}'_i = (\kappa_i, \gamma_i)$ and $\mathbf{d}_t = (1, t)'$. (Of course, we are taking here the most general case, for which (9.2) incorporates both a constant and a trend. If this equation only contained a constant, we would simply have the constant κ_i in (9.2) and if neither trend nor constant were present, the term $\boldsymbol{\alpha}'_i \mathbf{d}_t$ would disappear from this equation.)

As $\phi_i = \phi$ for all i, it follows that the null and alternative hypotheses underlying the LLC test are therefore

$$\begin{cases} H_0 : \phi = 0 \\ H_1 : \phi < 0. \end{cases}$$

The determination of the test statistic is based on a procedure which takes place in three stages.

Step 1: Perform the ADF Regressions and Generate the Orthogonalised Residuals

For each of the units i, we perform the augmented Dickey–Fuller regression (ADF)

$$\Delta y_{it} = \phi_i y_{i,t-1} + \sum_{j=1}^{k_i} \beta_{ij} \Delta y_{i,t-j} + \boldsymbol{\alpha}'_i \boldsymbol{d}_t + u_{it}. \tag{9.3}$$

The number of lags k_i can differ from one unit to another. To estimate it, we choose a maximum number $k_{\max}$ and we use the statistic t of $\widehat{\beta}_{ij}$ to see if a lower k is preferred. (Under $H_0 : \beta_{ij} = 0$, these statistics t follow a distribution $N(0, 1)$, both when $\phi_i = 0$ and $\phi_i < 0$.)

Once we have determined k_i, we then launch the two auxiliary regressions

$$\Delta y_{it} \text{ on } \Delta y_{i,t-j}, j = 1, \ldots, k_i, \text{ and } \mathbf{d}_t y_{i,t-1} \text{ on } \Delta y_{i,t-j}, j = 1, \ldots, k_i, \text{ and } \mathbf{d}_t,$$

which provides the respective orthogonalised residuals $\widehat{e}_{it}$ and $\widehat{v}_{i,t-1}$. We then standardise these residuals, so that

$$\widehat{e}_{it}/\widehat{\sigma}_{ui} \text{ and } \tilde{v}_{i,t-1} = \widehat{v}_{i,t-1}/\widehat{\sigma}_{ui},$$

with $\widehat{\sigma}_{ui}$ the standard error resulting from each of the ADF regressions for $i = 1, \ldots, N$.

Step 2: Estimate the Ratio Between Long-Term and Short-Term Standard Deviations

Under $H_0 : \phi = 0$, but after removing the trend, the long-term variance in (9.3), which is convergent, is estimated by

$$\sigma_{yi}^2 = \frac{1}{T-1}\sum_{t=2}^{T} \Delta y_{it}^2 + 2\sum_{j=1}^{\bar{K}} w_{\bar{K}j}\left(\frac{1}{T-1}\sum_{t=2+j}^{T} \Delta y_{it}\Delta y_{i,t-j}\right),$$

where $\bar{K}$ is a shift window used to truncate the number of delays taken into consideration and the $w_{\bar{K}j}$ are the corresponding weights (more explanations are given in "White's Estimator" section).

For each unit i, the sought ratio is estimated by $\hat{s}_i = \widehat{\sigma}_{yi}/\widehat{\sigma}_{ui}$, while the average of the standard deviations, denoted S_N, is estimated by

$$\widehat{S}_N = (1/N)\sum_{i=1}^{N} \hat{s}_i.$$

Step 3: Calculate the Test Statistic

We perform the grouped regression

$$\tilde{e}_{it} = \phi\tilde{v}_{i,t-1} + \tilde{u}_{it},$$

using the $N\tilde{T}$ observations where $\tilde{T} = T - \bar{k} - 1$ and $\bar{k}$ is the average number of lags obtained from the ADF regressions in step 1.

The t-statistic, under the null hypothesis of a unit root, thus for $\phi = 0$, is given by

$$t_\phi = \frac{\widehat{\phi}}{\widehat{\sigma}\left(\widehat{\phi}\right)},$$

where

$$\hat{\phi} = \sum_{i=1}^{N}\sum_{t=2+k_i}^{T} \tilde{v}_{i,\,t-1}\tilde{e}_{it} / \sum_{i=1}^{N}\sum_{t=2+k_i}^{T} \tilde{v}_{i,\,t-1}^2$$

$$\hat{\sigma}(\hat{\phi}) = \hat{\sigma}_u / \left(\sum_{i=1}^{N} \sum_{t=2+k_i}^{T} \tilde{v}_{i,\,t-1}^2 \right)^{1/2}.$$

Finally, we obtain the adjusted t-statistic

$$t_\phi^* = \frac{t_\phi - N\tilde{T}\hat{S}_N \hat{\sigma}_{\tilde{u}}^{-2} \hat{\sigma}(\hat{\phi}) \mu_{\tilde{T}}^*}{\sigma_{\tilde{T}}^*}, \tag{9.4}$$

in which $\mu_{\tilde{T}}^*$ and $\sigma_{\tilde{T}}^*$ are respectively the adjusted mean and standard deviation as they were established in Table 2 of Levin et al. (2002, p. 14), of which here is an excerpt (Table 9.1).

Table 9.1 Adjusted means and standard deviations

$\tilde{T}$	$\bar{K}$	$\mu_{\tilde{T}}^*$	$\sigma_{\tilde{T}}^*$
25	9	−0.703	1.003
50	12	−0.614	0.818
100	15	−0.566	0.695
250	20	−0.533	0.603
∞	–	−0.500	0.500

Our three authors then demonstrate—this is their theorem 5—that asymptotically, the statistic t_ϕ^* follows a **standard normal distribution**. From a practical point of view, this test requires that N be small relative to T, that is to say that $N \to +\infty$ in such a way that $N/T \to 0$.

Other Unit Root Tests

There are many other unit root tests in panels. In this section, we will limit ourselves to presenting three of them: those of Im, Pesaran and Shin, Breitung and Hadri.

One of the limitations of the LLC test is that it imposes on ϕ to be homogeneous for all units i. The Im, Pesaran and Shin (IPS) test does not suffer from this limitation, insofar as it allows this coefficient to be **heterogeneous,** to vary between units.

IPS are based on a model analogous to (9.1) and take an average individual tests. The null hypothesis is that each individual series contains a unit root, in other words

$$H_0 : \phi_i = 0, \forall i.$$

The alternative hypothesis allows some individual series, but not all, to be non-stationary:

$$H_1 : \begin{cases} \phi_i < 0 & \text{pour } i = 1, 2, \ldots, N_1 \\ \phi_i = 0 & \text{pour } i = N_1 + 1, \ldots, N. \end{cases} \tag{9.5}$$

The IPS statistic, denoted $\bar{t}$,, is defined as the average of the individual ADF statistics, i.e.

$$\bar{t} = (1/N) \sum_{i=1}^{N} t_{\phi_i}, \tag{9.6}$$

where t_{ϕ_i} is the individual t statistic intended to test the null hypothesis.

There are still some complications to implement this test, mainly due to the number of lags to incorporate in the ADF regressions. *Ultimately*, IPS show that the statistic $\bar{t}$, properly standardised, follows, asymptotically, a distribution $N(0, 1)$. The same convergence conditions as those seen for the LLC test apply, namely that N must be small relative to T.

The **Breitung** test abandons the common requirement to LLC and IPS that $N \to +\infty$, in such a way that $N/T \to 0$. This test therefore offers the possibility that N can be large relative to T.

The obtaining of the Breitung statistic follows the same steps as those seen for the determination of the LLC statistic, but with notable differences. Thus, during the first step, only $\Delta y_{i,t-j}$ is used to obtain the residuals $\hat{e}_{it}$ and $\hat{v}_{it-1}$. In the second, the residuals $\hat{e}_{it}$ are orthogonalised using the Arellano-Bover transformation.

$$e_{it}^* = \sqrt{\frac{T-1}{T-t+1}} \left(\tilde{e}_{it} - \frac{\tilde{e}_{i,t+1} + \cdots + \tilde{e}_{iT}}{T-t} \right).$$

Similarly, the residuals $\hat{v}_{i,t-1}$ are orthogonalised as follows:

$$v_{i,t-1}^* = \tilde{v}_{i,t-1} - \tilde{v}_{i1} - \frac{t-1}{T} \tilde{v}_{iT}.$$

Finally, in the third step, we perform the "pooled"regression

$$e_{it}^* = \phi v_{i,t-1}^* + u_{it}^*$$

and we then obtain the statistic t which, under H_0, follows, asymptotically, a **standardised normal distribution**.

The **Hadri** test is a generalisation of the KPSS designed for simple time series (see "Other Tests of Stationarity" section). As for the latter, stationarity is the null hypothesis for **all** units. More precisely, it is a Lagrange multiplier (ML) test—see "Presence of Autocorrelated Disturbances" section—based on the residuals of the regression, by least squares, of y_{it} on a constant and a deterministic trend

$$y_{it} = r_{it} + \beta_i t + u_{it}, \tag{9.7}$$

where $r_{it} = r_{i,t-1} + \varepsilon_{it}$ is a random walk. Furthermore, $u_{it} \sim IIN\left(0, \sigma_u^2\right)$ and $\varepsilon_{it} \sim IIN\left(0, \sigma_\varepsilon^2\right)$ are normally distributed random variables and mutually independent both on i and on t.

By substituting the random walk recursively into (9.7), we get:

$$y_{it} = r_{io} + \beta_i\, t + \sum_{j=1}^{t} \varepsilon_{ij} + u_{it} = r_{io} + \beta_i\, t + v_{it}, \tag{9.8}$$

with $v_{it} = \sum_{j=1}^{t} \varepsilon_{it} + u_{it}$. The null hypothesis is then

$$H_0 : \sigma_\varepsilon^2 = 0,$$

in which case $v_{it} = u_{it}$.

A first ML statistic is given by

$$M\,L_1 = N^{-1}\left(\sum_{i=1}^{N} T^{-2} \sum_{t=1}^{T} S_{it}^2\right)/\hat{\sigma}_u^2, \tag{9.9}$$

where the $S_{it} = \sum_{j=1}^{t} \hat{u}_{ij}$ are the cumulative sums of the residuals from (9.8) and $\widehat{\sigma}_u^2$ is a convergent estimator of σ_u^2 under H_0.

Hadri proposes a second ML test that takes into account possible heteroscedasticity among the *i's* and leads to

$$ML_2 = N^{-1}\left(\sum_{i=1}^{N}\left(T^{-2}\sum_{t=1}^{T} S_{it}^2/\hat{\sigma}_{ui}^2\right)\right). \tag{9.10}$$

He then demonstrates that the test statistic Z defined as follows, tends asymptotically towards a standardised normal distribution

$$Z = \frac{\sqrt{N}(ML - \xi)}{\chi} \rightarrow N(0, 1),$$

where $\xi = 1/15$ and $\chi = 11/6300$, when the model includes both a constant and a trend.

The Monte-Carlo simulations contained in Hadri (2000) show that the empirical level of the test is close to its nominal level of 0.05, at least when N and T are sufficiently large.

Testing Cross-Section Dependence

Almost all the tests described here are based on the assumption of independence of cross-sectional series. It is therefore important to check whether this assumption is verified or not.

From this point of view, Pesaran (2004) proposed a test of cross-section error dependence (*cross-section dependence*, CD), which is an extension of that of Breusch–Pagan. The latter—we showed it in "Lagrange Multiplier Test" section—is actually a Lagrange multiplier test.

The main difference between these two approaches is that Pesaran uses the pair correlations of the residuals obtained by estimating the ADF regressions on the units, rather than their squares.

Thus, the modified Breusch and Pagan test, denoted CD_{ML}, leads to the statistic

$$CD_{ML} = T \sum_{i=1}^{N-1} \sum_{j=i+1}^{N} \hat{\rho}_{ij}^2 \sim \chi^2(N(N-1)/2), \tag{9.11}$$

where the pair correlations are defined by

$$\hat{\rho}_{ij} = \hat{\rho}_{ji} = \frac{\sum_{t=1}^{T} \hat{u}_{it}\hat{u}_{jt}}{\left(\sum_{t=1}^{T} \hat{u}_{it}^2\right)^{1/2} \left(\sum_{t=1}^{T} \hat{u}_{jt}^2\right)^{1/2}}.$$

The Pesaran test uses the same pair correlations and results in the statistic *CD*, namely

$$CD = \sqrt{\frac{2T}{N(N-1)}} \left(\sum_{i=1}^{N-1} \sum_{j=i+1}^{N} \hat{\rho}_{ij} \right), \tag{9.12}$$

the difference with (9.11) coming, let's repeat, from the use of correlations $\widehat{\rho}_{ij}$ and not their squares.

Let's add that this statistic asymptotically follows a **standardised normal distribution** and that the test is particularly well suited to panels for which N is small and T large. Moreover, whatever the test, the null hypothesis is that of **independence** of cross-sectional errors.

Example 9.1 Bismans and Damette (2008) use the *intraday* data (24 November to 25 November 2004) on the bid and ask prices of four exchange rates, namely US dollar/euro, dollar/yen, dollar/sterling and dollar/Canadian dollar. Here, $N = 4$ and T is equal to several thousand.

One would suspect a priori that the exchange rates in question are not independent. This is confirmed by the tests presented in the following table:

	Statistic	Critical value (5%)
CD_{ML}	11.30	7.81
CD	−3.21	−1.96

In both cases, the null hypothesis of cross-sectional independence is rejected.

9.2 Second and Third-Generation Tests

When the results of the cross-sectional error dependence tests conclude with the effective presence of a form of correlation between the individual units, the previous unit root tests are biased.

Therefore, the use of these tests (Levin-Lin-Chu, IPS, Breitung) which postulate independence between individual units, is prohibited [see O'Connell (1998) and Westerlund (2005)]. Most of the macroeconomic series in panel concern countries, firms or individuals sharing however common underlying trends that generate inter-individual dependence.

The second-generation tests precisely relax the hypothesis of independence between the series and make the unit root tests in panel again applicable. Although their general structure consists in estimating a common factor to approach the dependence between individual units, there are several types depending on how to consider this factor.

The PANIC Test

Bai and Ng (2004, 2010) developed a test called PANIC (for *Panel Analysis of Non-stationarity in the Idiosyncratic and Common components*). This deals with the correlation between individual units by distinguishing the sources of non-stationarity that are present in all units (this is the common component) from those that are specific to a particular unit (called idiosyncratic).

The authors develop an aggregate test, in which the null hypothesis of non-stationarity follows the data generating process (DGP)

$$y_{i,t} = D_{i,t} + \lambda_i F_t + e_{i,t}, \tag{9.13}$$

with i and t the indices of individual units and time, while $D_{i,t}$ is a deterministic component and λ_i a country-specific adjustment coefficient (*loading factor*) or each country's specific elasticity to the common factor, source of dependencies between individuals.

This formalisation via a factorial model allows to capture the cross-sectional correlation within the common component F_t. The stationarity of this component, F_t can then be analysed separately from the stationarity of the idiosyncratic component

$e_{i,t}$. Thus, a variable $y_{i,t}$ will be non-stationary as soon as at least one of the common factors is non-stationary or the idiosyncratic term is non-stationary. It is quickly understood that F_t and $e_{i,t}$ can be integrated into different orders in practice and the combination of the two can prove difficult to establish. That's why the authors will separately test the existence of a unit root on F_t and $e_{i,t}$ rather than on $y_{i,t}$.

Let's rewrite the model in the following form where the deterministic component $D_{i,t}$ is replaced by a constant and a linear trend:

$$y_{i,t} = \alpha_i + \beta_{i,t} + \lambda_i F_t + e_{i,t}, \quad t = 1, ..., T \tag{9.14}$$

$$F_{m,t} = \tau_m F_{m,t-1} + \nu_{m,t}, \quad m = 1, ..., r \tag{9.15}$$

$$e_{i,t} = \rho_i e_{i,t-1} + \varepsilon_{i,t}, \quad i = 1, ...N. \tag{9.16}$$

The common factors $F_{m,t-1}$ are indexed by m. The m-th among them—there are r in total, which are assumed to be known—will be stationary if $\tau_m < 1$, whereas the idiosyncratic component $e_{i,t}$ will be stationary for the ith statistical individual if $\rho_i < 1$. These components not being observable, it is therefore necessary to estimate them in advance. As explained by the synthesis of Hurlin and Mignon (2007), the validity of the PANIC test relies on the fact that it is possible to estimate these components while preserving their degree of integration, whether the idiosyncratic component is stationary or not.

The estimation is carried out by a principal component analysis based on the series expressed in first differences:

$$\Delta y_{i,t} = \lambda_i \Delta F_t + \Delta e_{i,t}, \quad t = 1, ..., T \tag{9.17}$$

$$\Delta y_{i,t} = \lambda_i f_t + z_{i,t}, \quad t = 1, ..., T. \tag{9.18}$$

First, we estimate f_t and λ_i in model (9.18) by a principal component analysis, which allows us to obtain:

$$\widehat{z}_{i,t} = \Delta y_{i,t} - \widehat{\lambda}_i \widehat{f}_t. \tag{9.19}$$

In the second step, since we now know $\widehat{z}_{i,t}$ and $\widehat{f}_t$, it is possible to determine the variables of the cumulative components in the general form:

$$\widehat{F}_{m,t} = \sum_{s=2}^{t} \widehat{f}_{m,s} \text{ and } \widehat{e}_{i,t} = \sum_{s=2}^{t} \widehat{z}_{i,s}. \tag{9.20}$$

The non-stationarity of the idiosyncratic component is tested using the ADF test statistics calculated via the estimated components $\widehat{e}_{i,t}$ using the following model without a constant or deterministic trend:

$$\Delta\hat{e}_{i,t} = \delta_{i,0}\hat{e}_{i,t-1} + \delta_{i,1}\Delta\hat{e}_{i,t-1} + \ldots + \delta_{i,p}\Delta\hat{e}_{i,t-p} + \mu_{i,t}. \tag{9.21}$$

We denote $ADF_e^c(i)$ the t-statistic of the augmented Dickey–Fuller test of the idiosyncratic component of the ith individual. As the asymptotic distribution is standard (model without constant), it is then possible to apply an ADF test on each individual and each idiosyncratic component.

The dimension T may be lower in the case of panels, Bai and Ng use an average statistic of the IPS type. In this last case, the statistic is written as:

$$Z_e^c = \frac{1}{\sqrt{4N}}\left[-2\sum_{i=1}^{N}\log p^c(i) - 2N\right], \tag{9.22}$$

with $Z_e^c \sim N(0,1)$, regardless of the dimension N of the panel and under the assumption that the estimated idiosyncratic components are independent.

It remains to test the non-stationarity of the common factors. A standard ADF test can be used, when there is only one common factor, thus $r = 1$, among the individual units and this, on the basis of the following model:

$$\Delta\widehat{F}_{1,t} = c + \gamma_{i,0}F_{1,t-1} + \gamma_{i,1}\widehat{F}_{1,t-1} + \ldots + \gamma_{i,p}\Delta\widehat{F}_{1,t-p} + \nu_{i,t}. \tag{9.23}$$

This leads to the statistic $ADF_F^c(i)$ which, once again, follows the asymptotic distribution of a standard ADF test with a constant.

In the case where the number of common factors is larger, thus> 1, the authors test the number of common stochastic trends. They finally propose two statistics, MQ_f and MQ_c, based on these r estimated factors $\widehat{F}_{m,t}$ for $= 1, \ldots, r$.

As explained by Hurlin and Mignon (2007), these statistics originate from Stock and Watson (1986). Their purpose is to test whether the real part of the smallest eigenvalue of the AR coefficients matrix is equal to one and they rely on nested tests (or tests using sequences of hypotheses).

In this case, the null hypothesis is $H_0 : r_1 = m$; if it is rejected, the test is applied again by setting $m = m-1$. We then test the equality between the number of common stochastic trends thus detected and the number of common factors $r_1 = r$.

According to Bai and Ng, their test provides robust results in terms of size and power, even for moderately sized panels such as country macro-panels.

The Pesaran Test

Unlike previous tests, Pesaran (2007) suggests a unit root test, where the common factor is approximated by averages of individual units (or *cross sections*). This method proves simpler than the previous ones, insofar as the common factor is approximated by terms easy to obtain and where ordinary least squares are used for the estimation. This is probably why it is the most used by econometrics practitioners.

Indeed, instead of considering the unit root relative to deviations from the estimated common factors, it transforms the standard ADF test with the addition of an inter-individual mean term. More exactly, it is a matter of augmenting the ADF form test with averages of lags of level variables and first differences of individual series. The test includes two sets of statistics, depending on whether one considers a modified version of the ADF test or the IPS test: CADF (for *Cross-sectionally Augmented ADF*) or CIPS (for *Cross-sectionally Augmented IPS*).

Like the IPS test presented previously, the CIPS panel unit root test is based on averages of the Dickey–Fuller augmented statistics for the different individual units.

Consider a variable $y_{i,t}$ observable over $t = 1, ..., T$ time periods and $i = 1, ..., N$ individual units, which follow a process of the heterogeneous panel model type autoregressive of the following form:

$$y_{i,t} = (1 - \rho_i)\alpha_i + \rho_i y_{i,t-1} + u_{i,t}, \tag{9.24}$$

where $u_{i,t}$ is an error term that has a common factor f, that is

$$u_{i,t} = \gamma_i f_i + \varepsilon_{it}. \tag{9.25}$$

By construction, ε_{it} groups together a set of idiosyncratic shocks, that is, specific to the individual unit and not dependent on common factors. The variances of the common factor and the idiosyncratic term are assumed to be finite.

Equation (9.24) can be rewritten in the form:

$$\Delta y_{i,t} = \alpha_i + \rho_i y_{i,t-1} + \gamma_i f_i + u_{i,t}, \tag{9.26}$$

with $\Delta y_{i,t} = y_{i,t} - y_{i,t-1}$ the first difference of the variable of interest $y_{i,t}$, $\alpha_i = (1 - \phi_i)\mu_i$ and $\rho_i = \phi_i - 1$.

This allows us to define the null hypothesis as follows:

$$H_0 : \rho_i = 0 \text{ for all } i.$$

The alternative hypothesis is then written as:

$$H_1 : \rho_i < 0$$

According to Pesaran, it is possible to approximate the unobserved common factor f by the inter-individual averages of Δy_{it} and $y_{i,t-1}$. On this basis, he proposes the following CADF regression model (*Cross-sectionally Augmented Dickey–Fuller*):

$$\Delta y_{i,t} = \alpha_i + \rho_i y_{i,t-1} + c_i \overline{y}_{t-1} + d_i \Delta \overline{y}_t + v_{i,t}, \tag{9.27}$$

with $\bar{y}_{t-1} = (1/N)\sum_{i=1}^{N} y_{i,t-1}$ and $\Delta\bar{y}_t = (1/N)\sum_{i=1}^{N} \Delta y_{i,t}$.

Under the assumption that the residuals are not autocorrelated and that there is no collinearity between the residuals and the common factor (although this case was taken into account by Pesaran), we will have the proposition for the individual unit: if t_i designates the individual t-statistics—for example for each country in a panel—it is therefore possible to write their average as:

$$CIPS = \frac{1}{N}\sum_{i=1}^{N} t_i(N,T), \tag{9.28}$$

where t_i is the statistic of the individual unit root test obtained by simply estimating ρ_i by Ordinary Least Squares (OLS) in Eq. (9.27). Implicitly, we are therefore testing $\rho_i = 0$.

$CIPS(N,T)$ thus represents an augmented version of the inter-individual averages of the IPS test, hence its name as a statistic or CIPS test. Its asymptotic distribution being non-standard, it has been tabulated by its author (see Pesaran, 2007) for different sizes of N and T.

It should be noted that Pesaran also proposes a truncated statistic that we will denote $CIPS^*$ and which is based on the truncated individual statistics defined as follows:

$$CIPS^* = (1/N)\sum_{i=1}^{N} t_i^*(N,T), \tag{9.29}$$

with $t_i^*(N,T)$.

Although the statistics $CIPS$ and $CIPS^*$ do not show level distortion and have similar power characteristics, when the residuals are not autocorrelated, the truncated statistic $CIPS^*$ proves to be superior (lower-level distortions) in the presence of error autocorrelation.

Tests with Structural Breaks

Macroeconomic series are likely to be marked by structural breaks, especially in the long term. These breaks are likely to alter the distribution of the unit root statistic and can lead to erroneous conclusions regarding the rejection of the null hypothesis.

Carrion-I-Silvestre et al. (2005) develop a methodology that allows for the existence of one or more breaks in the dynamic process of panel series. The test is not a unit root test but a stationarity test, i.e. it postulates stationarity under the null hypothesis. The authors rely on the KPSS (Kwiatkowski-Phillips-Schmidt-Shin) test to construct a test whose null hypothesis is variance stationarity. They calculate the global stationarity test as an average of the univariate KPSS statistics.

The stationarity test of Hadri (2000)—see "Other Unit Root Tests" section—can be modified to include one or more structural breaks in the deterministic component of the model through *dummy* variables. Under the null hypothesis, the data generating process is written as:

$$y_{i,t} = \alpha_i + \sum_{k=1}^{m_i} \theta_{i,k} DU_{i,k,t} + \beta_{i,t} + \sum_{k=1}^{m_i} \gamma_{i,k} DT^*_{i,k,t} + \varepsilon_{i,t}, \tag{9.30}$$

with the dummy variable $DT^*_{i,k,t} = t - T^i_{b,k}$ for $t > T^i_{b,k}$ or 0 otherwise. The number of individual structural breaks m is such that $m_i \geq 1$ with $k = \{1, ..., m_i\}$.

These breaks induce a change in the average in the temporal trends. This is a transposition of analogous tests on time series. When $\beta_i = \gamma_{i,k} = 0$, the model (8.30) becomes the Perron and Vogelsang (1992) model on time series or model C of Perron (1989) if $\beta_i \neq \gamma_{i,k} \neq 0$.

The effects of structural breaks are therefore measured through $\theta_{i,k}$ and $\gamma_{i,k}$. Moreover, these breaks can exist at all periods, since there is no restriction here: $T^i_{b,k} = T_{b,k}$. In addition, each individual unit can be affected by a different number of breaks $m_i \neq m_j, \forall_i \neq j, \{i, j\} = \{1, ..., T\}$.

On this basis, the test follows in the footsteps of Hadri (2000) and the corresponding statistic is written, under the null hypothesis,

$$LM(\lambda) = N^{-1} \sum_{i=1}^{N} \left(\hat{\omega}_i^{-2} T^{-2} \sum_{t=1}^{T} S^2_{i,t} \right), \tag{9.31}$$

with $S_{i,t} = \sum_{j=1}^{t} \hat{\varepsilon}_{i,j}$ the sum of the estimated residuals resulting from the application of OLS to Eq. (8.30) and $\hat{\omega}_i^{-2}$ an estimate of the long-term variance of $\varepsilon_{i,t}$. The test's dependence on break periods is materialised by λ, so that for each unit i, there is a vector $\lambda_i = \left(\lambda_{i,1}, ..., \lambda_{i,m_i}\right)' = \left(T^i_{b,1}/T, ..., T^i_{b,m_i}/T\right)'$, which allows to indicate the relative positions of each break with respect to the entire sample T.

The estimation of the number of breaks m_i is carried out through the method of Bai and Perron, which consists of minimising the sum of the squares of the residuals (*SSR*). The optimal number of breaks is then selected based on the BIC information criteria, modified Schwarz (LWS) or by a sequential procedure using F tests.

An application of this general procedure was notably carried out on the health expenditures of OECD countries by Carrion-I-Silvestre (2005). The author demonstrates the existence of breaks both in level and in the slope of the series. He also shows that health expenditure series can be considered as stationary around a long-term deterministic trend characterised by a break.

The use of cointegration techniques is then not necessary to obtain converging regressions of the link between expenditures of health and GDP, a question widely discussed in the literature.

9.3 Spurious Regressions and Panel Model Estimation

We have observed—see notably "On Canonical Correlations" section—that there exist spurious regressions between time series. Such a proposition meant that econometric relationships could appear significant on a statistical level, even though this significance was due to the presence of non-stationarity in the variables in question. Similar phenomena also characterise regressions using panel data.

Spurious Regressions and Cointegration

Phillips and Moon (1999) proposed an asymptotic theory of non-stationary panel data, for which both T and N are large. In this specific case, panel regressions can behave very differently from usual time series regressions. To put it simply, their basic idea is that considering the double dimension, particularly the cross-sectional dimension, of panels leads to better asymptotic results than those obtained on time series.

The two authors are actually trying to build an asymptotics for the processes $\{\mathbf{X}_{T,N}\}$ doubly indexed, in this case by time T and the individual unit N. In other words, it is about analysing limits for such processes, when T and N tend towards infinity. Several approaches are possible in this regard, but here, we will limit ourselves to considering two of these approaches:

- we fix one of the indices, for example N, and let the other index T tend towards infinity, which gives an intermediate limit, after which N in turn tends towards infinity. In this way, we have constructed a theory of the **sequential limit**. We will denote this type of limit process in the form $(T, N \to +\infty)_{\text{seq}}$, while specifying that the order of the indices is of utmost importance.
- the alternative is to allow both indices to tend **simultaneously** towards infinity, which provides us with a theory of the **joint limit**. We can then write this limit process in the general form $(T, N \to +\infty)$.

Phillips and Moon (1999) use the results of these two theoretical approaches to study the asymptotic behaviour of estimators obtained in the following four main cases:

1. spurious regressions in panels, for which there is no cointegration of time series;
2. cointegration in heterogeneous panels, where each individual unit admits its own cointegration relationship;
3. cointegration in homogeneous panels, where individual units have the same cointegration relationship;
4. cointegration in quasi-homogeneous panels, characterised by the fact that individual units admit very close cointegration relationships determined by what the authors call a position parameter.

For the first of these cases, where we are dealing with purely spurious regressions, the authors show that by using pooled data, the least squares estimators of the long-term average coefficients are $(N)^{1/2}$—convergent and follow, asymptotically, a normal distribution.

Such a result differs greatly from that valid for purely temporal spurious regressions—see, particularly, the example analysed in "Test on Constrained Estimators" section—since then, the OLS estimators of the coefficients in question were non-degenerate random variables, functionals of Brownian motions.

We will now examine how Phillips and Moon apply this asymptotic theory to the estimation of the panel model incorporating specific individual effects—in short, the fixed effects model (see "A Brief History of VAR Models" section of the previous chapter).

Here is, first, an example that shows how to estimate the long-term coefficients of a simple cointegration relationship between two variables that are integrated of order one, thus $I(1)$.

Example 9.2 Let's assume, initially, that y_t and x_t are two temporal variables with a long-term variance–covariance matrix equal to

$$\boldsymbol{\Omega} = \begin{pmatrix} \Omega_{yy} & \Omega_{yx} \\ \Omega_{xy} & \Omega_{xx} \end{pmatrix}.$$

We can then interpret $\widehat{\beta} = \Omega_{yx}\Omega_{xx}^{-1}$ as the estimator of the long-term coefficient linking the two non-stationary variables.
In the case of panel series integrated of the same order, but characterised by the heterogeneity of individual units, the analogue of $\boldsymbol{\Omega}$ is, for all i, given by $\boldsymbol{\Omega}_i$. This last matrix is supposed to have been drawn from a population of average $\boldsymbol{\Omega} = E(\boldsymbol{\Omega}_i)$. Consequently, the regression coefficient associated with the average long-term variance–covariance matrix $\boldsymbol{\Omega}$ is given by

$$\widehat{\beta} = E(\Omega_{y_i x_i})E(\Omega_{x_i x_i})^{-1} = \Omega_{yx}\Omega_{xx}^{-1}.$$

Following Phillips and Moon, let's first consider the general panel model based on the integrated vector process

$$z_{it} = z_{i,\,t-1} + \boldsymbol{u}_{it}, \qquad t = 1, \ldots, T, \quad i = 1, \ldots, N. \tag{9.32}$$

Let's partition the m-vectors $\mathbf{z}_{it}$ and $\mathbf{u}_{it}$ from (9.32) into m_y and m_x components such that $z_i = (\mathbf{y}'_{it}, \mathbf{x}'_{it})'$ and $\boldsymbol{u}_{it} = (\boldsymbol{u}'_{y_i,t}, \boldsymbol{u}'_{x_i,t})'$, with $m_y + m_x = m$. Let then $\boldsymbol{\Omega}_i$ be the variance–covariance matrix of $\mathbf{z}_{it} = (\mathbf{y}_{it}, \mathbf{x}_{it})'$ conditional on an information σ-algebra—it is enough to remember that it is a conditional matrix—that is, based on example 9.2, that

$$\boldsymbol{\Omega}_i = \begin{pmatrix} \boldsymbol{\Omega}_{y_i y_i} & \boldsymbol{\Omega}_{y_i x_i} \\ \boldsymbol{\Omega}_{x_i y_i} & \boldsymbol{\Omega}_{x_i x_i} \end{pmatrix}.$$

The matrix of **long-term average variance–covariance** of $\mathbf{z}_{it}$ is then given by

$$\mathbf{\Omega} = E(\mathbf{\Omega}_i) = \begin{pmatrix} \Omega_{yy} & \Omega_{yx} \\ \Omega_{xy} & \Omega_{xx} \end{pmatrix}.$$

Let's now introduce, while adopting the previous notations, the simplified fixed effects model:

$$y_{it} = \alpha_i + \beta_i x_{it} + e_{it} x_{it} = x_{i,t-1} + u_{x_i,t}.$$

We already know that the fixed effect α_i can be eliminated—see "The Within Estimator" section—by taking the average of observations in each individual group i and subtracting it from the initial data. By then applying the least squares (OLS) to the thus modified model, we finally obtain the Within estimator of β.

We then have the following proposition (Phillips & Moon, 1999, p. 1090), this time general, using joint limits:

Proposition 9.1 *Under a set of three assumptions and given that the data generating process* (DGP) *is given by* (9.32), *it follows*

(i) *when* $(N, T \to +\infty)$, *we have* $\widehat{\boldsymbol{\beta}}_{NT} \xrightarrow{p} \boldsymbol{\beta}$.
(ii) *if* $(N, T \to +\infty)$ *and if* $N/T \to 0$,

$$\sqrt{N}\left(\widehat{\boldsymbol{\beta}}_{NT} - \boldsymbol{\beta}\right) \xrightarrow{L} N\left(0, 36(\mathbf{\Omega}_{xx}^{-1} \otimes \mathbf{I}_{m_y})\mathbf{\Theta}_f(\mathbf{\Omega}_{xx}^{-1} \otimes \mathbf{I}_{m_y})\right),$$

where the operator $\otimes$, *already introduced in section* ("Preliminaries" section), *denotes the* ***Kronecker product*** (*or outer product*) *of two matrices,* $\mathbf{I}_{m_y}$ *representing the identity matrix of dimension* $(m_y)^2$ *and* $\mathbf{\Theta}_f$ *is the long-term average variance–covariance matrix* (*average, because the data are expressed in deviation from the group mean*).

The conclusion to be drawn from this proposition is that the least squares estimators of the fixed effects panel model are asymptotically convergent and normally distributed.

Estimation of Cointegrated Panel Models

We discussed the problem of estimating temporal cointegrated relations in Chapter 4. In short, three main methods were available to estimate such relations: Fully Modified Ordinary Least Squares (FM-OLS, see "Alternative Methods" section); Dynamic Ordinary Least Squares (DOLS, also see "Alternative Methods" section); and finally, Johansen's multidimensional method, which differs from the previous two methods as it is based on the representation of a VAR model in error correction form (see "On Canonical Correlations"–"The Maximum Likelihood Estimator" sections).

All these estimation procedures find their counterpart for panel models. However, before detailing them, we will first examine the reasons why simple least squares are not suitable for estimating such models.

To do this, let's consider the fixed effects model

$$y_{it} = \alpha_i + \boldsymbol{\beta}'\boldsymbol{x}_{it} + u_{it}, \quad i = 1, \ldots, N, \quad t = 1, \ldots, T. \tag{9.33}$$

The meaning of the different symbols in (9.33) is known to us. It will suffice to add that u_{it} is a supposed stationary random variable and that $\{\mathbf{x}_{it}\}$ is a multivariate process $(K \times 1)$, supposed to be integrated of order one, so $I(1)$, for all i, with

$$\boldsymbol{x}_{it} = \boldsymbol{x}_{i,t-1} + \boldsymbol{\varepsilon}_{it}. \tag{9.34}$$

Under such specifications, (9.33) represents a system of cointegrated equations.

The least squares estimator of $\boldsymbol{\beta}$ is given by

$$\begin{aligned}\hat{\boldsymbol{\beta}}_{MCO} = & \left[\sum_{i=1}^{N}\sum_{t=1}^{T}(\boldsymbol{x}_{it} - \overline{\boldsymbol{x}}_i)(\boldsymbol{x}_{it} - \overline{\boldsymbol{x}}_i)'\right]^{-1} \\ & \times \left[\sum_{i=1}^{N}\sum_{t=1}^{T}(\boldsymbol{x}_{it} - \overline{\boldsymbol{x}}_i)\left(y_{it} - \overline{y}_i\right)\right].\end{aligned} \tag{9.35}$$

Subject to three assumptions, namely (i) the theory used is a sequential limit theory, $T \to +\infty$ being followed by $N \to +\infty$; (ii) the variance–covariance matrix of the vector $\left(u_{it}, \varepsilon'_{it}\right)'$ is equal to

$$\boldsymbol{\Omega} = \begin{pmatrix} \boldsymbol{\Omega}_u & \boldsymbol{\Omega}_{u\varepsilon} \\ \boldsymbol{\Omega}_{\varepsilon u} & \boldsymbol{\Omega}_{\varepsilon} \end{pmatrix};$$

(iii) the panel considered is homogeneous, that is to say that the variances are constant between the cross-sectional units and the corresponding long-term matrix is

$$\boldsymbol{\Delta} = \begin{pmatrix} \boldsymbol{\Delta}_u & \boldsymbol{\Delta}_{u\varepsilon} \\ \boldsymbol{\Delta}_{\varepsilon u} & \boldsymbol{\Delta}_{\varepsilon} \end{pmatrix},$$

Kao and Chiang (2000) then demonstrate the following proposition:

Proposition 9.2 $T(\hat{\boldsymbol{\beta}}_{MCO} - \boldsymbol{\beta}) \xrightarrow{P} -3\boldsymbol{\Omega}_{\varepsilon}^{-1}\boldsymbol{\Omega}_{\varepsilon u} + 6\boldsymbol{\Omega}_{\varepsilon}^{-1}\boldsymbol{\Delta}_{\varepsilon u}.$

It follows that the least squares estimator does not converge in probability towards the true value of $\boldsymbol{\beta}$. (This proposition remains valid if we consider heterogeneous panels.)

There is a second, practical reason why least squares do not yield optimal estimators of cointegration relationships in panels: indeed, Kao and Chiang (2000) have shown, using Monte Carlo simulations, that at finite distance, the OLS estimator showed a significant bias compared to its FM-OLS and DOLS competitors.

Let's now examine the FM-OLS estimation method, which finds its origins in the article by Phillips and Hansen (1990), but which has been adapted for panel cointegration regressions by Phillips and Moon (1999) and by Pedroni (2000).

The FM-OLS estimator is obtained by making corrections for endogeneity and autocorrelation to the OLS estimator (9.35):

$$\hat{\boldsymbol{\beta}}_{FM} = \left[\sum_{i=1}^{N}\sum_{t=1}^{T}(\boldsymbol{x}_{it}-\overline{\boldsymbol{x}}_i)(\boldsymbol{x}_{it}-\overline{\boldsymbol{x}}_i)'\right]^{-1} \times \left[\sum_{i=1}^{N}\sum_{t=1}^{T}(\boldsymbol{x}_{it}-\overline{\boldsymbol{x}}_i)\hat{y}_{it}^{+} - NT\,\hat{\boldsymbol{\Delta}}_{\varepsilon u}^{+}\right], \tag{9.36}$$

where $\hat{y}_{it}^{+} = y_{it} - \hat{\boldsymbol{\Omega}}_{u\varepsilon}\hat{\boldsymbol{\Omega}}_{\varepsilon}^{-1}\Delta\boldsymbol{x}_{it}$ and $\hat{\boldsymbol{\Delta}}_{\varepsilon u}^{+} = \hat{\boldsymbol{\Delta}}_{\varepsilon u} - \hat{\boldsymbol{\Delta}}_{\varepsilon}\hat{\boldsymbol{\Omega}}_{\varepsilon}^{-1}\hat{\boldsymbol{\Omega}}_{\varepsilon u}$, with $\widehat{\boldsymbol{\Delta}}_{\varepsilon u}$ and $\widehat{\boldsymbol{\Delta}}_{\varepsilon}$ being non-parametric estimators of $\boldsymbol{\Delta}_{\varepsilon u}$ and $\boldsymbol{\Delta}_{\varepsilon}$.

The corrected estimator $\widehat{\boldsymbol{\beta}}_{FM}$ is convergent. Moreover, it is asymptotically normal.

Proposition 9.3 *Under the assumptions detailed in the previous proposition, it follows*

$$\sqrt{NT}\,\hat{\boldsymbol{\beta}}_{FM} \to^{L} N(\boldsymbol{0},\ 6\boldsymbol{\Omega}_{\varepsilon}^{-1}\boldsymbol{\Omega}_{u\varepsilon}).$$

Let's consider again the model (9.33)–(9.34). We obtain the DOLS estimator of β, denoted β_D, by launching the regression

$$y_{it} = \alpha_i + \boldsymbol{\beta}'\boldsymbol{x}_{it} + \sum_{j=-q}^{q} c_{ij}\Delta\boldsymbol{x}_{i,t+j} + \dot{v}_{it},$$

where $\dot{v}_{it} = v_{it} + \sum_{|j|>q} c_{ij}\varepsilon_{i,t+j}$.

Given two additional assumptions, one on the spectral density of a correction matrix (this is therefore a non-parametric estimation method), the asymptotic distribution of $\boldsymbol{\beta}_D$ is identical to that of the FM-OLS estimator:

Proposition 9.4 $\sqrt{NT}\,\hat{\boldsymbol{\beta}}_{D} \to^{D} N(\boldsymbol{0},\ 6\boldsymbol{\Omega}_{\varepsilon}^{-1}\boldsymbol{\Omega}_{u\varepsilon})$.

The previous developments rely on the assumption of homogeneity of panel data in their cross-sectional dimension. The questioning of this assumption does not, in reality, pose a major difficulty. Indeed, it is possible to correct the FM-DOLS and DOLS estimators, so that they satisfy propositions 9.2 and 9.3 while incorporating the **cross-sectional heterogeneity** of the panel data used.

However, several criticisms can be made of these two approaches. First, they are based on the consideration of a single equation and therefore, implicitly, they presuppose that there is only one cointegration relationship between the selected variables. Then, these methods both use a non-parametric estimation approach, which, even if it is elegant on a theoretical level, does not meet the usual standards of the econometrics practitioner.

That's why a purely parametric and systemic approach (multiple cointegration relationships) is welcome. This is what Breitung (2005) achieved by considering a vector error correction model.

In short, this approach goes through two main stages:

(i) during the first, the adopted vector error correction model provides an estimator that is asymptotically equivalent to the Gaussian maximum likelihood estimator;
(ii) the second stage allows to obtain, using the Johansen MV estimator—see "Alternative Methods" and "The Maximum Likelihood Estimator" sections—a convergent estimator (when $T \rightarrow +\infty$) resulting from the estimation of separate models for the N cross-sectional units.

The final estimator obtained satisfies all the desired asymptotic properties, in particular those of convergence and normality.

9.4 Cointegration Tests

As with the analysis of time series, these tests fall into two main categories: on the one hand, cointegration tests focusing on estimation residuals; on the other hand, those based on vector autoregressive processes (VAR).

Kao Tests Based on Residuals

Kao (1999) proposes unit root tests of both the DF (Dickey Fuller) and ADF (Augmented Dickey–Fuller) types. The first is based on the use of residuals from the regression

$$y_{it} = \alpha_i + \boldsymbol{\beta}'\boldsymbol{x}_{it} + u_{it}, \quad i = 1, \ldots, N, \quad t = 1, \ldots, T, \tag{9.37}$$

where y_{it} and $\mathbf{x}_{it}$ are $I(1)$, but not cointegrated. These residuals $\hat{u}_{it}$ are used to estimate the relationship

$$\hat{u}_{it} = \rho\, \hat{u}_{i,t-1} + \varepsilon_{it}. \tag{9.38}$$

The ordinary least squares estimator of ρ in (9.38) is then given by

$$\hat{\rho} = \frac{\sum_{i=1}^{N}\sum_{t=2}^{T}\hat{u}_{it}\hat{u}_{i,t-1}}{\sum_{i=1}^{N}\sum_{t=2}^{T}\hat{u}_{i,t-1}^{2}},$$

and the statistic t by

$$t_\rho = \frac{(\hat{\rho}-1)\sqrt{\sum_{i=1}^{N}\sum_{t=2}^{T}\hat{u}_{i,t-1}^{*\,2}}}{s_u},$$

with $s_u^2 = (1/NT)\sum_{i=1}^{N}\sum_{t=2}^{T}(\hat{u}_{it} - \hat{\rho}\,\hat{u}_{i,t-1})^2$.

The null hypothesis of **no cointegration** is written as $H_0 : \rho = 1$. Kao then proposes four DF-type statistics:

$$DF_\rho = \frac{\sqrt{N}T(\hat{\rho}-1) + 3\sqrt{N}}{\sqrt{10,2}},$$
$$DF_t = \sqrt{1,25}\,t_\rho + \sqrt{1,875N},$$

and

$$DF_\rho^* = \frac{\sqrt{N}T(\hat{\rho}-1) + 3\sqrt{N}(\hat{\sigma}_\varepsilon^2/\hat{\sigma}_{0\varepsilon}^2)}{\sqrt{3 + (36\hat{\sigma}_\varepsilon^4/5\hat{\sigma}_{0\varepsilon}^4)}},$$
$$DF_t^* = \frac{t_\rho + \sqrt{6N}(\hat{\sigma}_\varepsilon/2\hat{\sigma}_{0\varepsilon})}{\sqrt{\frac{\hat{\sigma}_{0\varepsilon}^2}{2\hat{\sigma}_\varepsilon^2} + \frac{3\hat{\sigma}_\varepsilon^2}{10\hat{\sigma}_{0\varepsilon}^2}}},$$

where $\hat{\sigma}_\varepsilon^2 = \hat{\Sigma}_{yy} - \hat{\Sigma}_{yx}\hat{\Sigma}_{xx}^{-1}$ and $\hat{\sigma}_{0\varepsilon}^2 = \hat{\Omega}_{yy} - \hat{\Omega}_{yx}\hat{\Omega}_{xx}^{-1}$.

The ADF type test requires adding lagged variations of u_{it} in (9.38), let

$$\hat{u}_{it} = \rho\,\hat{u}_{i,t-1} + \sum_{j=1}^{p}\varphi_j\,\Delta\hat{u}_{i,t-j} + \varepsilon_{it,p}. \tag{9.39}$$

Under the null hypothesis of no cointegration, that is, the estimator of ρ in regression (8.39) is equal to one, the ADF statistic is defined by

$$ADF = \frac{t_{ADF} + \sqrt{6N}(\hat{\sigma}_\varepsilon/2\hat{\sigma}_{0\varepsilon})}{\sqrt{\frac{\hat{\sigma}_{0\varepsilon}^2}{2\hat{\sigma}_\varepsilon^2} + \frac{3\hat{\sigma}_\varepsilon^2}{10\hat{\sigma}_{0\varepsilon}^2}}}, \tag{9.40}$$

where t_{ADF} is the t-statistic of ρ in (9.39).

It can be shown, using the sequential limit theory, that these five statistics all follow, asymptotically, a standardised normal distribution.

Other Tests Based on the Residuals

Pedroni (1999, 2004) considers the panel regression model incorporating **heterogeneous** coefficients in the cross-sectional dimension

$$y_{it} = \alpha_i + \gamma_i\, t + \boldsymbol{\beta}'_i \boldsymbol{x}_{it} + u_{it}, \tag{9.41}$$

where y_{it} and $\mathbf{x}_{it}$ are respectively a scalar and a K-vector $I(1)$. It is also required that this last vector does not have cointegrated components. Such a model also allows the long-term variance of $\{\Delta y_{it}, \Delta \mathbf{x}_{it}\}$ to be heterogeneous.

Pedroni estimates model (9.41) separately for each i by least squares and then suggests seven tests using the residuals of (9.41). However, Monte Carlo simulations carried out by Gutierrez (2003) and by Pedroni (2004) himself indicate that a variant of the Im, Pesaran and Shin (IPS, see "Other Unit Root Tests" section) test is the most promising among the seven of them, insofar as it maintains nominal levels within acceptable limits, even for N small.

This variant is, in fact, a **normalised** sum of the Phillips and Ouliaris (1990) statistic, intended to test the null hypothesis of no cointegration, while using the residuals drawn from each regression operated on i in (9.41).

The requirement for normalisation requires the use of the mean and the variance of the Phillips and Ouliaris test, whose values are tabulated in Pedroni (2004, p. 606) based on 100,000 random draws.

Westerlund (2007) proposes a test of the null hypothesis of **no cointegration** in an error correction model. He then gives four test statistics, denoted G_α, G_τ, P_α, P_τ, which all asymptotically follow a normal distribution. To demonstrate this convergence in law, Westerlund uses the theory of sequential limits. Of these statistics, the simplest to calculate is G_τ which surpasses the other three, when the sample size is taken into account.

Let's now examine the McCoskey and Kao (1998) test. Its main feature is to adopt as the null hypothesis, unlike the previous tests, the **presence of cointegration** while using appropriate residuals. In this case, it is obviously essential to use an efficient estimation method, for example FM-OLS or DOLS (see "Estimation of Cointegrated Panel Models" section), which yield asymptotically normally distributed estimators with zero mean.

Specifically, let $\{y_{it}, \mathbf{x}_{it}\}$ be integrated processes of order one. Our two authors then consider the model

$$\begin{aligned} y_{it} &= \alpha_i + \boldsymbol{\beta}'_i \boldsymbol{x}_{it} + u_{it}, \\ x_{it} &= x_{i,t-1} + \varepsilon_{it}, \\ u_{it} &= \gamma_{it} + v_{it}, \end{aligned} \tag{9.42}$$

and

$$\gamma_{it} = \gamma_{i,t-1} + \theta\, v_{it},$$

where $\gamma_{it} = \gamma_{i,t-1} + \theta\, v_{it}$,

The null hypothesis of cointegration is $H_0 : \theta = 0$ against the alternative $H_1 : |\theta| \neq 0$. By repeatedly substituting the third equation of (9.42) into the first, it follows

$$\begin{aligned} y_{it} &= \alpha_i + \boldsymbol{\beta}'_i \boldsymbol{x}_{it} + \theta \sum_{j=1}^{t} v_{ij} + v_{it} \\ &= \alpha_i + \boldsymbol{\beta}'_i \boldsymbol{x}_{it} + e_{it}, \end{aligned} \tag{9.43}$$

with $e_{it} = \theta \sum_{j=1}^{t} v_{ij} + v_{it}$.

The test statistic is then

$$LM = \frac{\sum_{i=1}^{N} \sum_{t=1}^{T} S_{it}^2}{s^2}, \tag{9.44}$$

where S_{it} represents the partial sum of the residuals of (9.43)

$$\forall t,\ S_{it} = \sum_{j=1}^{t} \hat{e}_{it},$$

while s^2 is a convergent estimator of σ_v^2 under H_0, with

$$s^2 = (1/NT) \sum_{i=1}^{N} \sum_{t=1}^{T} \hat{e}_{it}^2.$$

McCoskey and Kao then show that the statistic (9.44) follows asymptotically a normal distribution with zero mean

$$\sqrt{N}(LM - \mu_r) \xrightarrow{L} N\left(0, \sigma_r^2\right),$$

where the moments μ_r and σ_r^2 are obtained by running stochastic simulations.

Tests Based on a VAR Model

Let the VAR(p) model be

$$\boldsymbol{y}_t = \boldsymbol{A}_1 \boldsymbol{y}_{t-1} + \boldsymbol{A}_2 \boldsymbol{y}_{t-2} + \cdots + \boldsymbol{A}_p \boldsymbol{y}_{t-p} + \boldsymbol{u}_t, \quad \boldsymbol{u}_t \sim IIN(\boldsymbol{0}, \boldsymbol{\Omega}), \tag{9.45}$$

where $\mathbf{u}_t$ is the k-column vector and $\boldsymbol{\Omega}$ is a positive definite matrix of dimension $k \times k$ such that $E\left(\mathbf{u}_s \mathbf{u}'_t\right) = 0$, $\forall s \neq t$.

This autoregressive model can still be written in first differences in the form (see "The VAR(p) Model" section):

$$\Delta \boldsymbol{y}_t = \sum_{j=1}^{p-1} \boldsymbol{\Phi}_j \, \Delta \boldsymbol{y}_{t-j} + \boldsymbol{\Pi} \, \boldsymbol{y}_{t-1} + \boldsymbol{u}_t, \tag{9.46}$$

with $\boldsymbol{\Pi} = -\left(\mathbf{I}_k - \mathbf{A}_1 - \cdots - \mathbf{A}_p\right)$ and $\boldsymbol{\Phi}_j = -\sum_{i=j+1}^{p} \mathbf{A}_i, j = 1, 2, \ldots, p-1$.

The matrix $\boldsymbol{\Pi}$ is of dimension $k \times k$. When its rank, let's denote it r, is less than k, then $(k - r)$ gives us the number of cointegration relationships. Johansen—see "Determining the Cointegration Rank" section—proposed a test based on the likelihood ratio.

Suppose, from this point of view that we test the null hypothesis

$$H_r: \ \ \text{rk}(\boldsymbol{\Pi}) = r < k$$

against the alternative

$$H_k: \ \ \text{rk}(\boldsymbol{\Pi}) = k,$$

where $r = 1, 2, \ldots, k-1$ represents the number of cointegration relations. The likelihood ratio statistic is then written as:

$$LR = 2\left[\ln L_c\left(\hat{\boldsymbol{\beta}}, k\right) - \ln L_c\left(\hat{\boldsymbol{\beta}}, r\right)\right].$$

Larsson et al. (2001) consider the analogue of (9.46) for panel data, namely the heterogeneous error correction vector model of order p_i :

$$\Delta \boldsymbol{y}_{it} = \sum_{j=1}^{p_i} \boldsymbol{\Phi}_{ij} \, \Delta \boldsymbol{y}_{i,t-j} + \boldsymbol{\Pi}_i \, \boldsymbol{y}_{i,t-1} + \boldsymbol{u}_{it}, \tag{9.47}$$

where $\mathbf{u}_{it} \sim IIN(0, \boldsymbol{\Omega}_i)$.

The authors assume that the cross-sectional units are independent and then consider the null hypothesis

$$H_0 : rg(\boldsymbol{\Pi}_i) = r_i \leq r < k, \quad \forall i = 1, \ldots, N$$

against the alternative

$$H_1 : rg(\boldsymbol{\Pi}_i) = k, \quad \forall i = 1, \ldots, N.$$

We now designate the Johansen likelihood ratio test (1988) for the individual unit i, by LR_i, and the corresponding cross-sectional average by $LRB_{NT} = (1/N) \sum LR_i$.

The test statistic is then given by

$$\lambda[H(r)|\,H(k)] = \frac{\sqrt{N}(LRB_{NT} - m_{LR})}{\sqrt{V_{LR}}},$$

where m_{LR} and V_{LR} are mean and variance correction factors.

This last statistic asymptotically follows a standard normal distribution $N(0, 1)$, when N and T tend towards infinity in such a way that $\sqrt{N}/T \to 0$.

One of the limitations of this test lies in the fact that it relies on the assumption of independence of individual units. Groen and Kleibergen (2003) attempted to lift this assumption, but the result is not entirely conclusive.

9.5 Application: Solar Energy Production

Our empirical application will consist of evaluating the determinants of renewable solar energy production in Europe. It directly relies on the use of the non-stationary panel methodology. Once we have determined the statistical properties and the order of the series through panel unit root tests, we will then study their potential long-term relationships.

The Data and the Equation to Be Estimated

The database used consists of a set of panel series composed of 20 European Union countries, each of them considered over the period 1990–2015.

However, the dimension T is limited, as some countries indeed could not develop a significant amount of photovoltaic infrastructure, given their reduced annual sunshine capacity. Moreover, the panel only started in 1990, due to the small share of renewables in total energy production before this date.

The series come from the ENERDATA platform, which is an expert and independent company, specialised in providing data and reports on energy.

We will use six variables in our modelling exercise:

- y_{it} is the percentage share of solar energy in total electricity production for each country, variable approximated by the installed capacity (measured in megawatt-hours, MWh) allowing the production of solar energy;
- GDP_{it} is the Gross Domestic Product *per capita*;
- $CO2_{it}$ refers to carbon dioxide emissions relative to GDP;
- $CONS_{it}$ is the domestic energy consumption of the twenty selected countries;
- BAL_{it} aggregates net electricity imports;
- OIL_{it} refers to the share of oil in the total electricity production.

Explicitly, the model, in which y_{it} constitutes the endogenous variable, is specified as follows:

$$y_{i,t} = \alpha_1 + \alpha_2 GDP_{it} + \alpha_3 CO2_{it} + \alpha_4 CONS_{it} + \alpha_5 BAL_{it} + \alpha_6 OIL_{it} + u_{it}. \quad (9.48)$$

Unit Root Tests in Panel

To avoid a potential spurious regression, we will implement a battery of unit root tests in panel—see "The LLC Test" and "Other Unit Root Tests" sections of this chapter—on each of the variables in our sample.

Initially, we will perform the classic so-called first-generation tests of Levin, Lin and Chu (LLC hereafter) and Im, Pesaran, and Shin (IPS hereafter). Both assume that the individual units—the countries in this case—are independent. Therefore, the rejection of the null hypothesis of the existence of a unit root does not necessarily imply that all these units are stationary.

Here are the results of applying these tests to our series (Table 9.2).

The Akaike criterion is used to determine the lags in both tests. Moreover, the specification without trend is retained for each of these tests.

Although the LLC test, applied to energy consumption and GDP, leads to the rejection of the non-stationarity of these two variables, all the other tests clearly conclude the non-rejection of the null hypothesis of unit root, the p-values converging towards unity in the vast majority of cases.

However, a question arises: are the 20 countries concerned really independent on a statistical level? A priori, one can suppose the existence of a certain convergence between European countries when implementing economic policies for the deployment of renewable energies, if only because these policies are based, at least in part, on common impulses, such as the Kyoto or COP21 Paris climate agreements for example.

We therefore need to carry out cross-dependency tests between individual units, referring to the *CD* statistic of Pesaran (see this chapter, "Testing Cross-Section Dependence" section). Table 9.3 presents the statistics in question and the associated *p*-values.

Table 9.2 First-generation tests

	LLC		IPS	
	Stat	*p*-values	Stat	*p*-values
y	6.406	1.00	8.410	1.00
GDP	−2.382	0.01	2.061	0.98
CO2	−0.052	0.48	5.30	1.00
CONS	−4.344	0.00	−0.807	0.21
BAL	2.529	0.99	6.587	1.00
OIL	0.437	0.67	2.868	0.99

Table 9.3 Cross-sectional dependency

	CD	*p*-value
y	51.356	0.000
GDP	56.394	0.000
CO2	56.667	0.000
CONS	57.007	0.000
BAL	66.707	0.000
OIL	20.688	0.000

This table shows that, regardless of the variable considered, the *CD* test by Pesaran leads to the rejection of the null hypothesis of independence between individual units.

The statistically significant dependence of all these variables can, of course, be explained by the common factor that represents the European public policies, climate agreements or even, through a certain synchronisation of indicators and macroeconomic performances, due to the concordance of economic cycles specific to the European Union, for example.

In light of these results, it is instructive to compare the conclusions provided by the first-generation tests with those that would result from the application of second-generation tests.

We will limit ourselves here to the Pesaran test—presented in "The Pesaran Test" section of this chapter. From a practical point of view, we have taken into consideration three time lags (Table 9.4).

The conclusions to be drawn from this test vary depending on the number of lags used. However, when one or two lags are retained, all series, with the exception of the energy balance, are integrated of order one, that is, non-stationary. The "energy balance" variable is clearly stationary.

Table 9.4 Pesaran test

	$y_{i,t}$	$k = 1$	$k = 2$
y	53.925 (0.000)	26.371 (0.952)	18.486 (0.999)
GDP	2.675 (0.996)	0.338 (0.651)	0.235 (0.593)
CO2	−2.184 (0.014)	−2.075 (0.019)	−0.618 (0.268)
CONS	−2.831 (0.002)	−1.245 (0.107)	−0.233 (0.408)
BAL	−5.410 (0.000)	−2.826 (0.000)	−3.120 (0.001)
OIL	26.886 (0.944)	21.727 (0.992)	19.327 (0.998)

Panel Cointegration Tests

In light of the results of the previous unit root tests, it seems reasonable to consider that the variables considered are I(1) or quasi I(1). As a result, we will analyse the existence of a cointegration relationship (null hypothesis of non-cointegration), through the use of Pedroni and Westerlund tests.

The Pedroni test—see this chapter, "Other Tests Based on the Residuals" section—is a so-called first-generation test, which was constructed from the analysis of the stationarity of the residuals from the potential cointegration relationship specified. In addition, Pedroni introduced a dose of heterogeneity in the cross-sectional dimension of the panel, which leads to "pooled" tests or heterogeneous, based on group averages.

Here are the seven statistics developed by Pedroni, with the associated p-values (remember that these statistics all follow a standard normal distribution and that the null hypothesis is the absence of a cointegration relationship) (Table 9.5).

The regression model incorporates a constant, but no trend.

From this table, it emerges above all that five out of the seven statistics programmed by Pedro do not lead to rejecting the null hypothesis of no cointegration. Taken as a whole, the results clearly lean towards the existence of a cointegration relationship between the six variables selected for the modelling.

Let's now move on to the Westerlund test. As we have highlighted—see "Non-stationary Panels" chapter, "Other Tests Based on the Residuals" section, this test proposes a test of the null hypothesis of no cointegration in an error correction model, while allowing a form of interindividual dependence.

It then provides a set of four test statistics, denoted respectively G_α, G_τ, P_τ, are asymptotically normally distributed. Table 9.6 groups statistics and p-values, while specifying that the application of the Akaike criterion leads to no shift being introduced in the estimated relationship.

Only the last two "panel" statistics, P_α and P_τ, lead to accepting the alternative hypothesis of cointegration, even though G_α and G_τ do not allow to reject the null hypothesis of non-cointegration.

Table 9.5 Pedroni's cointegration tests

	Stat	Standardised values and p-values
Pooled panel	$Z_{\nu_{N,T}}$	3.301 (0.001)
	$Z_{\rho_{N,T-1}}$	0.965 (0.83)
	$Z_{t_{N,T}}$	−2.525 (0.01)
	$Z^*_{t_{N,T}}$	−2.776 (0.00)
Heterogeneous Panel	$\tilde{Z}_{\rho_{N,T-1}}$	2.31 (0.99)
	$\tilde{Z}_{t_{N,T}}$	−2.44 (0 01)
	$\tilde{Z}^*_{t_{N,T}}$	−2.23 (0 01)

Table 9.6 Westerlund Test (inter-individual dependence)

	G_α	G_τ	P_α	P_τ
Statistic	3.373	3.455	−2.385	−2.385
p-value	1.000	1.000	0.009	0.006

Furthermore, we have no statistical indications suggesting the presence of structural breaks in the dynamics of solar energy production. Consequently, we did not resort to cointegration tests, known as "3rd-generation", integrating the presence of potential trend breaks in the dynamics of the stochastic processes at work.

Estimation of the Panel Cointegration Relationship

The cointegration tests that we conducted in the previous section confirmed the existence of a cointegration relationship between solar capacities and their determinants.

There is therefore potentially a panel equilibrium relationship that requires the use of adequate estimators (as we showed in "Estimation of Cointegrated Panel Models" section).

First, there is the FMOLS estimator, which modifies the least squares on pooled data through a semi-parametric correction of the long-term variance–covariance matrices. The DOLS estimator, then, corrects the least squares to take into account endogeneity and autocorrelation. Also note that it performs very well in the temporal dimension of the panel in finite samples.

The FMOLS and DOLS estimators can themselves be corrected to incorporate the cross-sectional heterogeneity of the panel data used.

Here are the parameter estimates of Eq. (9.48) obtained by each of the two methods (Table 9.7).

Table 9.7 FMOLS and DOLS estimators

	GDP	*CO2*	*CONS*	*BAL*	*OIL*
FMOLS	−14.693 (−0.60)	−38.698 (−0.06)	40.057 (0.76)	0.033 (9.114)	−26.37 (−1.27)
DOLS	−1.668 (−0.05)	−157.234 (−0.15)	−5.229 (−0.07)	0.054 (7.17)	−11.399 (−0.29)

Upon reading this table, three main remarks immediately come to mind:

1. only the coefficients of the variable *BAL*, i.e. the net electricity imports, are significantly different from zero. Given the observed sign, we conclude that these imports positively influence long-term solar energy production and, correspondingly, its deployment.

2. even if the coefficients of the variable *CONS* are not significant, they are of opposite sign depending on the estimation method used.
3. the coefficients of the variables *CO2* and *GDP* negatively impact long-term solar energy production, which is a result contrary to intuition.

In total, the poor statistical quality of our estimates could be explained by the fact that few countries have really developed significant electricity production from solar over the considered period (1990–2015), especially in the first part of this period (specifically, the nineties).

However, let's remember the result of the analysis related to the variable *BAL*, which is, we have established, positively correlated with solar energy production.

In other words, the more countries are energy dependent on the outside world through their net energy imports, the more they are also encouraged to deploy the necessary resources for solar energy production.

The energy transition is thus taking place before our eyes.

9.6 Exercises

9.6.1 Using your own words, explain what is the difference between the "first-generation", "the second-generation" and the "third-generation" panel unit root tests.

9.6.2 Identify the types of time series tests that the LLC, IPS, Breitung and Hadri tests are associated with.

9.6.3 In the same spirit, explain what is the difference between the Kao, Pedroni and Westerlund cointegration tests.

9.6.4 Answer the following questions about second and third-generation panel unit root tests.

1. Explain why traditional unit root tests like LLC, IPS and Breitung are biased in the presence of cross-sectional error dependence. Provide an example of a macroeconomic series that might exhibit this dependence.
2. Describe the main steps of the PANIC test. What is the role of the common and idiosyncratic components in this test?
3. How do Bai and Ng test the stationarity of the common and idiosyncratic components separately?
4. Outline the procedure for the CADF test proposed by Pesaran. How does it differ from the traditional ADF test? What is the CIPS statistic and how is it calculated?
5. Discuss how structural breaks can impact the results of unit root tests. Explain the methodology developed by Carrion-I-Silvestre (2005) for handling structural breaks in panel data. How is the Hadri (2000) test modified to include structural breaks?

9.6.5 Consider the estimation of a cointegrated panel model as discussed in "Estimation of Cointegrated Panel Models" section. Suppose we have a fixed-effects model of the form:

$$y_{it} = \alpha_i + \boldsymbol{x'}_{it}\boldsymbol{\beta} + u_{it}, \qquad i = 1, \dots, N; \qquad t = 1, \dots, T.$$

All variables included in $\boldsymbol{x'}_{it}$ are assumed to be integrated of order one, i.e. I(1).

1. Show why simple OLS is not suitable for estimating the parameters of this model.
2. Describe the FMOLS method for estimated cointegrated panel models. What corrections are made to the OLS estimator in this case?
3. Explain the DOLS method for estimating the same parameters. How did it differ from the previous estimator?
4. Discuss the Johansen method for estimating the same model in this context.
5. Discuss the implications of panel heterogeneity on the estimation methods discussed. How can FM-OLS and DOLS techniques be adapted to account for heterogeneity across cross-sectional units?

9.6.6 Using the data in the file "Forex.xls" from Bismans and Damette (2008), specifically intraday data from November 24 to November 25, 2004, on the bid and ask prices of four exchange rates, namely US dollar/euro, dollar/yen, dollar/sterling, and dollar/Canadian dollar, apply the panel unit root tests of first, second and third generations (if necessary) and compare the results. Along this way, find the results of cross-section dependence illustrated in Example 9.1.

References

Bai, J., & Ng, S. (2004). A PANIC attack on unit roots and cointegration. *Econometrica, 72*, 1127–1178.

Bai, J., & Ng, S. (2010). Panel unit root tests with cross-section dependence: A further investigation. *Econometric Theory, 26*, 1088–1114.

Bismans, F., & Damette, O. (2008). Currency transaction tax elasticity: An econometric evaluation. *Économie Internationale, 115*, 193–212.

Breitung, J. (2005). A parametric approach to the estimation of cointegration vectors in panel data. *Econometric Reviews, 24*, 151–173.

Carrion-I-Silvestre, J. (2005). Health care expenditure and GDP. *Journal of Health Economics, 24*, 839–854.

Groen, J. J. J., & Kleibergen, F. R. (2003). Likelihood-based cointegration analysis in panels of vector error correction models. *Journal of Business and Economic Statistics, 21*, 295–318.

Gutierrez, L. (2003). On the power of panel cointegration tests: A monte Carlo comparison, *Economic Letters, 80*, 105–111.

Hadri, K. (2000). Testing for stationarity in heterogenous panel data. *Econometrics Journal, 3*, 148–161.

Hurlin, C., & Mignon, V. (2007). Une synthèse des tests de racine unitaire sur données de panel. *Economie Et Prévision, 169–171*, 251–295.

Johansen, S. (1988). Statistical analysis of cointegration vectors. *Journal of Economic Dynamics and Control, 12*, 231–254.

Kao, C. (1999). Spurious regression and residual-based tests for cointegration in panel data. *Journal of Econometrics, 90*, 1–44.

Kao, C., & Chiang, M. H. (2000). On the estimation and inference of a cointegrated regression in panel data. *Advances in Econometrics, 15*, 179–222.

Larsson, R., Lyhagen, J., & Löthgren, M. (2001). Likelihood-based cointegration tests in heterogenous panels. *Econometrics Journal, 4*, 109–147.

Levin, A., Lin, C.-F., & Chu, C.-S.J. (2002). Unit root tests in panel data: Asymptotic and finite-sample properties. *Journal of Econometrics, 108*, 1–24.

McCoskey, S., & Kao, C. (1998). A residual-based test of the null of cointegration in panel data. *Econometric Reviews, 17*, 57–84.

O'Connell, P.G.J. (1998). The overvaluation of purchasing power parity, *Journal of International Economics, 44*, 1–19.

Perron, P. (1989), The great crash, the oil price shock and the unit root hypothesis, *Econometrica, 57*, 1361–1401.

Pedroni, P. (1999). Critical values for cointegration tests in heterogenous panels with multiple regressors, *Oxford Bulletin of Economics and Statistics, 61*, 653–678.

Pedroni, P. (2000). Fully modified OLS for heterogenous cointegrated panels. *Advances in Econometrics, 15*, 93–130.

Pedroni, P. (2004). Panel cointegration: asymptotic and finite sample properties of pooled time series tests with an application to the PPP hypothesis. *Econometric Theory, 20*, 597–625.

Pesaran, M. H. (2004). *General diagnostic tests for cross section dependence in panels*. Cambridge Working Papers in Economics, 435, University of Cambridge.

Pesaran, M. H. (2007). A simple panel unit root test in the presence of cross section dependence. *Journal of Applied Econometrics, 22*, 265–312.

Perron, P., & Vogelsang, T.J. (1992). Nonstationarity and level shifts with an application to purchasing power parity, *Journal of Business and Economic Statistics, 10*, 301–320.

Phillips, P. C. B., & Hansen, B. E. (1990). Statistical inference in instrumental variable regression with $I(1)$ processes. *Review of Economic Studies, 57*, 99–125.

Phillips, P. C. B., & Moon, H. R. (1999). Linear regression limit theory for nonstationary panel data. *Econometrica, 67*, 1057–1111.

Phillips, P. C. B., & Ouliaris, S. (1990). Asymptotic properties of residual based tests for cointegration. *Econometrica, 58*, 165–193.

Stock, J. M., & Watson, M. H. (1993). A simple estimator of cointegrating vectors in higher order integrated systems. *Econometrica, 61*, 783–820.

Westerlund, J. (2005). New simple tests for panel cointegration. *Econometric Reviews, 24*, 297–316.

Westerlund, J. (2007). Testing for error correction in panel data. *Oxford Bulletin of Economics and Statistics, 69*, 709–748.

Chapter 10
The Binary Qualitative Model

Statistical data sometimes relate to **qualitative characteristics**, such as whether people are working or registered as unemployed, or whether the economy is expanding or contracting. These qualitative characteristics can have $K + 1$ disjoint modalities, i.e. $k = 0, 1, \ldots, K$.

Any qualitative characteristic can be associated with a numerically quantified variable by coding.

Example 10.1 The variable "socio-professional category" is given by c and takes three forms: 0 = blue-collar worker; 1 = white-collar worker and 2 = manager. We then define the variable $\tilde{c}$ as

$$\begin{aligned}
\tilde{c} &= 0 \quad \text{if } c = \text{blue-collar worker} \\
\tilde{c} &= 1 \quad \text{if } c = \text{white-collar worker} \\
\tilde{c} &= 2 \quad \text{if } c = \text{manager.}
\end{aligned}$$

In this way, if we know the value taken by $\tilde{c}$, we also know the modality assumed by c.

In this chapter, we will restrict ourselves to qualitative variables which take $K + 1 = 2$ modalities. Such variables are said to be **dichotomous** and if we use the 0 and 1 coding, they are similar to the binary or dummy variables frequently used in previous chapters.

We will therefore refer to R_t, $t = 1, \ldots, T$, as the binary qualitative variable taking the value 0 or 1 over the T time periods considered.

Supplementary Information The online version contains supplementary material available at https://doi.org/10.1007/978-3-031-72910-2_10.

F. J. Bismans and O. Damette, *Dynamic Econometrics*,
https://doi.org/10.1007/978-3-031-72910-2_10

10.1 The Linear Regression Model

We begin by examining linear regression, in which the binary dependent variable is linked to a set of appropriate explanatory variables. We will then detail the shortcomings of this approach.

Linear Regression

Let's assume that we have $t = 1, \ldots, T$ observations on the qualitative variableR_t, while the values of the K explanatory variables are respectively $\boldsymbol{x}_t = (x_{t1}, \ldots, x_{tK})'$ with $t = 1, \ldots, T$.

The linear model is then written as

$$R_t = \boldsymbol{\beta}'\boldsymbol{x}_t + u_t, \qquad t = 1, \ldots, T, \tag{10.1}$$

where $\boldsymbol{\beta}'$ is a vector of K parameters and u_t is the error or disturbance associated with the t-th observation.

We will make just one assumption about the error term, namely that its mathematical expectation is zero for all t. The conditional expectation $E(R_t|\boldsymbol{x}_t)$, is equal to $\boldsymbol{\beta}'\boldsymbol{x}_t$ and can therefore be interpreted as the probability of the binary variable taking the value 1 or 0.

This formulation has a number of shortcomings, both logical and probabilistic.

Some Shortcomings of the Linear Model

Firstly, in (10.1), R_t is a qualitative variable and $\boldsymbol{\beta}'\boldsymbol{x}_t + u_t$ a quantitative variable, which makes little sense. It is true that the left-hand side takes the value 0 or 1, but this coding is arbitrary and other coding values would lead to a different estimate of $\hat{\boldsymbol{\beta}}$. As a result, the estimated values of $\hat{\boldsymbol{\beta}}$ are not interpretable.

From a probabilistic point of view, as R_t can only take the values 1 or 0, the disturbance u_t is necessarily dichotomous:

u_t takes the value $1 - \boldsymbol{\beta}'\boldsymbol{x}_t$ with probability p_t
u_t takes the value $-\boldsymbol{\beta}'\boldsymbol{x}_t$ with probability $1 - p_t$.

Consequently, the disturbance necessarily follows a discrete distribution, which precludes the formulation of the normality hypothesis for the residuals.

If, as we have done, we impose zero mean on the perturbation, the probability p_t is uniquely determined, since

$$E(u_t) = p_t(1 - \boldsymbol{\beta}'\boldsymbol{x}_t) - (1 - p_t)\boldsymbol{\beta}'\boldsymbol{x}_t = 0$$

and that

$$p_t = \boldsymbol{\beta}'\boldsymbol{x}_t.$$

The vector $\boldsymbol{\beta}$ of parameters cannot be arbitrary, given that

$$0 \leq p_t = \boldsymbol{\beta}'\boldsymbol{x}_t \leq 1, \qquad t = 1, \ldots, T.$$

In addition, these constraints may be incompatible and the estimated probabilities may fall outside the [0, 1], interval, in which case the model would make no sense.

Finally, the variance of the disturbances is given by

$$\begin{aligned} \operatorname{Var}(u_t) &= \boldsymbol{\beta}'\boldsymbol{x}_t(1-\boldsymbol{\beta}'\boldsymbol{x}_t)^2 + (1-\boldsymbol{\beta}'\boldsymbol{x}_t)(\boldsymbol{\beta}'\boldsymbol{x}_t)^2 \\ &= \boldsymbol{\beta}'\boldsymbol{x}_t(1-\boldsymbol{\beta}'\boldsymbol{x}_t)\big[[(1-\boldsymbol{\beta}'\boldsymbol{x}_t)+\boldsymbol{\beta}'\boldsymbol{x}_t]\big] \\ &= \boldsymbol{\beta}'\boldsymbol{x}_t(1-\boldsymbol{\beta}'\boldsymbol{x}_t). \end{aligned}$$

It follows that this variance is heteroscedastic. We could undoubtedly try to apply generalised least squares, but this proves impossible, given that the variance–covariance matrix of the disturbances depends on the unknown vector parameter $\boldsymbol{\beta}$.

10.2 Non-linear Binary Models

Since the linear model is inappropriate for regressing a binary qualitative variable on a set of explanatory variables, we have to turn to other models, these being non-linear.

In this respect, there are two main possible categories: probit models (also known as normit models; we will soon see why) and logit models, which are based on a specific probability distribution, the logistic distribution.

The Probit Model

The first alternative approach—which leads to the probit model—assumes the existence of an unobservable (and therefore latent) variable R_t^* defined by the regression

$$R_t^* = \boldsymbol{\beta}'\boldsymbol{x}_t + u_t, \qquad t = 1, \ldots, T. \tag{10.2}$$

The qualitative binary variable, on the other hand, is observed, but it is constructed from the latent variable according to the following scheme:

$$\begin{cases} R_t = 1 & \text{if } R_t^* > 0 \\ R_t = 0 & \text{otherwise.} \end{cases} \tag{10.3}$$

Given this latest formulation, $\boldsymbol{\beta}'\boldsymbol{x}_t$ is equal to $E(R_t^*|\boldsymbol{x}_t)$ and no longer $E(R_t|\boldsymbol{x}_t)$, as was the case for the usual linear model.

It follows from (10.2) and (10.3) that

$$\begin{aligned} \Pr(R_t = 1) &= \Pr(u_t > -\boldsymbol{\beta}'\boldsymbol{x}_t) \\ &= 1 - F(-\boldsymbol{\beta}'\boldsymbol{x}_t), \end{aligned} \tag{10.4}$$

where F *is* the cumulative distribution function of u.

The observed values of R are then realisations of a binomial process whose probabilities are given by (10.4) and vary as a function of $\boldsymbol{x}_t$. Consequently, the likelihood function is equal to

$$L = \prod_{R_t=0} F(-\boldsymbol{\beta}'\boldsymbol{x}_t) \prod_{R_t=1} F(1 - \boldsymbol{\beta}'\boldsymbol{x}_t). \tag{10.5}$$

If we now assume that $u_t \sim IIN(0, \sigma^2)$, the corresponding cumulative distribution function is that of a normal random variable, which leads, by reducing it, to

$$\Phi(-\boldsymbol{\beta}'\boldsymbol{x}_t) = \int_{-\infty}^{-\boldsymbol{\beta}'\boldsymbol{x}_t/\sigma} (2\pi)^{-1/2} e^{-t^2/2}\, dt. \tag{10.6}$$

It immediately follows from (10.6) that we simply need to estimate the ratio $\boldsymbol{\beta}/\sigma$ and not these parameters separately (In practice, we will take $\sigma = 1$).

When F in (10.4) is normal, we obtain the **probit** model (we should say **normit**, but the former name has been adopted).

The Logistic Function and the Logit Model

The logistic function is given by

$$y = \frac{N}{1 + b\,e^{-rt}},$$

where $N, b, r > 0$, N *is* the saturation value, an expression that is easy to understand when this function is represented graphically.

The logistic curve is as follows (Fig. 10.1):

If $t = 0, y = N/(1 + b)$ and for $t \to +\infty$, y tends towards N, which represents the saturation value. What's more, the logistic function has a form very close to that of the normal distribution function, provided we take $N = 1$.

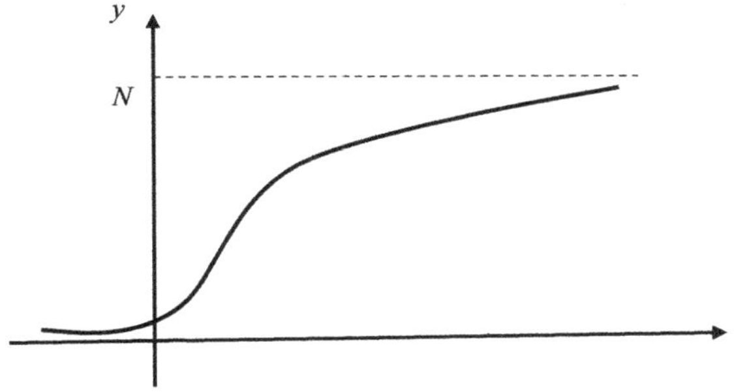

Fig. 10.1 Graph of the logistic function

This gives us the definition:

Definition 10.1 The random variable X follows a **logistic distribution** if its cumulative distribution function is given by.

$$\Lambda(x) = \frac{1}{1 + e^{-x}}. \tag{10.7}$$

The expectation of X *is* equal to 0, while its variance is equal to $\pi^2/3$.
Note that, as with the normal distribution, we have

$$\Lambda(-x) = 1 - \Lambda(x).$$

When the logistic distribution function (10.7) is introduced into (10.4), we obtain the **dichotomous logit model**.

10.3 Maximum Likelihood Estimation

There are several methods for estimating the simple dichotomous model. However, we prefer the maximum likelihood method, applying it first to the probit and then to the logit.

Likelihood Functions

Substituting (10.6) into Eq. (10.5), the likelihood function of the probit model is written:

$$L = \prod_{t=1}^{T} \left[\Phi(\boldsymbol{\beta}'\boldsymbol{x}_t)\right]^{R_t} \left[1 - \Phi(\boldsymbol{\beta}'\boldsymbol{x}_t)\right]^{1-R_t},$$

so that the log-likelihood is given by

$$\ln L = \sum_{t=1}^{T} R_t \ln \Phi(\boldsymbol{\beta}'\boldsymbol{x}_t) + \sum_{t=1}^{T} (1 - R_t) \ln[1 - \Phi(\boldsymbol{\beta}'\boldsymbol{x}_t)]. \tag{10.8}$$

Deriving (10.8) with respect to $\boldsymbol{\beta}$ and then cancelling this derivative, we obtain:

$$S(\boldsymbol{\beta}) = \frac{\partial \ln L}{\partial \boldsymbol{\beta}} = \sum_{t=1}^{T} \frac{\left[R_t - \Phi(\boldsymbol{\beta}'\boldsymbol{x}_t)\right]}{\Phi(\boldsymbol{\beta}'\boldsymbol{x}_t)\left[1 - \Phi(\boldsymbol{\beta}'\boldsymbol{x}_t)\right]} \phi(\boldsymbol{\beta}'\boldsymbol{x}_t)\boldsymbol{x}_t = \mathbf{0}, \tag{10.9}$$

where ϕ is the probability density of a standardised normal random variable.

By solving the system of non-linear equations in (10.9), which requires an iterative procedure, we obtain the maximum likelihood estimator $\boldsymbol{\beta}_{ML}$.

For the logit model, the calculations are simpler because it is not necessary to solve an integral to obtain the logistic distribution function. Using (10.7), we immediately have

$$\Lambda(-\boldsymbol{\beta}'\boldsymbol{x}_t) = \frac{1}{1 + e^{\boldsymbol{\beta}'\boldsymbol{x}_t}},$$

from which we deduce that

$$1 - \Lambda(-\boldsymbol{\beta}'\boldsymbol{x}_t) = \frac{e^{\boldsymbol{\beta}'\boldsymbol{x}_t}}{1 + e^{\boldsymbol{\beta}'\boldsymbol{x}_t}}.$$

Using these last two results, the likelihood corresponding to the logit model is given by

$$L = \prod_{t=1}^{T} \left[\frac{1}{1 + e^{\boldsymbol{\beta}'\boldsymbol{x}_t}}\right]^{1-R_t} \left[\frac{e^{\boldsymbol{\beta}'\boldsymbol{x}_t}}{1 + e^{\boldsymbol{\beta}'\boldsymbol{x}_t}}\right]^{R_t}$$

$$= \frac{e^{\boldsymbol{\beta}' \sum_{t=1}^{T} \boldsymbol{x}_t R_t}}{\prod_{t=1}^{T} \left[1 + e^{\boldsymbol{\beta}'\boldsymbol{x}_t}\right]}. \tag{10.10}$$

Note in (10.10) $\sum \boldsymbol{x}_t R_t = k$. The log-likelihood is then written as

$$\ln L = \boldsymbol{\beta}' k - \sum_{t=1}^{T} \ln\left(1 + e^{\boldsymbol{\beta}'\boldsymbol{x}_t}\right). \tag{10.11}$$

Cancelling the vector of first derivatives in (10.11) gives the system of likelihood equations:

$$S(\boldsymbol{\beta}) = \frac{\partial \ln L}{\partial \boldsymbol{\beta}} = \sum_{t=1}^{T} \frac{e^{\boldsymbol{\beta}'\mathbf{x}_t}}{1 + e^{\boldsymbol{\beta}'\mathbf{x}_t}} \mathbf{x}_t + k = \mathbf{0}. \tag{10.12}$$

Solving this system of non-linear equations requires the use of an iterative procedure—to which we will return shortly—exactly as in the case of estimating the parameters of the probit model.

First, however, we need to say a few words about the comparison of the maximum likelihood estimators of the two models under consideration. From this point of view, we have already pointed out that the normal and logistic distribution functions are very similar, except that the tails of the latter are thicker and the corresponding variance—see Definition 10.1—is equal to $\pi^2/3$.

It follows that the estimators obtained from the two models are not directly comparable. Given the difference in the values of the variance, the estimates of $\boldsymbol{\beta}$ resulting from the logit model must be multiplied by $\sqrt{3}/\pi$ in order to be compared with those obtained with the probit model (in the latter case, the variance and standard deviation are conventionally equal to unity).

Procedures for Solving Non-linear Systems

We have seen that solving systems (10.9) and (10.12) requires the use of an iterative numerical procedure to determine the maximum likelihood estimators. The maximum of the log-likelihood function must also be unique, which is guaranteed when this function is strictly concave or, equivalently, the cumulative distribution function used is also **strictly concave**. This condition is met for both the probit and logit models. Let's look at the latter example.

The first derivative of the logistic distribution function is equal to

$$\begin{aligned}\frac{d}{dx}[\ln \Lambda(x)] &= \frac{1}{\Lambda(x)}\frac{d}{dx}\Lambda(x) = (1 + e^{-x})[-(1 - e^{-x})^{-2}(e^{-x})\\ &= \frac{e^{-x}}{1 + e^{-x}} = 1 - \Lambda(x).\end{aligned}$$

The second derivative is given by

$$\begin{aligned}\frac{d^2}{dx^2}[\ln \Lambda(x)] &= \frac{d}{dx}\left(\frac{e^{-x}}{1 + e^{-x}}\right) = \frac{-e^{-x}}{(1 + e^{-x})^2} = -\Lambda(x)\frac{e^{-x}}{1 + e^{-x}}\\ &= -\Lambda(x)(1 - \Lambda(x)) < 0.\end{aligned}$$

As a result, this function is strictly concave. The same applies to $1 - \Lambda(x)$, since

$$\ln[1-\Lambda(x)] = \ln\left(\frac{e^{-x}}{1+e^{-x}}\right) = \ln\left(e^{-x}\frac{1}{1+e^{-x}}\right) = -x + \ln[\Lambda(x)].$$

It is therefore sufficient to check the first-order conditions and solve the corresponding systems of non-linear equations to obtain the maximum likelihood estimator of $\boldsymbol{\beta}$. It is at this precise point that the iterative solution procedures come into play, of which we will detail two examples: the Newton–Raphson algorithm and the score algorithm. In all cases, the problem comes down to finding a root of the vector equation $\partial \ln L/\partial \boldsymbol{\beta}$.

The **Newton–Raphson** procedure involves choosing an initial value $\hat{\boldsymbol{\beta}}_0$, and then searching for the tangent plane at $\hat{\boldsymbol{\beta}}_0$ using the function

$$\boldsymbol{\beta} \mapsto \boldsymbol{d} = \partial \ln L/\partial \boldsymbol{\beta}.$$

This plane, whose equation is

$$\boldsymbol{d} = \frac{\partial \ln L}{\partial \boldsymbol{\beta}}(\hat{\boldsymbol{\beta}}_0) + \frac{\partial^2 \ln L}{\partial \boldsymbol{\beta}\partial \boldsymbol{\beta}'}(\hat{\boldsymbol{\beta}}_0)(\boldsymbol{\beta} - \hat{\boldsymbol{\beta}}_0)$$

is an approximation of $\partial \ln L/\partial \boldsymbol{\beta}$. The root of $\partial \ln L/\partial \boldsymbol{\beta} = \mathbf{0}$ can therefore be approximated by the root of

$$\begin{aligned} &\frac{\partial \ln L}{\partial \boldsymbol{\beta}}(\hat{\boldsymbol{\beta}}_0) + \frac{\partial^2 \ln L}{\partial \boldsymbol{\beta}\partial \boldsymbol{\beta}'}(\hat{\boldsymbol{\beta}}_0)(\boldsymbol{\beta} - \hat{\boldsymbol{\beta}}_0) = \mathbf{0} \\ &\Leftrightarrow \frac{\partial^2 \ln L}{\partial \boldsymbol{\beta}\partial \boldsymbol{\beta}'}(\hat{\boldsymbol{\beta}}_0)\boldsymbol{\beta} = \frac{\partial^2 \ln L}{\partial \boldsymbol{\beta}\partial \boldsymbol{\beta}'}(\hat{\boldsymbol{\beta}}_0)(\hat{\boldsymbol{\beta}}_0) - \frac{\partial \ln L}{\partial \boldsymbol{\beta}}(\hat{\boldsymbol{\beta}}_0). \end{aligned} \quad (10.13)$$

The solution of (10.13) for $\boldsymbol{\beta} = \hat{\boldsymbol{\beta}}_1$ is

$$\hat{\boldsymbol{\beta}}_1 = \hat{\boldsymbol{\beta}}_0 - \left[\frac{\partial^2 \ln L}{\partial \boldsymbol{\beta}\partial \boldsymbol{\beta}'}(\hat{\boldsymbol{\beta}}_0)\right]^{-1} \frac{\partial \ln L}{\partial \boldsymbol{\beta}}(\hat{\boldsymbol{\beta}}_0).$$

A new iteration is then started, using $\hat{\boldsymbol{\beta}}_1$ as the new initial value and determining $\hat{\boldsymbol{\beta}}_2$, which is then taken as the new initial value, and so on.

The recurrence formula for the j-th iteration will therefore be given by

$$\hat{\boldsymbol{\beta}}_{j+1} = \hat{\boldsymbol{\beta}}_j - \left[\frac{\partial^2 \ln L}{\partial \boldsymbol{\beta}\partial \boldsymbol{\beta}'}(\hat{\boldsymbol{\beta}}_j)\right]^{-1} \frac{\partial \ln L}{\partial \boldsymbol{\beta}}(\hat{\boldsymbol{\beta}}_j). \quad (10.14)$$

If the sequence of $\hat{\boldsymbol{\beta}}_j$ s converges to a limit $\hat{\boldsymbol{\beta}}$, the latter is a solution of the likelihood equations, since

$$\hat{\boldsymbol{\beta}} = \lim_{j \to +\infty} \hat{\boldsymbol{\beta}}_j = \hat{\boldsymbol{\beta}} - \left[\frac{\partial^2 \ln L}{\partial \boldsymbol{\beta} \partial \boldsymbol{\beta}'}(\hat{\boldsymbol{\beta}})\right]^{-1} \frac{\partial \ln L}{\partial \boldsymbol{\beta}}(\hat{\boldsymbol{\beta}}) = \frac{\partial \ln L}{\partial \boldsymbol{\beta}}(\hat{\boldsymbol{\beta}}).$$

The **score algorithm**—which is, remember, the likelihood gradient—differs from the previous one in that, in (10.14), $(\partial^2 \ln L / \partial \hat{\boldsymbol{\beta}} \partial \hat{\boldsymbol{\beta}}')(\hat{\boldsymbol{\beta}}_j)$ is replaced by its mathematical expectation.

The recurrence formula is therefore as follows:

$$\hat{\boldsymbol{\beta}}_{j+1} = \hat{\boldsymbol{\beta}}_j - \left\{E\left[\frac{\partial^2 \ln L}{\partial \boldsymbol{\beta} \partial \boldsymbol{\beta}'}(\hat{\boldsymbol{\beta}}_j)\right]\right\}^{-1} \frac{\partial \ln L}{\partial \boldsymbol{\beta}}(\hat{\boldsymbol{\beta}}_j). \tag{10.15}$$

We could give an alternative formulation of (10.15), which is particularly interesting, but which would take us too far. Suffice it to say that this formulation, as indicated by Amemiya (1985, pp. 274–275), allows the score algorithm to be interpreted as a set of iterations applying *weighted non-linear least squares.*

Properties of Maximum Likelihood Estimators

We have just seen that iterative procedures lead to a global maximum of the log-likelihood, at least if the algorithm used converges. Unfortunately, this is not always the case, as the following example shows.

Example 10.2 Consider the logit model with a single explanatory variable, all of whose values are positive. The likelihood equation—see (10.12)—takes the form

$$\sum_{t=1}^{T} \frac{e^{\beta x_t}}{1 + e^{\beta x_t}} x_t = \sum_{t=1}^{T} x_t R_t.$$

The member on the left, when β varies from $-\infty$ to $+\infty$, takes all its values in the interval $]0, \sum x_t[$. On the other hand, the member on the right has its values between 0 (all R_t are zero) and $\sum x_t > 0$ (all R_t are equal to unity).

The solution to the likelihood equation therefore only exists if R_t takes some of its values equal to 0 and others equal to unity.

When the global maximum exists, it is still necessary to ensure that the conditions are met for the estimators of $\boldsymbol{\beta}$ to be **convergent and asymptotically normal (CAN).**

Before stating these conditions, we will need an additional definition (already implicitly given in "Lagrange Multiplier Test" section):

Definition 10.2 The Fisher information matrix is defined by

$$I(\boldsymbol{\beta}) = -E\big(\partial^2 \ln L(\boldsymbol{\beta})/\partial\boldsymbol{\beta}\partial\boldsymbol{\beta}'\big),$$

where E *is* the mathematical expectation.

Following Gouriéroux and Monfort (1981), if all the explanatory variables x_t^k are **certain** and **bounded**, i.e. if

$$\exists\ m > 0 \text{ and } M < +\infty \text{ such that } m < |x_t^k| < M, \qquad \forall t, k$$

and if, in addition, these variables are such that the variance–covariance matrix exists, then the maximum likelihood estimator $\boldsymbol{\beta}_{ML}$ exists asymptotically, converges towards the true value of the parameter $\boldsymbol{\beta}$ and asymptotically follows a **normal distribution** with an expectation vector of zero and a variance–covariance matrix equal to the **inverse of the Fisher information matrix**, i.e. $I^{-1}(\boldsymbol{\beta})$.

It follows that the asymptotic variance–covariance matrix, noted AVar and, moreover, dependent on the unknown parameters β, will, consequently, be estimated by

$$\text{AVar}(\hat{\boldsymbol{\beta}}) = \left[-E\left(\frac{\partial^2 \ln L(\hat{\boldsymbol{\beta}})}{\partial\boldsymbol{\beta}\partial\boldsymbol{\beta}'}\right)\right]^{-1}. \tag{10.16}$$

Using Eqs. (10.9) and (10.12), we obtain the respective information matrices of the probit and logit models, i.e.

$$I_p(\boldsymbol{\beta}) = E\left(-\frac{\partial^2 \ln L(\boldsymbol{\beta})}{\partial\boldsymbol{\beta}\partial\boldsymbol{\beta}'}\right) = \sum_{t=1}^{T} \frac{[\phi(\boldsymbol{\beta}'\boldsymbol{x}_t)]^2}{\Phi(\boldsymbol{\beta}'\boldsymbol{x}_t)[1-\Phi(\boldsymbol{\beta}'\boldsymbol{x}_t)]}$$

$$I_l(\boldsymbol{\beta}) = E\left(-\frac{\partial^2 \ln L(\boldsymbol{\beta})}{\partial\boldsymbol{\beta}\partial\boldsymbol{\beta}'}\right) = \sum_{t=1}^{T} \frac{e^{\boldsymbol{\beta}'\boldsymbol{x}_t}}{\left[1+e^{\boldsymbol{\beta}'\boldsymbol{x}_t}\right]}\boldsymbol{x}_t\boldsymbol{x}'_t.$$

All that remains is to substitute $\hat{\boldsymbol{\beta}}_{MV}$ for $\boldsymbol{\beta}'\boldsymbol{x}_t$ and invert the two matrices to obtain the estimators we are looking for.

10.4 Evaluating Dichotomous Models

We will first look at what measure of goodness of fit should be used for simple probit and logit models. We will then review the various tests available for this type of model.

Measuring Statistical Fit

R^2-type measures are not suitable for discrete dependent variables—see Maddala (1983, pp. 37–41). Other measures based on maximum likelihood have therefore been proposed. The most common of these is described below.

Let L^u be the maximum of the likelihood function when all the parameters of the model are taken into account, so unconstrained, and L^c the maximum when some of these parameters are constrained, i.e. $\beta_k = 0$, with $k = 1, \ldots, K$. Using these two maxima, we obtain Mc Fadden's **likelihood ratio indicator (LRI)**, also known as the **pseudo-R^2**

$$LRI = 1 - \frac{\ln L^u}{\ln L^c}. \tag{10.17}$$

This indicator is interesting in that it is reduced by 0 and increased by 1. Indeed, if all the marginal effects of the different independent variables are zero, then $\ln L^u$ and $\ln L^c$ are equal in (10.16) and $LRI = 0$.

For the Mc Fadden indicator to be equal to unity, the distribution function in the log-likelihood function must be $F_t = 1$ whenever $R_t = 1$ and $F_t = 1$, whenever $R_t = 0$. In this case, $\ln L = \ln 1 = 0$ and $LRI = 1$.

However, the latter possibility can only arise if the fit of the unconstrained model is perfect, in the precise sense that, for any t, $P_t = F(\hat{\boldsymbol{\beta}}'\boldsymbol{x}_t) = R_t$.

However, care should be taken not to use the Mc Fadden indicator to compare two different models. In the latter case, it is sufficient to compare the values of the likelihood functions.

Tests on a Set of Restrictions

The simplest method for testing a single restriction on the coefficients of a binary model is based on the usual t-tests and uses the standard errors from the Fisher information matrix.

Alternatively, given that $\hat{\boldsymbol{\beta}}$ asymptotically follows a normal distribution, we could use the table of the standardised normal distribution to determine the critical values of the test.

Obviously, when there is more than one restriction, things get a little more complicated. We then have to use one of the asymptotic "Holy Trinity" tests. (We have already developed this point in "Test on Constrained Estimators" section, but here we will take up the question again and apply it to the case of probit and logit models).

The problem is now to test a set of linear restrictions, namely

$$H_0 : \boldsymbol{R\beta} = \boldsymbol{r}$$
$$H_1 : \boldsymbol{R\beta} \neq \boldsymbol{r},$$

where $\mathbf{R}$ is a matrix of appropriate dimensions and $\mathbf{r}$ is a column vector. If we want to formulate a test analogous to the F-test in the usual regression model, then these restrictions are simply written as

$$H_0 : \boldsymbol{\beta} = \mathbf{0}.$$

For the **likelihood ratio**, the corresponding statistic is

$$LR = 2(\ln \hat{L}^{u} - \ln \hat{L}^{c}), \tag{10.18}$$

where $\hat{L}^{u}$ and $\hat{L}^{c}$ represent the likelihood function evaluated on the basis of the free or unrestricted estimator and the constrained estimator respectively. In practice, the assumption is that all the slopes of the probit and logit models are zero and that the constant term remains constrained.

The *constrained* log-likelihood is then identical for both probit and logit models and is given by

$$n\ [P \ln P + (1 - P) \ln(1 - P)], \tag{10.19}$$

where n is the number of observations and P is the proportion of observations on the dependent variable equal to unity in the total number of observations.

The statistic (10.19) follows a Chi-square distribution with r degrees of freedom, where r is the number of restrictions, in other words, in our case, the number of constrained coefficients.

If this statistic is greater than the critical value in the χ^2 table for r degrees of freedom, the null hypothesis is rejected and the alternative accepted, i.e. $\boldsymbol{\beta} \neq \mathbf{0}$.

The Lagrange multiplier test statistic is, in general (see "Asymptotic Distribution of the Three Tests" section), equal to

$$\begin{aligned} LM &= \boldsymbol{s}(\boldsymbol{\theta}_0)'\boldsymbol{I}(\boldsymbol{\theta}_0)^{-1}\boldsymbol{s}(\boldsymbol{\theta}_0) \\ &= \boldsymbol{s}(\boldsymbol{\theta}_0)'\hat{\boldsymbol{V}}(\boldsymbol{\theta}_0)\boldsymbol{s}(\boldsymbol{\theta}_0), \end{aligned}$$

where $\boldsymbol{s}(\boldsymbol{\theta}_0)$ is the vector of first partial derivatives of the log-likelihood evaluated in the vector of constrained parameters and $\hat{\boldsymbol{V}}(\boldsymbol{\theta}_0)$ the variance–covariance matrix of $\boldsymbol{\theta}$ evaluated again in $\boldsymbol{\theta}_0$.

In this case, given that Davidson and MacKinnon (1984) have shown that the best estimator of the matrix in question is represented by (10.16), the LM statistic can be written as follows

$$LM = \left(\sum_{t=1}^{T} s_t \boldsymbol{x}_t\right)' \left[\sum_{t=1}^{T} E(h_t)\boldsymbol{x}_t \boldsymbol{x}'_t\right] \left(\sum_{t=1}^{T} s_t \boldsymbol{x}_t\right), \tag{10.20}$$

with $h_t = d^2 \ln F_t / d(\boldsymbol{\beta}'\boldsymbol{x}_t)^2$.

For the logit model, if the assumption is that all slopes are zero, the corresponding statistic is

$$LM = T \cdot R^2,$$

where R^2 is the coefficient obtained by regressing $(R_t - P)$ on $\boldsymbol{x}_t$, with P the proportion of R_t equal to unity in the sample.

Model Specification

To select a probit or logit model, we can naturally refer to the three usual information criteria—see "The Akaike Criterion" and "Other Selection Criteria" sections—namely Akaike, Bayesian and Hannan-Quinn criteria. They are based on the use of likelihood, which is the most widely used estimation method.

As for the specification tests themselves, the problem can conveniently be presented as follows. If we want to apply a Wald test—see Eq. (4.26) in "Asymptotic Distribution of the Three Tests" section—we consider the set of constraints

$$\boldsymbol{C} = (\boldsymbol{0}, \boldsymbol{I}_L) = \boldsymbol{0},$$

where L is the number of restrictions equal to zero.

The Wald statistic is then given by

$$W = \hat{\boldsymbol{\beta}}'_L \boldsymbol{V}_L^{-1} \hat{\boldsymbol{\beta}}_L,$$

where **V is** the estimated asymptotic variance–covariance matrix of $\hat{\boldsymbol{\beta}}$.

We saw in the previous section how to implement the likelihood ratio test. However, for the Lagrange multiplier test (also known as the score test), we proceed as follows:

(i) use maximum likelihood to estimate the model under H_0, in order to obtain the vector of constrained coefficients;
(ii) if $\mathbf{x}$ is the unrestricted vector, the LM statistic is calculated using (10.20).

10.5 Application

We will use a set of data taken from a study by Spector and Mazzeo (1980), data also used by Greene (2012).

Series and Programme

The data, which we will be detailing shortly, will enable us to answer a central question: will a new method of teaching economics lead or not to an improvement in the marks of the students who are the subjects of this pedagogical innovation?

More precisely, the dependent variable in this case is GRADE, which takes the value 1 or 0 depending on whether or not the student has significantly improved his or her marks in an intermediate macroeconomics course compared with the previous "principles of economics" course. From this point of view, we have a sample of 32 observations divided between 21 zeros and 11 units.

There are three explanatory variables:

- GPA (*Grade Point Average*), i.e. the weighted average of the grades obtained in the various economics courses;
- TUCE represents the score obtained in a pre-test which indicates the degree of knowledge of the subject;
- PSI (*Personalised System of Instruction*) is a binary variable which indicates whether or not the student has been exposed to the new teaching method (in fact, fourteen of the thirty-two students in the sample have been).

These different series are cross-sectional series.

Here is the GRETL program, the script we say was used to estimate the two models:

Probit GRADE const GPA TUCE PSI	
Logit GRADE const GPA TUCE PSI	

The Probit Model

Before moving on to the analysis of the model estimation results, two points need to be made. Firstly, the parameter estimators will be obtained using the maximum likelihood method via the Newton–Raphson algorithm.

On the other hand, as the probit is non-linear, the estimated coefficients, unlike those of the linear model, do not represent the derivatives of the centred reduced normal distribution function. In fact, in our case, the latter are given by

$$\frac{\partial\left(\Phi\left(\hat{\boldsymbol{\beta}}'\boldsymbol{x}\right)\right)}{\partial\,\boldsymbol{x}} = \phi\left(\hat{\boldsymbol{\beta}}'\,\overline{\boldsymbol{x}}\right)\boldsymbol{\beta},$$

where ϕ is the standard normal density and $\overline{\boldsymbol{x}}$ is the vector of means of the observations on each of the explanatory variables.

The derivatives in question are called **average marginal effects** or **slopes** in the language of GRETL. They should be interpreted as probabilities resulting from the variation in the independent variables.

Here is a table summarising the results obtained for the probit model (Table 10.1): The Mac Fadden coefficient (or pseudo-R^2) is close to 0.38. All the coefficients estimated by ML are significantly different from zero, with the exception of the TUCE coefficient, which is also extremely low.

In addition, these coefficients must be multiplied by the probability of 0.328 (= 6th row of the table) to give the **average marginal effects** (or slopes in GRETL terminology) of each of the independent variables. Unsurprisingly, it is the GPA and PSI variables that have the highest probability of the binary variable GRADE taking the value 1, in other words that the students' grade has improved in subsequent tests.

The analysis can then be completed by carrying out two hypothesis tests:

1. the first is a likelihood ratio test, the null hypothesis of which is the joint nullity of the three coefficients of the explanatory variables. The corresponding statistic is

$$LR = 15.546 \sim \chi^2(3) \qquad p\text{-value} = 0.014.$$

   The null hypothesis must therefore be rejected.
2. As the TUCE variable is not significant, it is logical to carry out a test of nullity and therefore of omission of this variable. Here is the appropriate Wald statistic (with the proviso that we consider it as a finite-distance sample, i.e. following a Fisher-Snedecor F distribution):

$$W = 0.380 \sim F(1,\ 28) \qquad p--\text{value} = 0.542.$$

   We therefore have to accept the nullity of this variable.

Consequently, we will estimate a model containing only the GPA and PSI variables. From this point of view, we already know that the selection of a model requires reference to information criteria.

Table 10.2 shows a comparison of these criteria, with model 1 containing the three initial independent variables and model 2 excluding the TUCE variable.

Table 10.1 Probit estimation (32 observations)

Variable	Coefficient	Standard error	Stat t	Slopes
Const	– 7.452	2.542	– 2.931	–
GPA	1.626	2.343	2.343	0.533
TUCE	0.052	0.084	0.617	0.017
PSI	1.426	0.595	2.397	0.464
$\phi(\boldsymbol{\beta}'\boldsymbol{x})$ evaluated at the mean of the independent variables = 0.328				
Pseudo-R^2 = 0.377			Log-likelihood: – 12.819	

Table 10.2 Information criteria

	AIC	BIC	HQC
Model 1	33.638	39.501	35.581
Model 2	32.033	36.430	33.491

Table 10.3 Estimation of the reduced probit model

Variable	Coefficient	Standard error	Stat t	Slopes
Const	– 6.783	2.234	– 3.036	–
GPA	1.785	0.659	2.707	0.596
PSI	1.422	0.586	2.426	0.469
$\phi(\boldsymbol{\beta}'\boldsymbol{x})$ evaluated at the mean of the independent variables = 0.334				
Pseudo-R^2 = 0.368			Log-likelihood: – 13.017	

Examination of these different criteria shows that model 2, from which the TUCE variable has been excluded, improves each of the three information criteria by minimising them.

It follows that we are justified in estimating a model with only the explanatory variables GPA and PSI (Table 10.3).

A few comments on this table will suffice:

1. all coefficients, including the constant, are significantly different from zero;
2. the average marginal effects (obtained by multiplying the estimated coefficients by the probability 0.334 in the 5ème row of the table) are close to the probability of 0.5;
3. lastly, the likelihood ratio test, which includes the following elements

$$LR = 15.15 \sim \chi^2(2) \qquad p - \text{-value} = 0.028,$$

leads us to reject the nullity of the two GPA and PSI coefficients.

The Logit Model

From the outset, we will focus on the model that incorporates only the GPA and PSI variables. Here are the results of the estimation of this logit model.

Table 10.4 Estimation of the logit model

Variable	Coefficient	Standard error	Stat t	Slope
Const	– 11.60	4.213	– 2.754	–
GPA	3.063	1.223	2.505	0.594
PSI	2.338	1.041	2.246	0.457
$\lambda(\boldsymbol{\beta}'\boldsymbol{x})$ evaluated at the mean of the independent variables = 0.194				
Pseudo-R^2 = 0.363			Log-likelihood: – 13.126	

Table 10.4 leads us to a number of comments:

1. all the estimated coefficients are significantly different from zero;
2. the average marginal effects resulting from the use of the probability density λ of the logistic distribution are close to a probability equal to 0.5 or 0.6;
3. the likelihood ratio test, under the assumption of nullity of the two estimated coefficients, leads to the statistic

$$LR = 14.93 \sim \chi^2(2) \qquad p - \text{-value} = 0.0006.$$

It follows that the null hypothesis must be rejected.

It is now interesting to compare the estimates obtained in the two models. The table below provides the relevant information (Table 10.5).

Table 10.5 Comparison of probit and logit models

Variable	Probit		Logit	
	Coefficient	Slope	Coefficient	Slope
Const	− 6.783	–	− 11.60	–
GPA	1.785	0.596	3.063	0.594
PSI	1.422	0.469	2.338	0.457
$f(\hat{\boldsymbol{\beta}}'\overline{x})$	0.334		0.194	

A number of points need to be made when examining the data in this table. Firstly, the average marginal effects are very close to each other, whatever the model considered.

On the other hand, the coefficients estimated by maximum likelihood differ greatly depending on whether we are dealing with a probit or a logit. On this subject, we have already pointed out—see "Likelihood Functions" section—that the variance of the standard normal distribution is equal to unity and that that of the logistic distribution is equal to $\pi^2/3$ It follows that the estimators of $\boldsymbol{\beta}$ in the probit model should be multiplied by $\pi/\sqrt{3} \approx 1.8$ to make them comparable with those obtained in the logit model. In reality, the multiplication factor is 1.71 for the GPA variable and 1.64 for the PSI variable. This confirms Amemiya's assertion (1981, p. 1487) that the factor in question is in fact close to 1.6.

Finally, it is also worth noting the very different, but logical, values for normal and logistic densities.

10.6 Dynamic Probit

For the last fifty years at least, economic activity has had a marked cyclical profile, with phases of economic expansion succeeding phases of contraction, the latter also being referred to as phases of recession.

The following graph provides a schematic representation of a typical cyclical profile (Fig. 10.2).

The high points of the cycle are called peaks and the low points are called troughs. A complete cycle goes from peak to peak or trough to trough and therefore includes an expansion phase and a recession phase. (The chart shows a complete cycle from peak to peak).

We are particularly interested in two questions: (1) how can we measure the cycle? (2) is it possible to predict cyclical downturns, or in our case, recessions?

We will defer answering the first question. On the other hand, the answer to the second question leads us to say that we will use a dichotomous probit model to forecast recessions.

Formally, let's call R_t the binary variable that takes the value 1 during recessions and the value 0 during expansions, with, of course, $t = 1, \ldots, T$. We then introduce a vector $\boldsymbol{x}_t$ of explanatory variables for recessions, specifying that this vector also includes lagged values of these independent variables.

The probit model developed in the previous sections, R_t is linked to the regressors by

$$E(R_t \mid \boldsymbol{x}_t) = \Pr(R_t = 1 \mid \boldsymbol{x}_t) = \Phi\left(\boldsymbol{\beta}'\boldsymbol{x}_t\right) = \Phi(z_t), \tag{10.21}$$

where $E(\cdot)$ denotes the mathematical expectation operator and $\Phi(\cdot)$ represents the cumulative distribution function of a standardised normal random variable.

The representation (10.21) is, in fact, that of the **static probit** model. We will now develop its dynamic version.

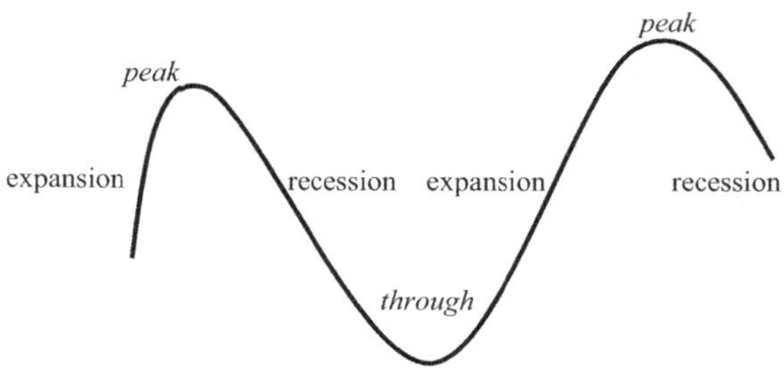

Fig. 10.2 A typical cyclical profile

Making the Model Dynamic

To introduce dynamics into (10.21), we begin by defining the information set

$$I(t) = \{R_t, \ldots, R_1, \boldsymbol{x}_t, \ldots, \boldsymbol{x}_1\}.$$

The dynamic analogue of (10.21) is then given by

$$E(R_t|\, I_{t-1}) = \Pr(R_t = 1|\, I_{t-1}) = \Phi(z_t), \tag{10.22}$$

where

$$z_t = \sum_{j=1}^{p} \rho_j R_{t-j} + \boldsymbol{\beta}'\boldsymbol{x}_t, \qquad t = 1, \ldots, T. \tag{10.23}$$

The model defined by (10.22) and (10.23) represents the **dynamic probit**. It is a simplified form of the *Binary Autoregressive model* (BAR(p)).

Before estimating it, however, we need to introduce a modification into the formulation of the model. Given that our objective is to forecast R_t, recessions, the explanatory variables are, in this case, predictors. They are therefore necessarily lagged with respect to the period t envisaged and in fact represent **predetermined variables.**

Therefore, if we consider that there are, for example, two predetermined variables, the model to be estimated will be as follows:

$$\Pr(R_t = 1) = \Phi(\gamma_0 + \sum_{j=1}^{p} \rho_j R_{t-j} + \gamma_1 x_{t-k_1} + \gamma_2 x_{t-k_2} + u_t), \tag{10.24}$$

where, as before, R_t and $\Phi(\cdot)$ represents respectively the binary dependent variable and the cumulative distribution function of a standardised normal random variable, while k_1, k_2 denotes the parameters of the forecast horizons associated with the predetermined variables and u_t is a disturbance term whose errors may be autocorrelated, but also heteroskedastic.

Let $\boldsymbol{\beta} = (\rho_1, \ldots, \rho_p, \gamma_0, \gamma_1, \gamma_2)$, be the vector of true parameters of the probit model, $\boldsymbol{\rho} = (\rho_1, \ldots, \rho_p) \in \mathbb{R}^p$ and $\boldsymbol{\gamma} = (\gamma_1, \ldots, \gamma_3) \in \mathbb{R}^3$. Then let $\boldsymbol{R}_{-p}$ be the vector such that $\boldsymbol{R}_{-p} = (R_{t-1}, \ldots, R_{t-p}) \in \mathbb{R}^p$ and $\boldsymbol{x}_{-1}$ the vector such that $\boldsymbol{x}_{-1} = (1, x_{t-k_1}, x_{t-k_2})$.

It is also assumed that the disturbances are normally distributed, while specifying that the following condition must be satisfied for maximum likelihood estimation to be possible:

$$E(R_t|\, \boldsymbol{x}_{-1}, R_{t-1}, R_{t-2}, \ldots) = \Phi\left(\sum_{t=1}^{p} \rho_j R_{t-j} + \boldsymbol{\gamma}'\boldsymbol{x}_{-1}\right).$$

The log-likelihood function conditional on the first p observations $R_1, \ldots, R_p$ is then equal to

$$\begin{aligned}\ln L(\hat{\boldsymbol{\beta}}_T) &= (T-p)^{-1} \sum_{t=p+1}^{T} \ln L_t(\hat{\boldsymbol{\beta}}) \\ &= (T-p)^{-1}[R_t \ln(\Phi(\boldsymbol{\rho}'\boldsymbol{R}_{-p} + \boldsymbol{\gamma}'\boldsymbol{x}_{-1}) \\ &\quad + (1-R_t)\ln(1-\Phi(\boldsymbol{\rho}'\boldsymbol{R}_{-p} + \boldsymbol{\gamma}'\boldsymbol{x}_{-1})]. \end{aligned} \tag{10.25}$$

This highly non-linear log-likelihood function is optimised using one of the algorithms detailed in "Procedures for Solving Non-linear Systems" section of this chapter.

Let us now look at the asymptotic properties of the estimators obtained in this way, based on the contribution of De Jong and Woutersen (2011). Under the assumptions developed by these two authors, in particular, that of *near epoch dependence* of the random variables R_t, we have the following proposition:

Proposition 10.1 *The maximum likelihood estimator of* $\hat{\boldsymbol{\beta}}_T$ ***converges in probability*** *to the true value of* $\boldsymbol{\beta}$, *so that*

$$\hat{\boldsymbol{\beta}}_T \to^P \boldsymbol{\beta}.$$

Furthermore, under the additional hypothesis that

$$u_t|\ (x_t, R_{t-1}), (x_{t-1}, R_{t-2}), \ldots \sim IIN(0, 1),$$

comes the proposal:

Proposition 10.2 *Under the assumptions adopted, the vector* $\hat{\boldsymbol{\beta}}_T$ *converges* ***asymptotically*** *to a* ***normal distribution,*** *so* that

$$T^{1/2}(\hat{\boldsymbol{\beta}}_T - \boldsymbol{\beta}) \to^D N(\mathbf{0}, \boldsymbol{I}^{-1}),$$

where $\boldsymbol{I}^{-1}$ *is the inverse of the* ***Fisher information matrix***.

These two propositions are totally reassuring because the properties of the estimators of the parameters of the dynamic probit are analogous to those of a purely linear model.

Forecasting Methods

In this section, we describe the forecasting method specifically adapted to the dynamic probit model. The problem is to forecast the value of the binary variable R_t

over a horizon of h quarters—or any other period of time, but here we will restrict ourselves to quarters—given the information available in $t - h$.

A well-known statistical result states that the estimator of R_t conditional on the information set I_{t-h} minimising the mean square error is given by the conditional expectation of R_t, i.e. by

$$E(R_t | I_{t-h}).$$

This expectation therefore makes the forecast optimal, i.e. minimises the mean squared error, given the information available h quarters earlier.

Considering the dynamic nature of the probit model defined by (10.22) and (10.23), the conditional expectation of R_t can be written using the law of iterated expectations:

$$E_{t-h}(R_t) = E_{t-h}\left[\Pr(R_t = 1 | I_{t-h})\right], \tag{10.26}$$

where $E_{t-h}(\cdot)$ is the conditional expectation with respect to I_{t-h}.

It follows, on the basis of (10.26) and the dynamic probit formulation as represented by (10.22) and (10.23), that the optimal forecast of R_t becomes

$$E_{t-h}(R_t) = E_{t-h}(\Phi(z)) = E_{t-h}\left[\Phi\left(\sum_{j=1}^{p} \rho_j R_{t-j} + \boldsymbol{\beta}'\boldsymbol{x}_{-1}\right)\right], \tag{10.27}$$

with $\Phi(\cdot)$ representing the distribution function of a standardised normal random variable.

As indicated by Marcellino et al (2006), R_t can be forecast in two ways: either directly or iteratively. The method we will be using is fully in line with the latter approach.

More specifically, the recession probabilities for each quarter t are generated using the appropriate values of the model's predictors. Using the convention that the economy will be considered to be in recession when these probabilities are greater than 0.5, the process provides as a by-product the predicted value of R_t, i.e. 0 or 1.

This is one of the major advantages of the forecasting method we have chosen since it allows us to get around the problem of the late publication of data on Gross Domestic Product. Not to mention the fact that this data is provisional and therefore subject to revision!

10.7 Application: Forecasting French Recessions

Our aim now is to estimate a dynamic probit model based on the French data and then use it to forecast the recession—worldwide, with the exception of China and India—observed in 2008.

We will begin, however, by addressing a question posed in the introduction to "Dynamic Probit" section, but whose answer we had deferred: how can we date the French economic cycle?

Cycle Dating Procedure

Basically, there are two main methods available for determining the cycle. The first is to isolate the cyclical component of a series using a filter, such as the Hodrick-Prescott or Baxter-King filter. These filters have two major shortcomings: (1) they can generate spurious fluctuations that do not exist in reality; (2) there are as many different cycles as there are filters, and even more if the parameters specific to each filter are varied.

The second method uses computer algorithms to select the **high and low points in the cycle**. The best known is that of Bry-Boschan (1971), designed to analyse monthly data. Harding and Pagan (2002) developed a quarterly version (known by the acronym BBQ, or *Bry-Boschan Quarterly*).

This algorithmic quarterly version that we will adopt is based on several foundations. To do this, consider the logarithm of the quarterly series of Gross Domestic Product, i.e. $y_t = \ln Y_t$. If this series were continuous, its peaks and troughs would be the local maxima and minima of this chronicle, respectively.

To transpose this optimisation criterion to a necessarily discrete series, let's introduce the indicator function $\mathbf{1}_A$ which takes the value 1 when event A occurs and the value 0 when it does not. Therefore, if P_t et Cr_t are dummy variables equal to unity when a peak and a trough are reached, respectively, and equal to 0 when a peak and a trough are not reached, then

$$\begin{aligned} P_t &= \mathbf{1}_{\{Y_t > Y_{t+j},\, j=1,\ldots,k\}} \\ Cr_t &= \mathbf{1}_{\{Y_t < Y_{t+j},\, j=1,\ldots,k\}}. \end{aligned}$$

Using the quarterly GDP series $y_t = \ln Y_t$, the BBQ algorithm runs through three steps:

1. Identify a set of potential peaks and troughs using the *turning point* rule: *a local peak (respectively a local trough) is reached at time t if and only if for* $k = 1, 2$, *i.e. two quarters,*

$$\{y_t > y_{t\pm k}\} \qquad (resp.\ \{y_t < y_{t\pm k}\}).$$

2. Impose the condition that peaks and troughs must alternate by selecting the higher of the two consecutive peaks, alternating with the lower of the two successive troughs.

3. Eliminate turning points according to **censorship rules** relating to the duration of phases and cycles:
    - an expansion or contraction phase must last at least two quarters;
    - a complete cycle (i.e. from peak to peak or trough to trough) must last at least five half-years.

The following table provides a summary of the peaks and troughs affecting the quarterly series of French GDP between the 1er quarter of 1979 and the 4ème quarter of 2010, using the BBQ algorithm (Table 10.6).

Table 10.6 Cyclical profile of the French economy (1979–2010)

Peaks	Troughs
1980-T1	1980-T4
1992-T1	1993-T1
2008-T1	2009-T1

During the reference period, there were three recessions (1980Q2–1980Q4; 1992Q2–1993Q1 and 2008Q2–2009Q1). The longest of these was the recession that began in the 2nd quarter of 2008, linked to the US *subprime* crisis, which began in the first half of 2007. It should also be noted that the fifteen years preceding this latest recession were years of economic calm!

Estimation of the Dynamic Probit Model

On the basis of the previous dating of the peaks and troughs in the logarithm of real GDP series, we can easily construct the R_t, series, which takes the value 1 during the three identified phases of contraction and 0 during all the other quarters. This is our binary dependent variable.

As far as recession predictors are concerned, following Bismans and Majetti (2013), we will retain two of them, namely:

1. the interest rate ***spread***, denoted S and defined as the difference between the ten-year government bond rate and the three-month PIBOR (*Paris Interbank Offered Rate*) rate;
2. the spot euro-dollar ***exchange rate***, in other words, the rate of one euro in terms of the US dollar; this exchange rate will be denoted C.

However, we still need to determine the appropriate lag for each of these variables, which is a crucial step, both because it affects the selection of the forecast horizon and the accuracy of the forecast itself.

To choose the number of lags associated with each of the two variables, we will first take a forecast horizon ranging successively from one to six quarters, after which we will estimate the two models separately:

$$\begin{cases} M1: \Pr(R_t = 1) = \Phi(\gamma_0 + \gamma_1 S_{t-k_1}), & k_1 = 1, 2, \ldots, 6 \\ M2: \Pr(R_t = 1) = \Phi(\gamma_0 + \gamma_2 C_{t-k_2}), & k_2 = 1, 2, \ldots, 6. \end{cases} \tag{10.28}$$

Finally, we will use two criteria to assess the predictive power of the variables selected: firstly, McFadden's pseudo-R^2, already presented in "Measuring Statistical Fit" section and applied on several occasions; secondly, the t-statistics established on the basis of robust standard errors.

Table 10.7 shows these two measures derived from estimates of the equations in (10.28).

Examination of the table shows that using two lags for each of the variables gives the best statistical fit. In other words, $k_1 = 2$ and $k_2 = 2$, should be used for the interest rate spread and the EUR/USD exchange rate respectively.

It now remains to determine the number p of lags to take into account for the binary variable R_{t-j}, $j = 1, \ldots, p$. Since the number of regressors in the dynamic model increases with p and there may consequently be an over-fitting problem (see Estrella & Mishkin, 1998), the pseudo-R^2 is no longer appropriate.

We are thus led to use the information criteria (Akaike, Bayesian and Hannan-Quinn) to assess the value of the various adjustments made on the basis of the dynamic equations

$$\Pr(R_t = 1) = \Phi\left(\gamma_0 + \sum_{j=1}^{4} \rho_j R_{t-j} + \gamma_1 S_{t-2} + \gamma_2 C_{t-2}\right),$$

where p varies from one to four quarters.

The results of these adjustments will not be reproduced here, but they lead to the conclusion that the best model is the one with $p = 1$.

Consequently, the final empirical model to be estimated is of the form

$$\Pr(R_t = 1) = \Phi(\gamma_0 + \rho R_{t-1} + \gamma_1 S_{t-2} + \gamma_2 C_{t-2}). \tag{10.29}$$

Here are the results of the estimation of (10.29), taking into account the various lags that appear in it. In addition, for the sake of comparison, we will add those of the estimation of the static probit model. It should also be noted that we use the sample of observations covering the entire period, i.e. 1979Q1–2010Q4 (Table 10.8).

Table 10.7 Selecting the number of lags

		1	2	3	4	5	6
M1	Pseudo-R^2	0.196	0.226	0.195	0.117	0.039	0.024
	Stat t	− 3.13	− 3.38	− 3.44	− 2.78	− 1.68	− 1.34
M2	Pseudo-R^2	0.381	0.42	0.292	0.203	0.183	0.16
	Stat t	4.56	4.28	3.79	3.60	3.66	3.62

Table 10.8 Estimation of probit models

	Static probit	Dynamic probit
Const	– 20.8*** (4.63)	– 20.56** (0.298)
R_{t-1}	–	1.88*** (0.703)
S_{t-2}	– 0.78*** (0.177)	– 0.73*** (0.230)
C_{t-2}	15.01*** (3.402)	14.3** (5.940)
Log-likelihood	– 13.253	– 10.039
Pseudo-R^2	0.645	0.731
AIC	32.505	28.078
BIC	41.014	39.423
HQC	35.962	32.687

Robust standard errors are shown in brackets in this table. *, ** and *** indicate the significance of the coefficients at 10%, 5% and 1% respectively.

The first observation to emerge from the table is that the dynamic probit model is to be preferred to its static analogue, given that the pseudo-R^2 of the former is higher than that of the latter. Similarly, the three information criteria (Akaike, Bayesian and Hannan-Quinn) are minimised in the case of the dynamic probit.

Furthermore, whatever the model, the estimated coefficients of S_t and C_t are fully significant and have the expected sign, negative for the interest rate spread and positive for the EUR/USD exchange rate. The economic interpretation of these coefficients is therefore quite clear: on the one hand, a flattening or inversion of the yield curve increases the probability of a recession; on the other hand, an appreciation of the euro relative to the US dollar leads to a higher probability of the occurrence of a recession in France. It should also be noted that the coefficient associated with the exchange rate (15.01) is, in absolute terms, much higher than that of the interest rate spread (0.78).

Finally, we obtain a positive and statistically significant estimator at 1% of the coefficient of the one-period lagged binary variable. Here again, the interpretation is immediate: the probability of a recession in the current quarter is much higher if the economy was already in recession in the previous quarter.

The following graph associated with the two probit models illustrates the value of these models (Fig. 10.3). Note that the graph includes both actual recession periods (shaded) and the probability of recession in each quarter t of the 1979Q1–2010Q4 time interval.

Two comments immediately spring to mind:

1. the estimated probabilities adjust particularly well to recessionary phases, which are preceded by an increase in the said probabilities and followed by their decrease even before the return to growth.
2. the adjustment obtained using the dynamic probit model is much better than that resulting from the estimation of its static analogue. From this point of view, the

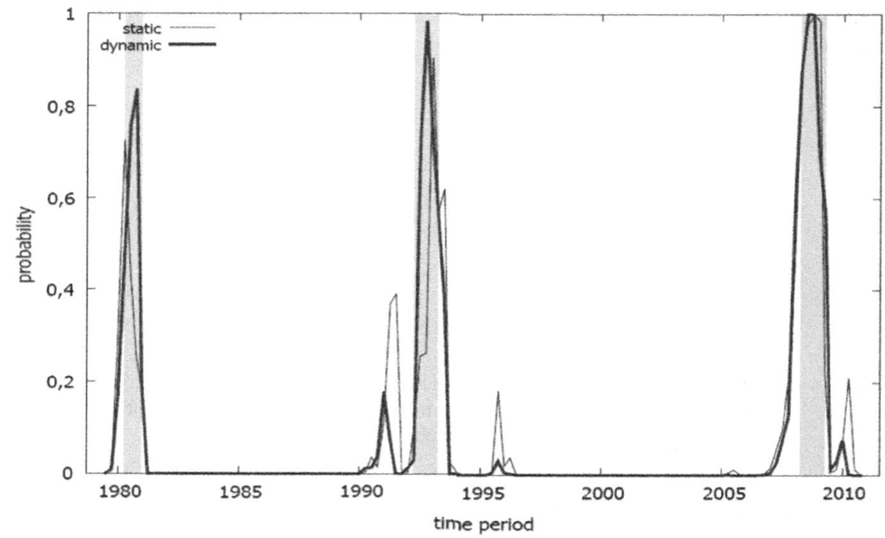

Fig. 10.3 Estimation of two probit models

dynamic model induces higher signals for the first two recessions while reducing spurious signals, such as those observed in 1991, 1996 and 2010.

Forecasting the 2008 recession

The forecasting procedure described in "Forecasting Methods" section of this chapter will be implemented in three stages:

(i) estimate the dynamic probit model as formulated in (10.29), with its two predictors S_{t-2} and C_{t-2}, and the binary variable R lagged by one quarter, all over the sample period 1979Q1–2005Q4;
(ii) generate the recession probabilities for the forecast period selected, i.e. 2006Q1–2012Q4;
(iii) assess the quality of the statistical adjustment obtained by comparing the forecasts for the period under consideration with the actual values of the binary variable.

Figure 10.4 shows the results of the three-stage forecasting procedure described above. However, it provides an additional piece of information, in that we have reproduced the dynamic forecasts obtained using S_{t-2} and C_{t-2} separately as predictors, in order to assess the role and specific influence of each of them.

The first conclusion to be drawn from Fig. 10.4 is that the dynamic model, with its two predictors and the lagged binary variable, behaves in such a way that it detects the occurrence of the 2008 recession very well. The second is that the two predictors

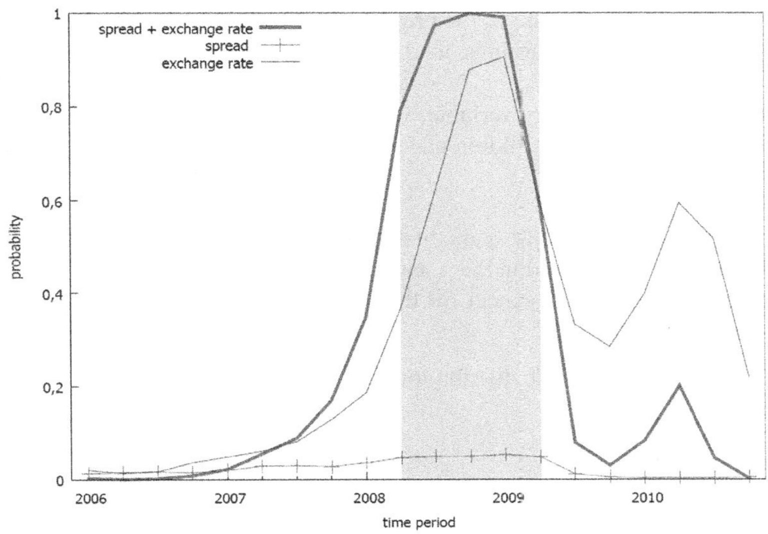

Fig. 10.4 Forecasting the 2008 recession

in question have a differentiated influence on the quality of the forecast, given that the EUR/USD exchange rate plays a primordial role in this respect, compared to the interest rate spread.

10.8 Exercises

10.8.1 Consider the linear (qualitative) model

$$R_t = \boldsymbol{\beta}'\boldsymbol{x}_t + u_t, \qquad t = 1, \ldots, T.$$

If you had the information that the vector $\boldsymbol{\beta}$ was close to the known value $\boldsymbol{\beta}_1$, could you justify using this linear model?

10.8.2 Mroz (1987), based on a sample of 753 women, 428 of whom were actually employed, estimated a linear equation, in which the dependent variable is ln *EARNINGS* which represents the hourly wage times hours worked. These earnings are a function of the following variables:

- *EDUCATION* refers to the number of years of schooling for the women in question.
- *AGE* is a variable introduced to take account of the fact that participants to the labour market have a higher income when they are older than when they are younger.

- AGE^2, which is the square of the previous explanatory variable, takes in consideration the fact that earnings tend to grow faster in the last few years of employment.
- Finally, *KIDS* is a binary variable, which takes the value 1 when the woman's family has one or more children under the age of 18 and the value 0 otherwise.

The question is twofold:

(i) how would you specify a qualitative binary equation that incorporates the different variables used in the linear regression?
(ii) what sign would you expect for the effect of the four variables in the probit model specified in (i)?

10.8.3 If X follows a logistic distribution, proof that.

(i) $E(X) = 0$
(ii) $Var(X) = \pi^2/3$.

10.8.4 Let be the log-likelihood function for the general qualitative model with binary variable (probit or logit):

$$\ln L(R, \boldsymbol{\beta}) = \sum_{t=1}^{T} \left[F(\boldsymbol{\beta}'\boldsymbol{x}_t)^{R_t} (1 - F(\boldsymbol{\beta}'\boldsymbol{x}_t))^{1-R_t} \right],$$

where F is a cumulative distribution function (CDF) with mathematical expectation equal to zero.

Proof that this log-likelihood function is strictly concave and therefore has a unique maximum (if it exists).

10.8.5 Consider the dynamic qualitative probit model given by

$$\Pr(R_t = 1) = \Phi\left(\gamma_0 + \sum_{j=1}^{4} \rho R_{t-j} + \gamma_1 S_{t-2} + \gamma_2 C_{t-2} \right),$$

where R_t is the binary indicator variable for French recessions, S_t denotes the interest rate spread and C_t the spot euro-dollar exchange rate.

Using the file used in "Estimation of the Dynamic Probit Model" section and the different information criteria, show that $j = 1$ in the previous equation.

10.8.6 You will now use the Germany2.gdt file, which contains the following quarterly series covering the period 1973Q1–2013Q4:

- *REC* takes the one during the contractions of the German economy and the value zero during the expansions.
- *SPREAD*: the yield spread between 10-year German Government bonds and 3-month EURIBOR.
- *ER* denotes the spot exchange rate EUR/USD (1 euro = x dollar).

- *SPI* is an index of total share prices for all German shares (quarterly growth rate).
- *M*1 denotes the monetary stock for Eurozone (semi-annual growth rate).

The aim of this exercise is to build a dynamic probit model that predicts German recessions. We begin with an initial function of the following form

$$\Phi(REC_t = 1) = \alpha_0 + \rho REC_{t-1} + \sum_{j=1}^{6} \beta_j \, SPREAD_{t-j} + \sum_{j=1}^{6} \gamma_j \, ER_{t-j} + \sum_{j=1}^{6} \delta_j \, SPI_{t-j} + \sum_{j=1}^{6} \phi_j M\,1_{t-j} + u_t, \tag{10.30}$$

where Φ is the standard normal CDF, and j is the number of quarters taken into account.

Based on (10.30), you are asked to determine the best model for predicting recessions in the German economy. To do this, you need to proceed in two stages:

(i) first, you will estimate the best predictor, if any, for each of the variables selected, using the significance of the associated coefficient and the pseudo-R^2.
(ii) then apply the GETS methodology to obtain the best and most parsimonious model.

10.8.7 Using the same data from the file germany2.gdt, we estimated a dynamic probit model over the period 1973Q1–2005Q4 (see Bismans et al., 2016, p. 79 sq). We then carried out a real-time forecast of the two recessions that affected the German economy between 2006 and 2014.

The figure below shows the two recessions and their duration, as well as the predictive probabilities obtained.

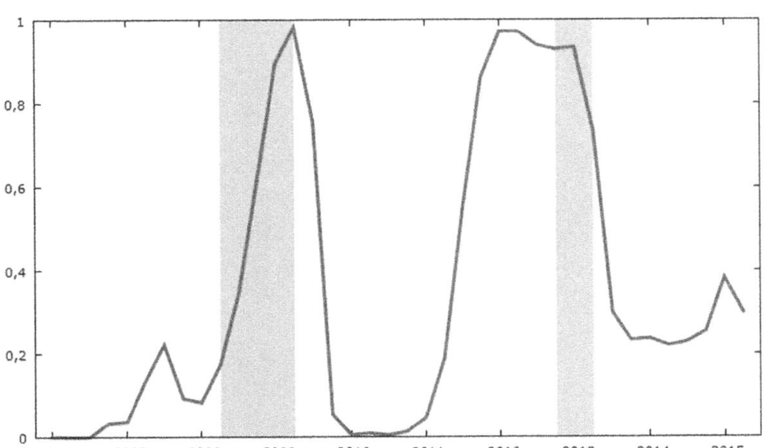

What conclusions do you draw from examining this figure?

References

Amemiya, T. (1981). Qualitative response models: A survey. *Journal of Economic Literature, 19*, 1481–1536.

Amemiya, T. (1985). *Advanced econometrics*. Harvard University Press.

Bismans, F., Aimar, T., & Diebolt, C. (2016). *Business cycles in the run of history*. Springer.

Bismans, F., & Majetti, R. (2013). Forecasting recessions using financial variables: The French case. *Empirical Economics, 44*, 419–433.

Bry, G., & Boschan, C. (1971). *Cyclical analysis of time series: Selected procedures and computer programs*, Technical Paper 20, National Bureau of Economic Research.

Davidson, R., & MacKinnon, J. G. (1984). Convenient specification tests for logit and probit models. *Journal of Econometrics, 25*, 241–262.

De Jong, R. M., & Woutersen, T. (2011). Dynamic time series binary choice. *Econometric Theory, 27*, 673–702.

Estrella, A., & Mishkin, F. (1998). Predicting US recessions: Financial variables as leading indicators. *Review of Economics and Statistics, 80*, 45–61.

Greene, W. H. (2012). *Econometric analysis* (7th ed.). Pearson: New York.

Gouriéroux, C., & Monfort, A. (1981). Asymptotic properties of the maximum likelihood estimator in dichotomous logit models, *Journal of Econometrics, 17*, 83–97.

Harding, D., & Pagan, A. (2002). Dissecting the cycle: A methodological investigation, *Journal of Monetary Economics, 49*, 365–381.

Maddala, G. S. (1983). *Limited-dependent and qualitative variables in econometrics*. Cambridge University Press.

Marcellino, M., Stock, J. H., & Watson, M. W. (2006). A comparison of direct and iterated multistep AR methods for forecasting macroeconomic series. *Journal of Econometrics, 135*, 499–526.

Mroz, T. (1987). The sensitivity of an empirical model of married women's hours of work to economic and statistical assumptions. *Econometrica, 55*, 765–799.

Spector, L., & Mazzeo, M. (1980). Probit analysis and economic education. *Journal of Economic Education, 11*, 37–44.

Appendix A
Softwares Used

The purpose of this appendix is limited: it simply aims to give a few details about the two software packages that have been used throughout this manual: GRETL in particular and Microfit, less frequently.

GRETL Software

GRETL is an acronym for *GNU Regression, Econometrics and Time-series Library*. (It's also a reference to the Grimm brothers' fairy tale, in which two children, Hansel and Gretel, have been lost by their parents in the forest and find themselves in the clutches of an antropophagous witch).

This is a multiplatform econometrics software package written in the C programming language. Its origins date back to the code written by Ramu Ramanathan (2002) for his ESL programme. It is also free software, which means that anyone can redistribute it and/or modify it under the terms of the GNU General Public License (GPL) published by the *Free Software Foundation*.

Its creators are Allin Cottrell, Wake Forest University, and Riccardo "Jack" Lucchetti, Università Politecnica delle Marche. The software can be downloaded with Debian GNU/Linux, Mac OS X and MS Windows. (The latest version was published on 21 May 2024).

GRETL supports a wide range of data dimensions. In addition to Excel.xlsx and CVS files, the following are of particular interest,

- Stata.dta files;
- SPSS.sav files;
- Eviews workfiles;
- JMulTi data files;
- RATS 4 databases;
- PC-Give databases.

F. J. Bismans and O. Damette, *Dynamic Econometrics*,
https://doi.org/10.1007/978-3-031-72910-2

From an analytical point of view, GRETL includes a number of econometric tools, including:

- a wide range of estimators: least squares, maximum likelihood, generalised method of moments (GMM), method of estimating a system of equations, LASSO, Ridge regressions, etc.;
- GRETL is particularly well equipped for the analysis of time series: ARIMA, GARCH-type models, VAR (including structural VAR) and VECM models, unit root and cointegration tests, Kalman filter;
- Qualitative or restricted dependent variables: logit, probit, tobit, models for count or duration data;
- GRETL also supports panel data estimators, including instrumental variable estimators, dynamic probit panel models and those based on GMM estimation.

Numerous *add-on* modules complement the available tools and add new functionalities, in particular for estimating structural VAR models, univariate GARCH models, Intra and Inter estimators on panel data, etc.

As GRETL is free software, it also includes a wide range of *packages* written by contributors. There are no fewer than 156 of these, which can be freely downloaded from the following link https://gretl.sourceforge.net/cgi-bin/gretldata.cgi?opt=SHOW_FUNCS.

Finally, GRETL has a powerful built-in scripting language, called HANSL, which provides a wide variety of programming tools and operations on matrices.

HANSL is in fact an acronym for "Hansl's A Neat Scripting Language". Incidentally, this is another allusion to Gretel's little brother, Hansel, in the Grimm brothers' fairy tale.

Cottrel and Lucchetti have put an introduction to HANSL online under the explicit title *A HANSL Primer. The Scripting Language of GRETL in 55 Minutes.* Here's the link to this little book, which has 15 chapters and 64 pages: file:///C:/Users/User/Downloads/hansl-primer%20(2).pdf.

MICROFIT Software

The latest version available is Microfit 5.5 (dated 3 September 2018), which includes a number of advances over its previous versions. It should also be noted that this is a free copy of Microfit exclusively for teaching and research. It easily can be downloaded by following the link https://www.econ.cam.ac.uk/people-files/emeritus/mhp1/Microfit/Microfit_5.5_Freeware_Setup.EXE.

The software was developed by Pesaran, Bahram and M. Hashem. It is accompanied by a manual entitled *Time Series Econometrics using Microfit 5.0*—see Pesaran and Pesaran (2009). It is also used extensively in the book that stands as a veritable econometric treatise, over a thousand pages long, *Time Series and Panel Data Econometrics* by Pesaran (2015).

Technically, the software can run regressions with up to 102 regressors on time series made up of five million observations.

Microfit has been specifically designed for modelling provenat time series data. It includes the following econometric tools:

- linear regression analysis;
- modelling one-dimensional time series;
- estimation of non-linear models;
- estimating probit and logit models;
- VAR-type models;
- cointegration analysis;
- VARX modelling and trend/cycle decomposition;
- univariate and multivariate GARCH models;
- estimation of multiple equation models.

However, Microfit does not allow panel data to be modelled.

Appendix B
Matrices and Linear Algebra

This appendix contains a set of results relating to matrix calculus which the author felt would be useful for understanding this book. It is presented in the form of definitions and theorems but does not include demonstrations or examples.

Notion of Matrix

1. Definition *A **matrix** is a real-valued function defined on*

$$\{(i, j)|\ i = 1, \ldots, m,\ j = 1, \ldots, n\}.$$

Informally, a matrix is a rectangular array of m rows and n columns of real numbers, written as

$$\boldsymbol{A}_{m\times n} = (a_{i,j}) = \begin{pmatrix} a_{11} & \ldots & a_{1n} \\ \vdots & & \vdots \\ a_{m1} & \ldots & a_{mn} \end{pmatrix}.$$

The element a_{ij} at the intersection of the i-th row and the j-th column is called the **general** or **typical element** of the matrix **A.** The latter, which has m rows and n columns, is said to have the **dimension** $m \times n$. (Read m times n.)

2. Definitions *A **diagonal** matrix* $\boldsymbol{D} = (d_{ij}) = \text{diag}(a_{11}, \ldots, d_{nn})$ *is a matrix such that* $d_{ij} = 0$ *if* $i \neq j$. *The **null** matrix is the matrix composed exclusively of zeros, in other words:* $\mathbf{0} = (0)$.

3. Definition *Let the **Kronecker delta be** given by*

F. J. Bismans and O. Damette, *Dynamic Econometrics*,
https://doi.org/10.1007/978-3-031-72910-2

$$\delta_{ij} = \begin{cases} 1 \text{ si } i = j \\ 0 \text{ si } i \neq j, \end{cases}$$

where *i and j are integers. Then the matrix of dimension $n \times n$ defined by $\boldsymbol{I}_n = (\delta_{ij})$ is called the* ***identity matrix****.*

Since the identity matrix has n rows and n columns, it is said to be **square** or ***n-square*****.** As we shall see, it represents the neutral element for matrix multiplication.
4. Definitions *A* ***scalar*** *is a matrix 1×1, i.e. a real number. A* ***vector*** *is a matrix with only one row or column. The* ***column vector*** *with n components* **x is** *written as*

$$\boldsymbol{x} = \begin{pmatrix} x_1 \\ \vdots \\ x_n \end{pmatrix}$$

and the ***row vector*** *with n components*

$$\boldsymbol{x}' = (x_1, \ldots, x_n).$$

(As a general rule, line vectors will be topped by the "prime" sign, a convention that will suffer almost no exceptions. As we shall soon see, the prime sign attached to a row vector also means that $\boldsymbol{x}'$ is the transpose of the column vector **x**).

Operations on Matrices

5. Definition *Let be the matrices $\boldsymbol{A} = (a_{ij})$* and *$\boldsymbol{B} = (b_{ij})$. If these matrices have the same number of rows and the same number of columns, then their* ***sum****, $\boldsymbol{A} + \boldsymbol{B}$, is the matrix $\boldsymbol{C} = (c_{ij})$ such that $c_{ij} = a_{ij} + b_{ij}$ for all i and all j.*
6. Definition *If $\boldsymbol{A}_{m\times n} = (a_{ij})$ and $\boldsymbol{B}_{n\times k} = (b_{ij})$, then their* ***product*** *is the matrix $\boldsymbol{C}_{m\times k} = \boldsymbol{AB}$ such that $c_{ij} = \sum_{r=1}^{n} a_{ik} b_{kj}$ for all i and all j.* **A** *is then said to* ***premultiply*** **B** *or, alternatively,* **B** ***postmultiplies*** **A.** *For matrix multiplication to be possible, the number of columns in the first matrix must be equal to the number of rows in the second matrix; otherwise, the product is not* ***defined****.*

In general, the matrix product is not commutative, which means that **AB** is not, in most cases, equal to **BA,** which may even be undefined.
7. Theorem *Let be the matrices $\boldsymbol{A}_{m\times n}$, $\boldsymbol{B}_{n\times k}$, $\boldsymbol{C}_{k\times p}$ and $\boldsymbol{D}_{n\times k}$. Then*
Associativity: $(\boldsymbol{AB})\boldsymbol{C} = \boldsymbol{A}(\boldsymbol{BC})$.
Distributivity with respect to addition: $\boldsymbol{A}(\boldsymbol{B} + \boldsymbol{D}) = \boldsymbol{AB} + \boldsymbol{AD}$; $(\boldsymbol{B} + \boldsymbol{D})\boldsymbol{C} = \boldsymbol{BC} + \boldsymbol{DC}$.
8. Theorem *If $\boldsymbol{I}_n$ is the identity matrix of dimension $n \times n$, then for any matrix $\boldsymbol{B}_{n\times k}$ and any matrix $\boldsymbol{A}_{m\times n}$,$\boldsymbol{I}_n\boldsymbol{B} = \boldsymbol{B}$* and *$\boldsymbol{A}\boldsymbol{I}_n = \boldsymbol{A}$.*
9. Definition *The* ***transpose*** *of the matrix $\boldsymbol{A}_{m\times n} = (a_{ij})$ is the matrix $\boldsymbol{A}'_{n\times m} = (a_{ji})$.*

10. Definition *Let be the n-square matrix* $\mathbf{A} = (a_{ij})$. *Then* **A** *is said to be* ***symmetrical*** *if* $\mathbf{A} = \mathbf{A}'$, *i.e. if for any* $i \neq j$, $a_{ij} = a_{ji}$.
11. Theorem *For any matrix* **A**, $(\mathbf{A}')' = \mathbf{A}$.
12. Theorem *Let* **A** *and* **B** *be two matrices such that* **AB** *is defined. Then*

$$(\mathbf{AB})' = \mathbf{B}'\mathbf{A}'.$$

The theorem can be generalised to a finite number of matrices. For example: if the product **ABC** *is defined, then*

$$(\mathbf{ABC})' = \mathbf{C}'\mathbf{B}'\mathbf{A}'.$$

13. Theorem *If* **A** *is any matrix, then* $\mathbf{A}'\mathbf{A}$ *and* $\mathbf{A}\mathbf{A}'$ *are* ***symmetric***. *Moreover*, the ***diagonal*** *elements of these two matrices are* ***non-negative***.

Trace and Determinant of a Matrix

14. Definition *The* ***trace*** *of a matrix* $\mathbf{A}_{n\times n}$, *denoted* $\text{tr}\,(\mathbf{A})$, *is the sum of the diagonal elements of* **A**:

$$\text{tr}\,(\mathbf{A}) = \sum_{i=1}^{n} a_{ii}.$$

15. Theorem *If* **A** *is an n-squared matrix, then*

$$\text{tr}\,(\mathbf{A}) = \text{tr}\,(\mathbf{A}').$$

16. Theorem *If* **A** *and* **B** *are n-squared matrices, then*

(i) $\text{tr}\,(\mathbf{A} \pm \mathbf{B}) = tr\,(\mathbf{A}) \pm \text{tr}\,(\mathbf{B})$;
(ii) $\text{tr}\,(\mathbf{AB}) = \text{tr}\,(\mathbf{BA})$.

17. Definition *If* $\mathbf{A} = (a_{ij})$ *is an n-squared matrix, then its* ***determinant***, *denoted* $\det\,(\mathbf{A})$,, *is the number defined by recurrence, for* ***any*** *fixed i, as*

$$\det\,(\mathbf{A}) = \begin{cases} A & n = 1, \\ \sum_{j=1}^{n} (-1)^{i+j} a_{ij} \det\,(\mathbf{A}_{ij}) & n > 1, \end{cases}$$

where $\mathbf{A}_{ij}$ *is the submatrix obtained by deleting the i-th row and j-th column of* **A.** *By proceeding in the latter way, we have* ***expanded*** **A** *along its i-th line.*
18. Theorem *Let be the n-squared matrix* **A** *and k a scalar. Then*

(i) *if two rows (columns) of the matrix* $\mathbf{A}$ *are swapped, the determinant changes sign;*
(ii) *if any multiple of the corresponding elements of a row (column) of* $\mathbf{A}$ *is added to the elements of another row (column) of* $\mathbf{A}$*, the determinant remains unchanged;*
(iii) $\det(\boldsymbol{A}) = \det(\boldsymbol{A}')$;
(iv) *if two rows (columns) of a matrix* $\mathbf{A}$ *are identical,* $\det(\boldsymbol{A}) = 0$;
(v) *if* $\boldsymbol{B} = k\boldsymbol{A}, \det(\boldsymbol{B}) = \det(k\boldsymbol{A}) = k^n \det(\boldsymbol{A})$..

19. Theorem *If* $\mathbf{A}$ *and* $\mathbf{B}$ *are two n-square matrices whose product is well defined, then*

$$\det(\boldsymbol{AB}) = \det(\boldsymbol{A}) \cdot det(\boldsymbol{B}).$$

Inverse and Rank of a Matrix

20. Definitions *If* $\mathbf{A}$ *is an* $n \times n$ *matrix and if the i-th row and the j-th column of this matrix are deleted, then* $\det(\boldsymbol{A}_{ij})$ *is called the* ***minor*** *of* a_{ij}. *The* ***cofactor*** *of the element* a_{ij}, *denoted* C_{ij}, *of* $\mathbf{A}$ *is such that*

$$C_{ij} = (-1)^{i+j} \det(\boldsymbol{A}_{ij}).$$

21. Definition *If* $A = (a_{ij})$ *is an n-square matrix,* the *matrix obtained by replacing each element* a_{ij} *of* $\mathbf{A}$ *by its cofactor* C_{ij}, *is called the* ***adjoint matrix*** *of* $\mathbf{A}$, *denoted* $\mathrm{Adj}\,\boldsymbol{A}$:

$$\mathrm{Adj}\,\boldsymbol{A} = \begin{pmatrix} C_{11} & \dots & C_{1n} \\ \vdots & \ddots & \vdots \\ C_{n1} & \dots & C_{nn} \end{pmatrix}.$$

22. Definition *Let* $\mathbf{A}$ *and* $\mathbf{B}$ *be two square matrices of dimension* $n \times n$. *Then* $\mathbf{B}$ *is the inverse of* $\mathbf{A}$ *if*

$$\boldsymbol{AB} = \boldsymbol{BA} = \boldsymbol{I}_n.$$

23. Theorem *If* A *has an inverse, this inverse is* ***unique*** *and is therefore denoted* A^{-1}.

24. Theorem *The* ***inverse*** *of the matrix* $\mathbf{A}$ *exists if the determinant of* $\mathbf{A}$ *is* ***non-zero***:

$$\det(\boldsymbol{A}) \neq 0.$$

The matrix $\mathbf{A}$ *is then said to be* ***non-singular***.

24. Definition *Let be the n-squared matrix* **A** *of non-zero determinant. Then the* ***inverse*** *of* **A** *is equal to the* ***transpose of the adjoint*** *of* **A** ***divided*** *by its* ***determinant***:

$$A^{-1} = \frac{1}{\det(A)}(\mathrm{Adj}\,A)'.$$

25. Theorem *Let A and B be two* ***non-singular*** *n-square matrices. Then*

(i) $\left(A^{-1}\right)^{-1} = A$;
(ii) $(A')^{-1} = (A^{-1})'$;
(iii) $(AB)^{-1} = B^{-1}A^{-1}$;
(iv) *if*, in addition, **A** *is* ***symmetric***, *then* A^{-1} *is also* ***symmetric***.

26. Theorem *If* **A** *is* ***non-singular***, *then*

$$\det\left(A^{-1}\right) = (\det(A))^{-1}.$$

27. Definitions *The* ***rank*** *of a matrix* **A**, denoted $\mathrm{rk}(A)$, *is the* ***maximum*** *number of* ***linearly independent*** *rows or columns in* **A**. *The* ***row rank*** *of* **A**, *denoted* $\mathrm{rk}_r(A)$, *is the number of* ***linearly independent*** *rows of* **A**. *The* ***column rank*** *of* **A**, *denoted* $\mathrm{rk}_c(A)$, *is the number of* ***linearly independent*** *columns of* **A**.
28. Theorem *Let be the matrices* $A_{m\times n}$ *and* $B_{m\times m}$, **B** *being non-singular. Then*

(i) $\mathrm{rk}_r(A) = \mathrm{rk}_c(A) = \mathrm{rk}(A)$.
(ii) $\mathrm{rk}(A') = \mathrm{rk}(A)$;
(iii) $\mathrm{rk}(BA) = \mathrm{rk}(A)$.

Systems of Linear Equations

Consider the system of m linear equations with n variables:

$$\begin{cases} a_{11}x_1 + a_{12}x_2 + \cdots + a_{1n}x_n = b_1 \\ a_{21}x_1 + a_{22}x_2 + \cdots + a_{2n}x_n = b_2 \\ \vdots \\ a_{m1}x_1 + a_{m2}x_2 + \cdots + a_{mn}x_n = b_m \end{cases}.$$

This system can still be written in a compact way:

$$Ax = b,$$

where **A** is the matrix of coefficients of dimension $m \times n$, **x** is the column vector of n variables and **b** is the column vector of m constants.
29. Definition *Let the system be* $Ax = b$. *If* $b = 0$, *the system is said to be* ***homogeneous***. On the *other hand, if* $b \neq 0$, *it is said to be* ***non-homogeneous***.

30. Definition *A system of linear equations* $\boldsymbol{Ax} = \boldsymbol{b}$ *is* ***compatible*** *if there is* ***at least*** *one set of n values* $x_1^*, x_2^*, \ldots, x_n^*$, *that simultaneously satisfies the m equations of the system.*
31. Definition *Let be the system* $\boldsymbol{Ax} = \boldsymbol{b}$. *The* ***augmented matrix*** *of dimension* $m \times (n+1)$ *is the matrix* $(\boldsymbol{A}, \boldsymbol{b})$ *such that*

$$(\boldsymbol{A}, \boldsymbol{b}) = \left(\begin{array}{cccc|c} a_{11} & a_{12} & \ldots & a_{1n} & b_1 \\ a_{21} & a_{22} & \ldots & a_{2n} & b_2 \\ \vdots & \vdots & \ddots & \vdots & \vdots \\ a_{m1} & a_{m2} & \ldots & a_{mn} & b_m \end{array}\right).$$

32. Theorem *The system* $\boldsymbol{Ax} = \boldsymbol{b}$ *is* ***compatible*** *if and only if the matrix of coefficients* **A** *and the augmented matrix* $(\boldsymbol{A}, \boldsymbol{b})$ *have the* ***same rank***.
33. Theorem *Let be the* ***non-homogeneous*** *system of m equations with n unknowns* $\boldsymbol{Ax} = \boldsymbol{b}$. *There are three mutually exclusive and collectively exhaustive possibilities*:

(i) *if* $\text{rk}(\boldsymbol{A}) = \text{rk}(\boldsymbol{A}, \boldsymbol{b})$, *the system is* ***incompatible***;
(ii) *if* $\text{rk}(\boldsymbol{A}) = \text{rk}(\boldsymbol{A}, \boldsymbol{b}) = n$, *the system has a* ***unique solution*** *given by*

$$x^* = \boldsymbol{A}^{-1}\, \boldsymbol{b}.$$

(iii) *if* $\text{rk}(\boldsymbol{A}) = \text{rk}(\boldsymbol{A}, \boldsymbol{b}) < n$, *the system has* ***infinitely many solutions***.

34. Theorem *Since the* ***homogeneous*** *system of m equations with n unknowns* $\boldsymbol{Ax} = \boldsymbol{0}$ *is always* ***compatible***, *there are only two possible cases, mutually exclusive and collectively exhaustive*:

(i) *if* $\text{rk}(\boldsymbol{A}) = n$, *the system has a* ***unique (trivial) solution*** $x^* = \boldsymbol{0}$;
(ii) *if* $\text{rk}(\boldsymbol{A}) < n$, *the system has* ***infinitely many solutions***.

Matrix Derivation

Let's briefly introduce a few notations and consider a numerical function f, differentiable up to order 2, of n variables $\boldsymbol{x} = (x_1, \ldots, x_n)'$.

We can group together all the partial derivatives of f in a column vector or a row vector, called the **gradient** of f, such that

$$\frac{\partial f(\boldsymbol{x})}{\partial \boldsymbol{x}} = \begin{pmatrix} \frac{\partial f(\boldsymbol{x})}{\partial x_1} \\ \vdots \\ \frac{\partial f(\boldsymbol{x})}{\partial x_1} \end{pmatrix} \quad \text{and} \quad \frac{\partial f(\boldsymbol{x})}{\partial \boldsymbol{x}'} = \left(\frac{\partial f(\boldsymbol{x})}{\partial x_1}, \ldots, \frac{\partial f(\boldsymbol{x})}{\partial x_n}\right).$$

In the same way, we can collect all the second partial derivatives of f in a matrix of dimension $n \times n$, called the **Hessian matrix** or simply the **Hessian** of f:

$$\frac{\partial^2 f(x)}{\partial x \partial x'} = \begin{pmatrix} \frac{\partial^2 f(x)}{\partial x_1 \partial x_1} & \cdots & \frac{\partial^2 f(x)}{\partial x_1 \partial x_n} \\ \vdots & \ddots & \vdots \\ \frac{\partial^2 f(x)}{\partial x_n \partial x_1} & \cdots & \frac{\partial^2 f(x)}{\partial x_n \partial x_n} \end{pmatrix}.$$

35. Definition *Let* $a' = (a_1, \ldots, a_n)$ *and* $x = (x_1, \ldots, x_n)'$. *The* ***linear form*** *in* **x** *is defined as the inner product*

$$a'x = \sum_{i=1}^{n} a_i x_i.$$

36. Definition *Let* $A_{n \times n}$ *be a symmetric real matrix. The function* $Q(x) : \mathbb{R}^n \to \mathbb{R}$ *defined by* $Q(x) = x'Ax$ *is said to be a* ***quadratic form*** *in* **x**.
37. Theorem *The vector derivative of the linear form* $a'x$ *in the vector* **x** *is given by*

$$\frac{\partial\, a'x}{\partial x} = a \text{ and } \frac{\partial\, a'x}{\partial x'} = a'.$$

38. Theorem *The derivative of the* ***linear mapping*** Ax *is equal to the matrix of the mapping*:

$$\frac{\partial Ax}{\partial x} = A.$$

39. Theorem *The vector derivative of the* ***quadratic form*** $Q = x'Ax$ *is given by*

$$\frac{\partial x'Ax}{\partial x} = 2Ax.$$

Partitioned Matrices

In general, a matrix **A** of dimension $m \times n$ can be divided into several blocks. For example, suppose it is structured as follows:

$$A = \begin{pmatrix} A_{11} & A_{12} \\ A_{21} & A_{22} \end{pmatrix}, \tag{B1}$$

where A_{11} has the dimension $m_1 \times n_1$, A_{12} has the dimension $m_1 \times n_2$, A_{21} has the dimension $m_2 \times n_1$ and A_{22} has the dimension $m_2 \times n_2$, with $m_1 + m_2 = m$ and $n_1 + n_2 = n$. **In this way, A** is said to be put into **partitioned form** or, more simply, to form a **partitioned matrix.**

It is of course possible to apply the usual matrix operations to the sub-matrices of **A,** in other words, to add them, multiply them and so on. For example, if **A** is shared like (B1) and the same is true for **B**, also of dimension $m \times n$, then the addition of these two shared matrices is defined by

$$A + B = \begin{pmatrix} A_{11} + B_{11} & A_{12} + B_{12} \\ A_{21} + B_{21} & A_{22} + B_{22} \end{pmatrix}.$$

However, we are mainly interested in the inverse of a shared matrix and its determinant.

40. Theorem *Let the matrix* **A** **be** *partitioned as follows*:

$$A = \begin{pmatrix} A_{11} & A_{12} \\ A_{21} & A_{22} \end{pmatrix},$$

where A_{11} and A_{22} ***are non-singular square*** *matrices. Then the* ***inverse*** *of* **A** *is given by*

$$A^{-1} = \begin{pmatrix} (A_{11} - A_{12}A_{22}^{-1}A_{21})^{-1} & -(A_{11} - A_{12}A_{22}^{-1}A_{21})^{-1}\Sigma_{12}\Sigma_{22}^{-1} \\ -A_{22}^{-1}A_{21}(A_{11} - A_{12}A_{22}^{-1}A_{21})^{-1} & A_{22}^{-1} + A_{22}^{-1}A_{21}(A_{11} - A_{12}A_{22}^{-1}A_{21})^{-1}A_{12}A_{22}^{-1} \end{pmatrix}.$$

41. Theorem *Let the partitioned matrix* **A** *of dimension* $m \times n$

$$A = \begin{pmatrix} A_{11} & A_{12} \\ A_{21} & A_{22} \end{pmatrix},$$

where A_{11} *has the dimension* $n_1 \times n_1, A_{12}$ *has the dimension* $n_1 \times n_2$, etc., *with* $n_1 + n_2 = n$. *Then if* A_{11} *is* ***non-singular***,

$$\det(A) = \det(A_{11}) \cdot \det(A_{22} - A_{21}A_{11}^{-1}A_{12}).$$

Alternatively, if A_{22} *is* ***non-singular***, *then*

$$\det(A) = \det(A_{22}) \cdot \det(A_{11} - A_{12}A_{22}^{-1}A_{21}).$$

Quadratic Forms and Definitability of Matrices

42. Definition *Let be the quadratic form* $Q(x) = x'Ax$ *and* $A_{n\times n}$ *the* ***symmetric*** *matrix associated with* $Q(x)$. *Then,* **A** *is said to be*

(i) ***positive definite*** *if for any* $x \neq 0, x'Ax > 0$;
(ii) ***positive semidefinite*** *if for any* $x \in \mathbb{R}^n$, $x'Ax \geq 0$;
(iii) ***negative definite*** *if for any* $x \neq 0$, $x'Ax < 0$;

(iv) ***negative semidefinite*** *if for any* $\boldsymbol{x} \in \mathbb{R}^n$, $\boldsymbol{x}'\boldsymbol{A}\boldsymbol{x} \leq 0$;
(v) ***indefinite*** *in all other cases.*

43. Theorem *The matrix* **A** *is* ***positive definite*** *if and only if there exists a* ***non-singular*** *matrix* $\boldsymbol{P}_{n \times n}$ *such that*

$$A = \boldsymbol{P}'\boldsymbol{P}.$$

44. Theorem *Given a matrix* **B** *of dimension* $m \times n$. *Then the* **symmetric** *matrix* $\boldsymbol{B}'\boldsymbol{B}$ *of dimension* $n \times n$ *is*

(i) positive semidefinite;
(ii) ***positive definite*** *if and only if* $\text{rk}(\boldsymbol{B}) = n$.

45. Theorem *If* **A** *is a* ***positive definite*** *matrix, then* A^{-1} *is also* ***positive definite***.
46. Definition *The* ***dominant principal minor*** *of order* $k \in \{1, 2, \ldots, n\}$ *of the n-square matrix* **A** *is the determinant of the submatrix of dimension* $k \times k$ *obtained by eliminating the last* $n - k$ *rows and the last* $n - k$ *columns of* **A**. *These are noted successively, for* $k = 1, \ldots, n$,

$$\det(\boldsymbol{A}_1) = |a_{11}|,\ \det(\boldsymbol{A}_2) = \begin{vmatrix} a_{11} & a_{12} \\ a_{21} & a_{22} \end{vmatrix}, \ldots,\ det(\boldsymbol{A}_n) = \begin{vmatrix} a_{11} & \ldots & a_{1n} \\ \vdots & \ddots & \vdots \\ a_{n1} & \ldots & a_{nn} \end{vmatrix}.$$

47. Theorem *Let* $A = (a_{ij})$ *be a* ***symmetric*** *n-square matrix. The matrix* **A** *is* ***positive definite*** *if and only if the dominant principal minors of* **A** *are* ***all positive***:

$$\det(\boldsymbol{A}_1) > 0,\quad \det(\boldsymbol{A}_2) > 0, \ldots,\ det(\boldsymbol{A}_n) > 0.$$

48. Theorem *Let* $A = (a_{ij})$ *be a* ***symmetric*** *n-square matrix. The matrix* **A** *is* ***negative definite*** *if and only if the dominant principal minors of* **A** ***alternate in sign***, *the* ***first*** *of them being negative*:

$$\det(\boldsymbol{A}_1) < 0,\quad \det(\boldsymbol{A}_2) > 0, \ldots,\ (-1)^n det(\boldsymbol{A}_n) > 0.$$

Eigenvalues and Eigenvectors

Consider an n-square matrix **A**.
49. Definition *An* ***eigenvalue*** *of* **A** *is a scalar* $\lambda \in \mathbb{R}$, *associated with a* ***non-zero*** $\boldsymbol{x} \in \mathbb{R}^n$, *such that*

$$\boldsymbol{A}\boldsymbol{x} = \lambda\boldsymbol{x}. \tag{B2}$$

The vector **x is** *said to be an* ***eigenvector*** *of* **A**.

Equation (B2) defining the eigenvalues and eigenvectors of **A** can be written as

$$(\boldsymbol{A} - \lambda \boldsymbol{I}_n)\boldsymbol{x} = \boldsymbol{0}. \tag{B3}$$

Equation (B3) represents a system of homogeneous linear equations. Following Theorem B.5.34, if we want to obtain a non-trivial solution, the rank of **A** must be less than n, alternatively the determinant of $(\boldsymbol{A} - \lambda \boldsymbol{I}_n)$ must be 0.
50. Theorem *A scalar* λ *is an* ***eigenvalue*** *of* **A** *if and only if*

$$\det\,(\boldsymbol{A} - \lambda \boldsymbol{I}_n) = 0. \tag{B4}$$

51. Definition *Equation* (B4) *is called the* ***characteristic equation*** *of* **A**. *The* ***roots*** *of this equation are the n* ***eigenvalues*** *of* **A**.
52. Theorem *If* **A is** *a* ***symmetric*** *matrix, its eigenvalues are all* ***real.***
53. Definition *An n-square matrix* **R** *is said to be* ***orthogonal*** *if*

$$\boldsymbol{R}'\boldsymbol{R} = \boldsymbol{I}_n$$

or alternatively,

$$\boldsymbol{R}' = \boldsymbol{R}^{-1}.$$

54. Theorem *For any* ***symmetric*** *n-square matrix* **A**, *there exists an* ***orthogonal matrix*** **R** *such that*

$$\boldsymbol{R}'\boldsymbol{A}\boldsymbol{R} = \text{diag}\,(\lambda_1, \lambda_2, \ldots, \lambda_n),$$

where λ_i *are the eigenvalues of* **A**.
55. Theorem *Let be the* ***symmetric*** *n-square matrix* **A** *and its eigenvalues* $\lambda_1, \lambda_2, \ldots, \lambda_n$. *Then*

$$\det\,(\boldsymbol{A}) = \prod_{i=1}^{n} \lambda_i.$$

56. Definition *The square matrix* **A** *is said to be* ***idempotent*** *if*

$$\boldsymbol{A}\boldsymbol{A} = \boldsymbol{A}.$$

57. Theorem *The* ***eigenvalues*** *of an* ***idempotent*** *matrix are equal to 0 or 1.*
58. Theorem *If the n-square matrix* **A** *is* ***symmetric***, ***idempotent*** *and of* ***rank*** *r, then*

$$\text{rk}(\boldsymbol{A}) + \text{rk}(\boldsymbol{I}_n - \boldsymbol{A}) = n.$$

Furthermore, if the matrices $\mathbf{A_1}$ and $\mathbf{A_2}$ are symmetrical, idempotent n-square matrices of rank r_1 and r_2 respectively, such that $\boldsymbol{A}_1\boldsymbol{A}_2 = \mathbf{0}$, then

$$\mathrm{rk}(\boldsymbol{A}_1) + \mathrm{rk}(\boldsymbol{A}_2) + \mathrm{rk}(\boldsymbol{I}_n - \boldsymbol{A}_1 - \boldsymbol{A}_2) = n.$$

59. Theorem *The only **non-singular idempotent** matrix is the **identity matrix**.*
60. Theorem *Let be the quadratic form $Q = \mathbf{x}'\mathbf{A}\mathbf{x}$, where $\mathbf{A}$ is n-squared and of rank r. There is an **orthogonal** matrix $\mathbf{B}$ such that if $\mathbf{y} = \boldsymbol{B}^{-1}\mathbf{x}$, then*

$$Q = \sum_{j=1}^{m} \lambda_j y_j^2,$$

*where $\lambda_1, \ldots, \lambda_m$ are the m **non-zero** eigenvalues of $\mathbf{A}$.*
61. Theorem *Let be the **quadratic form** $\mathbf{x}'\mathbf{A}\mathbf{x}$, where $\mathbf{x}$ is a column vector with n components and $\mathbf{A}$ is a **symmetric** n-squared matrix. Then*

(i) ***x′Ax*** *is **positive definite*** (resp. ***negative definite***) *if and only if **all the** eigenvalues of* **A** *are positive.*
(ii) ***x′Ax*** *is **positive semidefinite*** (resp. ***negative semidefinite***) *if and only if **all** the eigenvalues of* **A** *are **non-negative*** (resp. ***non-positive***).

References

Amemiya, H. (1971). The estimation of the variances in a variance-components model, *International Economic Review, 12*, 1–13.

Amemiya, T. (1981). Qualitative response models: A survey. *Journal of Economic Literature, 19*, 1481–1536.

Amemiya, T. (1985). *Advanced econometrics*. Harvard University Press.

Arellano, M., & Bond, S. (1991). Some tests of specification for panel data: Monte Carlo evidence and an application to employment equations, *Review of Economic Studies, 58*, 277–297.

Bai, J., & Ng, S. (2004). A PANIC attack on unit roots and cointegration. *Econometrica, 72*, 1127–1178.

Bai, J., & Ng, S. (2010). Panel unit root tests with cross-section dependence: A further investigation. *Econometric Theory, 26*, 1088–1114.

Bismans, F. (2016). *Probabilités et statistique inférentielle. Prélude à l'économétrie*. Ellipses.

Bismans, F., & Damette, O. (2008). Currency transaction tax elasticity: An econometric evaluation. *Économie Internationale, 115*, 193–212.

Bismans, F., & Majetti, R. (2013). Forecasting recessions using financial variables: The French case. *Empirical Economics, 44*, 419–433.

Bismans, F., Aimar, T., & Diebolt, C. (2016). *Business cycles in the run of history*. Springer.

Bollerslev, T. (1986), Generalized autoregressive conditional heteroscedasticity, *Journal of Econometrics, 31*, 307–327.

Box, G. E. P., & Jenkins, G. (1976). *Time series analysis: Forecasting and control* (Revised). Prentice-Hall.

Breitung, J. (2005). A parametric approach to the estimation of cointegration vectors in panel data. *Econometric Reviews, 24*, 151–173.

Breusch, T. S. (1987). Maximum likelihood estimation of random effects models, *Journal of Econometrics, 36*, 383–389.

Bry, G., & Boschan, C. (1971). *Cyclical analysis of time series: Selected procedures and computer programs*, Technical Paper 20, National Bureau of Economic Research.

Carrion-I-Silvestre, J. (2005). Health care expenditure and GDP. *Journal of Health Economics, 24*, 839–854.

Chung, K. L. (1979). *Elementary probability theory with stochastic processes* (3rd ed.). New York: Springer.

Davison, A.C. (2001), Biometrika centenary: Theory and general methodology, *Biometrika, 88*, 13–52.

F. J. Bismans and O. Damette, *Dynamic Econometrics*,
https://doi.org/10.1007/978-3-031-72910-2

Davidson, R., & MacKinnon, J. G. (1984). Convenient specification tests for logit and probit models. *Journal of Econometrics, 25*, 241–262.

de Jong, R. M., & Woutersen, T. (2011). Dynamic time series binary choice. *Econometric Theory, 27*, 673–702.

Dickey, D., & Fuller, W. (1979), Distribution of the estimators for autoregressive time series with a unit root, *Journal of the American Statistical Association, 74*, 427–431.

Elliot, G., Rothenberg, T.J., & Stock, J.H. (1996), Efficient tests for an autoregressive unit root, *Econometrica, 64*, 813–836.

Enders, W. (2010). *Applied econometric time series* (3rd ed.). Hoboken: Wiley.

Engle, R. F., & Granger, C. W. J. (1987). Co-integration and error-correction: Representation, estimation and testing. *Econometrica, 55*, 251–276.

Estrella, A., & Mishkin, F. (1998). Predicting US recessions: Financial variables as leading indicators. *Review of Economics and Statistics, 80*, 45–61.

Frisch, R. (1933). Propagation problems and impulse problems in dynamic economics. In *Economic essays in honour of Gustav Cassel* (pp. 171–205). George Allen & Unwin.

Fuller, W. A., & Battese, G. E. (1973). Transformations for estimation of linear models with nested error structure, *Journal of the American Statistical Association, 68*, 626–632.

Fuller, W. A., & Battese, G. E. (1974). Estimation of linear models with cross-error structure, *Journal of Econometrics, 2*, 67–78.

Geweke, J., Koop, G., & van Dijck, H. (Eds.). (2013). *The Oxford handbook of bayesian econometrics*. Oxford University Press.

Gouriéroux, C., & Monfort, A. (1981). Asymptotic properties of the maximum likelihood estimator in dichotomous logit models. *Journal of Econometrics, 17*, 83–97.

Gourieroux, C., & Monfort, A. (1995). *Séries Temporelles et Modèles dynamiques* (2nd ed.). Paris: Economica.

Granger, C. W. J. (1969). Investigating causal relations by econometric models and cross-spectral methods. *Econometrica, 37*, 424–438.

Granger, C. W. J. (1983). *Cointegrated variables and error-correcting models*. USCD Discussion Paper, 83–13.

Granger, C. W. J. (1991). Developments in the study of cointegrated economic variables. In R.F. Engle & C. W. J.Granger (Eds.), *Long-run economic relationships. Readings in cointegration* (Chapter 4). Oxford University Press.

Granger, C. W. J., & Newbold, P. (1974). Spurious regressions in econometrics. *Journal of Econometrics, 2*, 111–120.

Greene, W. H. (2012). *Econometric analysis* (7th ed.). New York: Pearson.

Groen, J. J. J., & Kleibergen, F. R. (2003). Likelihood-based cointegration analysis in panels of vector error correction models. *Journal of Business and Economic Statistics, 21*, 295–318.

Gutierrez, L. (2003). On the power of panel cointegration tests: A monte Carlo comparison, *Economic Letters, 80*, 105–111.

Hadri, K. (2000). Testing for stationarity in heterogenous panel data. *Econometrics Journal, 3*, 148–161.

Hamilton, J. D. (1994). *Time series analysis*. Princeton: Princeton University Press.

Harding, D., & Pagan, A. (2002). Dissecting the cycle: A methodological investigation, *Journal of Monetary Economics, 49*, 365–381.

Harvey, A. C. (1993). *Time series models* (2nd ed.). London: Harvester-Wheatsheaf.

Hendry, D. F., & Doornik, J. A. (2014). *Empirical model discovery and theory evaluation. Automatic selection methods in econometrics*. The MIT Press.

Hiemstra, C., & Jones, J. D. (1994). Testing for linear and nonlinear Granger causality in the price-volume relation. *Journal of Finance, 49*, 1639–1664.

Hurlin, C., & Mignon, V. (2007). Une synthèse des tests de racine unitaire sur données de panel. *Economie Et Prévision, 169–171*, 251–295.

Johansen, S. (1988). Statistical analysis of cointegration vectors. *Journal of Economic Dynamics and Control, 12*, 231–254.

Johansen, S. (1995). *Likelihood-based inference in cointegrated vector autoregressive models.* Oxford: Oxford University Press.

Johnston, J. (1963, 1972). *Econometric methods*, first and second edn. McGraw-Hill.

Jorda, O. (2005). Estimation and inference of impulse responses by local projections. *The American Economic Review, 95*, 161–182.

Juselius, K. (2006). *The cointegrated VAR model: Methodology and applications*. Oxford: Oxford University Press.

Kao, C. (1999). Spurious regression and residual-based tests for cointegration in panel data. *Journal of Econometrics, 90*, 1–44.

Kao, C., & Chiang, M. H. (2000). On the estimation and inference of a cointegrated regression in panel data. *Advances in Econometrics, 15*, 179–222.

Keynes, J.M. (1936), *The general theory of employment, interest and money, in Collected Writings of John Maynard Keynes*, Vol VII, Macmillan.

Kilian, L., & Lütkepohl, H. (2017). *Structural vector autoregressive analysis*. Cambridge University Press.

Kwiatkowski, D.P., Phillips, P.C.B., Schmidt, P., & Shin, Y. (1992), Testing the null hypothesis of stationarity against the alternative of a unit root: How sure are we that economic time series have a unit root? *Econometrica, 54*, 159–178.

Larsson, R., & Lyhagen, J. (1999). *Likelihood-based inference in multivariate panel cointegration models, Stockholm school of economics*. Working Paper Series in Economics and Finance, n° 331.

Larsson, R., Lyhagen, J., & Löthgren, M. (2001). Likelihood-based cointegration tests in heterogenous panels. *Econometrics Journal, 4*, 109–147.

Levin, A., Lin, C.-F., & Chu, C.-S.J. (2002). Unit root tests in panel data: Asymptotic and finite-sample properties. *Journal of Econometrics, 108*, 1–24.

Lütkepohl, H. (2005). *New introduction to multiple time series analysis*. Berlin: Springer.

Mackinnon, J. G. (1991). Critical values for cointegration tests. In R. F. Engle & C. W. J. Granger (Eds.), *Long-run economic relationship. Readings in cointegration*. Oxford: Oxford University Press.

Maddala, G. S. (1971). The use of variance components models in pooling cross section and time series data. *Econometrica, 39*, 341–358.

Maddala, G. S. (1983). *Limited-dependent and qualitative variables in econometrics*. London: Cambridge University Press.

Marcellino, M., Stock, J. H., & Watson, M. W. (2006). A comparison of direct and iterated multistep AR methods for forecasting macroeconomic series. *Journal of Econometrics, 135*, 499–526.

McCoskey, S., & Kao, C. (1998). A residual-based test of the null of cointegration in panel data. *Econometric Reviews, 17*, 57–84.

Mroz, T. (1987). The sensitivity of an empirical model of married women's hours of work to economic and statistical assumptions. *Econometrica, 55*, 765–799.

Nelson, C. R. (1972). The prediction performance of the FRB-MIT-PENN model of the US economy. *American Economic Review, 62*, 902–917.

Nerlove, M. (1971). A note on error components models, *Econometrica, 39*, 383–396.

Newey, W., & West, K. (1987). A simple positive semi-definite, heteroskedastic and autocorrelation consistent covariance matrix. *Econometrica, 55*, 703–708.

O'Connell, P.G.J. (1998). The overvaluation of purchasing power parity, *Journal of International Economics, 44*, 1–19.

Perron, P. (1989), The great crash, the oil price shock and the unit root hypothesis, *Econometrica, 57*, 1361–1401.

Pedroni, P. (1999). Critical values for cointegration tests in heterogenous panels with multiple regressors, *Oxford Bulletin of Economics and Statistics, 61*, 653–678.

Perron, P., & Vogelsang, T.J. (1992). Nonstationarity and level shifts with an application to purchasing power parity, *Journal of Business and Economic Statistics, 10*, 301–320.

Pedroni, P. (2000). Fully modified OLS for heterogenous cointegrated panels. *Advances in Econometrics, 15*, 93–130.

Pedroni, P. (2004). Panel cointegration: Asymptotic and finite sample properties of pooled time series tests with an application to the PPP hypothesis. *Econometric Theory, 20*, 597–625.

Pesaran, M. H. (2004). *General diagnostic tests for cross section dependence in panels.* Cambridge Working Papers in Economics, 435, University of Cambridge.

Pesaran, M. H. (2007). A simple panel unit root test in the presence of cross section dependence. *Journal of Applied Econometrics, 22*, 265–312.

Pesaran, M. H. (2015). *Time series and panel data econometrics.* Oxford: Oxford University Press.

Pesaran, B., & Pesaran, M. H. (2009). *Times series econometrics using microfit 5.0.* Oxford University Press.

Pesaran, M. H., Shin, Y., & Smith, R. J. (2001). Bounds testing approaches to the analysis of level relationships. *Journal of Applied Econometrics, 16*, 289–326.

Phillips, P. C. B. (1986). Understanding spurious regressions in econometrics. *Journal of Econometrics, 33*, 311–340.

Phillips, P. C. B. (1995). Fully modified least squares and vector autoregressions. *Econometrica, 63*, 1023–1078.

Phillips, P. C. B., & Hansen, B. E. (1990). Statistical inference in instrumental variable regression with $I(1)$ processes. *Review of Economic Studies, 57*, 99–125.

Phillips, P. C. B., & Moon, H. R. (1999). Linear regression limit theory for nonstationary panel data. *Econometrica, 67*, 1057–1111.

Phillips, P. C. B., & Ouliaris, S. (1990). Asymptotic properties of residual based tests for cointegration. *Econometrica, 58*, 165–193.

Press, S. J. (2005). *Applied multivariate analysis using Bayesian and frequentist methods of inference.* Dover Publications.

Ramanathan, R. (2002). *Introductory econometrics with applications* (5th ed.). Mason: South-Western.

Roos, C.F. (1933), Constitution of the Econometric Society, *Econometrica, 1*, 106–108.

Shiryaev, A. N. (1999). *Essentials of stochastic finance. Facts, models, theory.* Singapore: World Scientific.

Sims, C. (1980). Macroeconomics and reality. *Econometrica, 48*, 1–48.

Slutsky, E. (1927, 1937). The summation of random causes as the source of cyclic processes. *Econometrica, 5*, 105–146.

Solow, R. M. (1956). A contribution to the theory of economic growth, *The Quarterly Journal of Economics, 70*, 65–94.

Spector, L., & Mazzeo, M. (1980). Probit analysis and economic education. *Journal of Economic Education, 11*, 37–44.

Stock, J. M., & Watson, M. H. (1993). A simple estimator of cointegrating vectors in higher order integrated systems. *Econometrica, 61*, 783–820.

Stock, J. M., & Watson, M. H. (2007). *Introduction to econometrics* (2nd ed.): Pearson Education.

Swamy, P. A. V. B., & Arora, S. S. (1972). The exact finite sample properties of the estimators of coefficients in the error components regression models. *Econometrica, 40*, 261–275.

Swan, T. (1956). Economic growth and capital accumulation, *Economic Record, 32*, 334–361.

Triacca, U. (1998). Non-causality: The role of the omitted variables. *Economic Letters, 60*, 317–320.

Triacca, U. (2001). On the use of Granger causality to investigate the human influence on climate. *Theoretical and Applied Climatology, 69*, 137–138.

Toda, H. Y., & Yamamoto, T. (1995). Statistical inference in vector autoregressions with possibly integrated processes. *Journal of Econometrics, 66*, 225–250.

Wallace, T. D., & Hussain, A. (1969). The use of error components models in combining cross-section and time-series data, *Econometrica, 37*, 55–72.

Wansbeek, T. J, & Kapteyn, A. (1982). A class of decompositions of the variance-covariance matrix of a generalized error components model, *Econometrica, 50*, 713–724.

Westerlund, J. (2005). New simple tests for panel cointegration. *Econometric Reviews, 24*, 297–316.

Westerlund, J. (2007). Testing for error correction in panel data. *Oxford Bulletin of Economics and Statistics, 69*, 709–748.

White, H. (1980). A heteroskedasticity-consistent covariance matrix estimator and a direct test for heteroskedasticity. *Econometrica, 48*, 827–838.

Wold, H. (1954). *A study in the analysis of stationary time series* (2nd ed.), Almqvist & Wicksell.

Zellner, A. (1962). An efficient method of estimating seemingly unrelated regressions and tests of aggregation bias. *Journal of the American Statistical Association, 57*, 348–368.

Zellner, A., & Palm, F. (1974). Time series analysis and simultaneous equation econometric models. *Journal of Econometrics, 2*, 17–54.

Index

F. J. Bismans and O. Damette, *Dynamic Econometrics*,
https://doi.org/10.1007/978-3-031-72910-2

Made in the USA
Monee, IL
14 September 2026

62976164R10208